MICROPOROUS MEDIA

SURFACTANT SCIENCE SERIES

1. Nonionic Surfactants, *edited by Martin J. Schick* (see also Volumes 19, 23, and 60)
2. Solvent Properties of Surfactant Solutions, *edited by Kozo Shinoda* (see Volume 55)
3. Surfactant Biodegradation, *R. D. Swisher* (see Volume 18)
4. Cationic Surfactants, *edited by Eric Jungermann* (see also Volumes 34, 37, and 53)
5. Detergency: Theory and Test Methods (in three parts), *edited by W. G. Cutler and R. C. Davis* (see also Volume 20)
6. Emulsions and Emulsion Technology (in three parts), *edited by Kenneth J. Lissant*
7. Anionic Surfactants (in two parts), *edited by Warner M. Linfield* (see Volume 56)
8. Anionic Surfactants: Chemical Analysis, *edited by John Cross*
9. Stabilization of Colloidal Dispersions by Polymer Adsorption, *Tatsuo Sato and Richard Ruch*
10. Anionic Surfactants: Biochemistry, Toxicology, Dermatology, *edited by Christian Gloxhuber* (see Volume 43)
11. Anionic Surfactants: Physical Chemistry of Surfactant Action, *edited by E. H. Lucassen-Reynders*
12. Amphoteric Surfactants, *edited by B. R. Bluestein and Clifford L. Hilton* (see Volume 59)
13. Demulsification: Industrial Applications, *Kenneth J. Lissant*
14. Surfactants in Textile Processing, *Arved Datyner*
15. Electrical Phenomena at Interfaces: Fundamentals, Measurements, and Applications, *edited by Ayao Kitahara and Akira Watanabe*
16. Surfactants in Cosmetics, *edited by Martin M. Rieger* (see Volume 68)
17. Interfacial Phenomena: Equilibrium and Dynamic Effects, *Clarence A. Miller and P. Neogi*
18. Surfactant Biodegradation: Second Edition, Revised and Expanded, *R. D. Swisher*
19. Nonionic Surfactants: Chemical Analysis, *edited by John Cross*
20. Detergency: Theory and Technology, *edited by W. Gale Cutler and Erik Kissa*
21. Interfacial Phenomena in Apolar Media, *edited by Hans-Friedrich Eicke and Geoffrey D. Parfitt*
22. Surfactant Solutions: New Methods of Investigation, *edited by Raoul Zana*
23. Nonionic Surfactants: Physical Chemistry, *edited by Martin J. Schick*
24. Microemulsion Systems, *edited by Henri L. Rosano and Marc Clausse*
25. Biosurfactants and Biotechnology, *edited by Naim Kosaric, W. L. Cairns, and Neil C. C. Gray*
26. Surfactants in Emerging Technologies, *edited by Milton J. Rosen*
27. Reagents in Mineral Technology, *edited by P. Somasundaran and Brij M. Moudgil*
28. Surfactants in Chemical/Process Engineering, *edited by Darsh T. Wasan, Martin E. Ginn, and Dinesh O. Shah*
29. Thin Liquid Films, *edited by I. B. Ivanov*
30. Microemulsions and Related Systems: Formulation, Solvency, and Physical Properties, *edited by Maurice Bourrel and Robert S. Schechter*
31. Crystallization and Polymorphism of Fats and Fatty Acids, *edited by Nissim Garti and Kiyotaka Sato*

32. Interfacial Phenomena in Coal Technology, *edited by Gregory D. Botsaris and Yuli M. Glazman*
33. Surfactant-Based Separation Processes, *edited by John F. Scamehorn and Jeffrey H. Harwell*
34. Cationic Surfactants: Organic Chemistry, *edited by James M. Richmond*
35. Alkylene Oxides and Their Polymers, *F. E. Bailey, Jr., and Joseph V. Koleske*
36. Interfacial Phenomena in Petroleum Recovery, *edited by Norman R. Morrow*
37. Cationic Surfactants: Physical Chemistry, *edited by Donn N. Rubingh and Paul M. Holland*
38. Kinetics and Catalysis in Microheterogeneous Systems, *edited by M. Grätzel and K. Kalyanasundaram*
39. Interfacial Phenomena in Biological Systems, *edited by Max Bender*
40. Analysis of Surfactants, *Thomas M. Schmitt* (see Volume 96)
41. Light Scattering by Liquid Surfaces and Complementary Techniques, *edited by Dominique Langevin*
42. Polymeric Surfactants, *Irja Piirma*
43. Anionic Surfactants: Biochemistry, Toxicology, Dermatology. Second Edition, Revised and Expanded, *edited by Christian Gloxhuber and Klaus Künstler*
44. Organized Solutions: Surfactants in Science and Technology, *edited by Stig E. Friberg and Björn Lindman*
45. Defoaming: Theory and Industrial Applications, *edited by P. R. Garrett*
46. Mixed Surfactant Systems, *edited by Keizo Ogino and Masahiko Abe*
47. Coagulation and Flocculation: Theory and Applications, *edited by Bohuslav Dobiáš*
48. Biosurfactants: Production • Properties • Applications, *edited by Naim Kosaric*
49. Wettability, *edited by John C. Berg*
50. Fluorinated Surfactants: Synthesis • Properties • Applications, *Erik Kissa*
51. Surface and Colloid Chemistry in Advanced Ceramics Processing, *edited by Robert J. Pugh and Lennart Bergström*
52. Technological Applications of Dispersions, *edited by Robert B. McKay*
53. Cationic Surfactants: Analytical and Biological Evaluation, *edited by John Cross and Edward J. Singer*
54. Surfactants in Agrochemicals, *Tharwat F. Tadros*
55. Solubilization in Surfactant Aggregates, *edited by Sherril D. Christian and John F. Scamehorn*
56. Anionic Surfactants: Organic Chemistry, *edited by Helmut W. Stache*
57. Foams: Theory, Measurements, and Applications, *edited by Robert K. Prud'homme and Saad A. Khan*
58. The Preparation of Dispersions in Liquids, *H. N. Stein*
59. Amphoteric Surfactants: Second Edition, *edited by Eric G. Lomax*
60. Nonionic Surfactants: Polyoxyalkylene Block Copolymers, *edited by Vaughn M. Nace*
61. Emulsions and Emulsion Stability, *edited by Johan Sjöblom*
62. Vesicles, *edited by Morton Rosoff*
63. Applied Surface Thermodynamics, *edited by A. W. Neumann and Jan K. Spelt*
64. Surfactants in Solution, *edited by Arun K. Chattopadhyay and K. L. Mittal*
65. Detergents in the Environment, *edited by Milan Johann Schwuger*

66. Industrial Applications of Microemulsions, *edited by Conxita Solans and Hironobu Kunieda*
67. Liquid Detergents, *edited by Kuo-Yann Lai*
68. Surfactants in Cosmetics: Second Edition, Revised and Expanded, *edited by Martin M. Rieger and Linda D. Rhein*
69. Enzymes in Detergency, *edited by Jan H. van Ee, Onno Misset, and Erik J. Baas*
70. Structure–Performance Relationships in Surfactants, *edited by Kunio Esumi and Minoru Ueno*
71. Powdered Detergents, *edited by Michael S. Showell*
72. Nonionic Surfactants: Organic Chemistry, *edited by Nico M. van Os*
73. Anionic Surfactants: Analytical Chemistry, Second Edition, Revised and Expanded, *edited by John Cross*
74. Novel Surfactants: Preparation, Applications, and Biodegradability, *edited by Krister Holmberg*
75. Biopolymers at Interfaces, *edited by Martin Malmsten*
76. Electrical Phenomena at Interfaces: Fundamentals, Measurements, and Applications, Second Edition, Revised and Expanded, *edited by Hiroyuki Ohshima and Kunio Furusawa*
77. Polymer-Surfactant Systems, *edited by Jan C. T. Kwak*
78. Surfaces of Nanoparticles and Porous Materials, *edited by James A. Schwarz and Cristian I. Contescu*
79. Surface Chemistry and Electrochemistry of Membranes, *edited by Torben Smith Sørensen*
80. Interfacial Phenomena in Chromatography, *edited by Emile Pefferkorn*
81. Solid–Liquid Dispersions, *Bohuslav Dobiáš, Xueping Qiu, and Wolfgang von Rybinski*
82. Handbook of Detergents, *editor in chief: Uri Zoller*
    Part A: Properties, *edited by Guy Broze*
83. Modern Characterization Methods of Surfactant Systems, *edited by Bernard P. Binks*
84. Dispersions: Characterization, Testing, and Measurement, *Erik Kissa*
85. Interfacial Forces and Fields: Theory and Applications, *edited by Jyh-Ping Hsu*
86. Silicone Surfactants, *edited by Randal M. Hill*
87. Surface Characterization Methods: Principles, Techniques, and Applications, *edited by Andrew J. Milling*
88. Interfacial Dynamics, *edited by Nikola Kallay*
89. Computational Methods in Surface and Colloid Science, *edited by Małgorzata Borówko*
90. Adsorption on Silica Surfaces, *edited by Eugène Papirer*
91. Nonionic Surfactants: Alkyl Polyglucosides, *edited by Dieter Balzer and Harald Lüders*
92. Fine Particles: Synthesis, Characterization, and Mechanisms of Growth, *edited by Tadao Sugimoto*
93. Thermal Behavior of Dispersed Systems, *edited by Nissim Garti*
94. Surface Characteristics of Fibers and Textiles, *edited by Christopher M. Pastore and Paul Kiekens*
95. Liquid Interfaces in Chemical, Biological, and Pharmaceutical Applications, *edited by Alexander G. Volkov*

96. Analysis of Surfactants: Second Edition, Revised and Expanded, *Thomas M. Schmitt*
97. Fluorinated Surfactants and Repellents: Second Edition, Revised and Expanded, *Erik Kissa*
98. Detergency of Specialty Surfactants, *edited by Floyd E. Friedli*
99. Physical Chemistry of Polyelectrolytes, *edited by Tsetska Radeva*
100. Reactions and Synthesis in Surfactant Systems, *edited by John Texter*
101. Protein-Based Surfactants: Synthesis, Physicochemical Properties, and Applications, *edited by Ifendu A. Nnanna and Jiding Xia*
102. Chemical Properties of Material Surfaces, *Marek Kosmulski*
103. Oxide Surfaces, *edited by James A. Wingrave*
104. Polymers in Particulate Systems: Properties and Applications, *edited by Vincent A. Hackley, P. Somasundaran, and Jennifer A. Lewis*
105. Colloid and Surface Properties of Clays and Related Minerals, *Rossman F. Giese and Carel J. van Oss*
106. Interfacial Electrokinetics and Electrophoresis, *edited by Ángel V. Delgado*
107. Adsorption: Theory, Modeling, and Analysis, *edited by József Tóth*
108. Interfacial Applications in Environmental Engineering, *edited by Mark A. Keane*
109. Adsorption and Aggregation of Surfactants in Solution, *edited by K. L. Mittal and Dinesh O. Shah*
110. Biopolymers at Interfaces: Second Edition, Revised and Expanded, *edited by Martin Malmsten*
111. Biomolecular Films: Design, Function, and Applications, *edited by James F. Rusling*
112. Structure–Performance Relationships in Surfactants: Second Edition, Revised and Expanded, *edited by Kunio Esumi and Minoru Ueno*
113. Liquid Interfacial Systems: Oscillations and Instability, *Rudolph V. Birikh, Vladimir A. Briskman, Manuel G. Velarde, and Jean-Claude Legros*
114. Novel Surfactants: Preparation, Applications, and Biodegradability: Second Edition, Revised and Expanded, *edited by Krister Holmberg*
115. Colloidal Polymers: Synthesis and Characterization, *edited by Abdelhamid Elaissari*
116. Colloidal Biomolecules, Biomaterials, and Biomedical Applications, *edited by Abdelhamid Elaissari*
117. Gemini Surfactants: Synthesis, Interfacial and Solution-Phase Behavior, and Applications, *edited by Raoul Zana and Jiding Xia*
118. Colloidal Science of Flotation, *Anh V. Nguyen and Hans Joachim Schulze*
119. Surface and Interfacial Tension: Measurement, Theory, and Applications, *edited by Stanley Hartland*
120. Microporous Media: Synthesis, Properties, and Modeling, *Freddy Romm*
121. Handbook of Detergents, *editor in chief: Uri Zoller*
    Part B: Environmental Impact, *edited by Uri Zoller*

ADDITIONAL VOLUMES IN PREPARATION

Luminous Chemical Vapor Deposition and Interface Engineering, *Hirotsugu Yasuda*

MICROPOROUS MEDIA

Synthesis, Properties, and Modeling

Freddy Romm

Haifa College of Management
Technion–Israel Institute of Technology
Haifa, Israel

MARCEL DEKKER, INC. NEW YORK • BASEL

Library of Congress Cataloging-in-Publication Data
A catalog record for this book is available from the Library of Congress.

ISBN: 0-8247-5567-7

This book is printed on acid-free paper.

Headquarters
Marcel Dekker, Inc., 270 Madison Avenue, New York, NY 10016, U.S.A.
tel: 212-696-9000; fax: 212-685-4540

Distribution and Customer Service
Marcel Dekker, Inc., Cimarron Road, Monticello, New York 12701, U.S.A.
tel: 800-228-1160; fax: 845-796-1772

Eastern Hemisphere Distribution
Marcel Dekker AG, Hutgasse 4, Postfach 812, CH-4001 Basel, Switzerland
tel: 41-61-260-6300; fax: 41-61-260-6333

World Wide Web
http://www.dekker.com

The publisher offers discounts on this book when ordered in bulk quantities. For more information, write to Special Sales/Professional Marketing at the headquarters address above.

Current printing (last digit):

10 9 8 7 6 5 4 3 2 1

PRINTED IN THE UNITED STATES OF AMERICA

To my mother, Tamara Slavkin

Preface

This book is directed not only to specialists in interface science and porosity but also to graduate students and scientists having a strong background in natural sciences and interested in more knowledge about porosity and especially microporosity. Each chapter starts with a brief overview of well-known facts and concepts, then more significant matters are analyzed with recommendations and conclusions, and then we present examples of original, specific scientific and technical ideas. Thus, the reader can find in this book all kinds of information related to microporosity—from the general to very specific ideas. Of course, most references are not discussed in detail, because this is not a handbook.

WHAT IS A PORE?

A *pore* can be understood as a void inside a solid structure or solid material. A solid material containing pores is called *porous material.*

Porous structures are widespread around us. What solid materials are porous? This question needs too long an answer. It is much easier to list solid materials that cannot be considered porous: most crystals (comprising metals and their alloys, diamond, ionic salts, etc.), some special plastics, and special ceramics. Hence, it is not an error to claim that most existing materials are porous.

Let us list some widespread porous materials:

All noncrystalline solids found in nature
Concretes and most ceramics and composites

Rubbers and other polymers
All foams
Active carbon (also called *activated carbon*), zeolites, silica gel, alumina gel, and most other heterogeneous catalysts
Glasses
Metallic foams

Is the role of pores in our life positive or negative? That depends on the situation. High porosity may catastrophically reduce the mechanical stability of constructive materials, their resistance to corrosion and erosion. On the other hand, this fact allows the regulation of mechanical and chemical resistance of a material without any change in its chemical composition. Pores intensify all heat- and mass-transfer processes in solid phase. Even our body's tissue is porous; otherwise, it could not provide the variety of heat- and mass-transfer processes needed for our life.

The preceding analysis is relevant for visible pores, i.e., macropores, the properties of which are determined first of all by two factors: the total volume fraction of voids (porosity of the structure ξ) and their total surface (internal surface of a porous material). However, for very small pores (nanopores, the characteristic size of which is below 500 Å = 50 nm), not only porosity and internal surface are important for the determination of properties. For such small pores, quantum effects related to the interactions with the pore walls become so significant that they increase the total internal energy of the solid phase. This causes serious changes in properties of the porous media such as chemical reactions, adsorption, specific capillary effects, etc. Thus, for the case of nanopores we need to take into account not only porosity and internal surface but also the total internal energy and energy distribution of nanopores in the considered structure.

Nevertheless, even nanoporosity is not the limiting size for the variety of properties of the internal volume of solid structures. In nanopores below 20 Å (2 nm) a very strange phenomenon occurs: their internal surface is not as important as their internal volume—the volume of very narrow nanopores, named *micropores*. The energy of micropores is so high that many measurable properties of solid materials are determined first by the energy of the microporous substructure. For example, it is well known that carbon is hydrophobic ("repulses" water). Active carbon has approximately the same chemical composition as ordinary carbon; hence, active carbon should also be hydrophobic, obviously. However, due to the high energy of micropores, active carbon has the ability to adsorb water—as if this material was hydrophilic ("attracting" water).

Thus, we obtain a paradox: measurable properties of many existing materials can be determined by micropores—invisible empty fragments of solid structures. Two solid samples having the same chemical composition may seem absolutely identical except for differences in their measured density, but differ in their properties—due only to micropores!

The exceptional properties of microporous materials determine their uses. Let us note some of them.

PROPERTIES AND USES OF PORES

Electrical and thermal insulation. Most existing microporous materials are very effective as electrical and thermal insulators. If necessary, their electrical and thermal resistance can be increased up to that of air, and even further, because air can eventually be ionized (which makes it electrically conductive) or have mechanical motion (which provides convective heat transfer).

Let us notice that in some cases we need conductive microporous materials, which is an unusual situation, but such materials can be synthesized (e.g., polymer electrolytes for lithium batteries).

Separation processes. All microporous materials can be used for certain processes of fluid mixture separation. Inside micropores there are two principal mechanisms allowing separation: adsorption and selective separation, both related to the exceptional properties of micropores.

Adsorption is a process of preferable fixation, in which gas or liquid molecules are captured by micropores; after the rest of the fluid is removed, the adsorbed substance can be delivered from the solid phase by heating and/or pressure decrease. For example, if we want to separate a water–oil mixture, it is enough to ensure its contact with active carbon; there the oil is adsorbed (you may be reminded that, as was mentioned earlier, water is also adsorbed; this is true, but if carbon micropores have a "choice," they preferably adsorb the organic phase), the water is separated, the carbon is heated, and oil purified from water is delivered.

A special sequence of the adsorptive properties of microporous materials is involved in their *catalytic activity*. Adsorbing some substances, the internal surface of microporous catalysts becomes the zone of complex chemical reactions. Such *heterogeneous catalysis* is the base of modern chemical technology. In several situations, adsorption of reagents in micropores may even change the result of the reaction.

Selective separation takes place in very special microporous structures (molecular sieves), the pore size of which is below 7 Å (so-called ultramicropores), in such materials as zeolites and some kinds of active carbon, silica, or alumina gel. The selection is due to several factors, including the effective size of the separated particles and their interactions with the material of the micropore walls. The simplest example is an organic polymer, which can be removed from water in a zeolite due to not only the small size of molecules of water but also because of the preferable interaction of water with zeolite, whereas macromolecules, so big for ultramicropores, are repulsed by the zeolite material. The selective properties of ultramicropores are the basis for membranes used for water purification and other separation processes.

Military uses. The opportunity to regulate the mechanical properties of solids offers some military applications. It is well known that multilayer protection of "steel + microporous ceramics + steel + microporous ceramics..." is more effective against explosions than simple steel of the same weight. That is due to the specific interactions of microporous structure with the explosion wave.

Sanitary and medical uses. The ability of microporous materials to selectively adsorb organics finds applications in health protection. For example, such microporous materials as active carbon and silica gel are used in gas masks for protection from poison and aggressive gases and vapors. The undesired gas is adsorbed by the microporous material and cannot harm anyone.

A similar application of the same materials is found in the pharmacy. Tablets of active carbon and/or silica gel adsorb nonappropriate products of metabolism in the human organism, which are then easily removed.

FORECASTING PROPERTIES OF MICROPOROUS STRUCTURES

Since micropores are so important for us, it seems very attractive to deduce all the characteristics related to microporosity from several predictable factors, such as conditions of material preparation (synthesis). This is the main purpose of this book: the investigation of the relationship between the conditions of synthesis of a porous (microporous) material, its structural parameters, and its measurable properties—properties such as adsorption, percolation, permeability, and mechanical resistance. Of course, the conditions of the synthesis determine the structure of the product (microporous material), whereas its structure determines all properties that could interest the researchers.

In the light of our main objective, we analyze the existing models of micropores and microporosity-related phenomena, particularly their usefulness for forecasting the technical merit of new materials—meaning materials just designed in laboratories.

METHODOLOGY OF MODELING MICROPOROSITY

Modeling of microporosity is extremely complicated because of the combination of quantum effects inside the micropores with the rich variety of shapes and large spectrum of characteristic sizes of micropores. Therefore, traditional methods of modeling the condensed phase are not applicable to microporosity. A possible solution is the use of modified statistical nonequilibrium thermodynamics. Such an approach allows the solution of some problems related to microporosity. Such a modified thermodynamic approach is applicable not only to modeling microporous structures but also to the simulation of very special phenomena, such as the beginning of the universe by the Big Bang and the beginning of life (this aspect is analyzed in Appendix 2).

STRUCTURE OF THE BOOK

The goals outlined above determine the structure of this book.

Chapter 1 presents general concepts regarding microporosity and the most important definitions. The classification of pores and micropores is given. Several methodological problems (classification of models of porosity, the similarity and divergences between *micropore* and *defect in crystal*, similarity and differences between adsorption in macro-, meso-, micropores and absorption, etc.) are analyzed. Most of the important concepts are defined and explained.

Chapter 2 focuses on the analysis of experimental methods for the study of microporous media: techniques used, problems solved, and errors of measurement and secondary problems that appear because of "intervention" in the microporous structure.

Chapter 3 presents the theoretical aspect of pore formation. Typical processes for porous material synthesis, together with their mathematical models, are considered and compared. The relationship between preparation conditions and the characteristics of the obtained microporous structure is always the central point. Such problems as "chemical" properties of pores, fractal formation, and branched structure formation are discussed. We analyze the specific features of the methodology of microporous media, in comparison with traditional methods of study of condensed matter.

Chapter 4 contains the analysis of properties of microporous structures: adsorption and desorption, percolation, permeability, and mechanical resistance. These properties are analyzed without specifying the structural parameters.

Chapter 5 discusses the existing models of microporous structure. Existing models are classified, compared, and analyzed. Most attention is paid to the fractal model, the thermodynamic model, and the polymeric model: their applicability and shortcomings, the opportunities they offer for combination, etc. The relationship of synthesis–structure–measurable properties is analyzed throughout.

Chapter 6 presents the engineering applications of models of microporosity. The theoretical results discussed in Chapters 3 to 5 are transformed to calculative techniques available for engineers, with energetic, technical, and economic estimations.

Chapter 7 offers the perspectives for the further development of the concept of microporosity, analyzing experimental research, and linking it to the relevant theoretical models. The opportunities for the development of theoretical models are considered. Some recommendations regarding investment policy in porous material studies are presented.

ACKNOWLEDGMENT

This book could not have been prepared without the technical help of the College of Management in Haifa, Israel (Haviv Grave, Academic Manager) and the Department of Chemical Engineering of Technion in Haifa, Israel (Professor Ishi Talmon, Dean).

Freddy Romm

Contents

1
Concepts and Definitions

I. CLASSIFICATION OF PORES

A. Methods of Classifications of Pores

The notion of *pore* usually means a void inside a solid structure, especially if this void is connected to the surface. Pores and related porous structures can be classified according to [1]

1. Their characteristic size
2. Their connections to the surface
3. Their regularity of randomness
4. Their homogeneity or heterogeneity

Now, let us consider all above classifications.

B. Classification by the Characteristic Size of Pores

According to IUPAC [2], pores are classified by their size. Pores having size over 500 Å (50 nm) are defined as *macropores*; pores with size from 20 to 500 Angstrom (Å), *mesopores*; and pores with size below 20 Å, *micropores* [2]. Soviet researcher Dubinin recommended subdividing micropores into two groups usually termed *ultramicropores* (below 7 Å) and supermicropores (from 7 to 20 Å) [2–5]. The reasonability of such subdivision is related to the fact that ultramicropores have so specific properties as, for example, selective separation of molecules, and the energy of ultramicropores (per unit of volume) is enormous [2,6].

The IUPAC classification of pores would be very simple and logical if we were always able to measure pores, but that is the problem. Even macropores may have such varied shapes that the correctness of their size measurement (e.g., by using microscope) is too doubtable. For mesopores and especially micropores, observation of which is impossible without very precious electronic microscopes, the same problem seems absolutely critical. Nevertheless, we can introduce the notion of *characteristic size* that allows us to apply the IUPAC's definitions with the least logical error.

Since the shapes of pores can be absolutely various, it is very difficult to talk about pore size, radius, etc. However, we may assume that each pore is more or less characterized by its maximal size r_{max} (maximal distance between the closest opposite walls in the pore), minimal size r_{min} (minimal distance between the closest opposite walls in the pore), and the distribution of the internal surface (or internal volume) on the variable value of size. Let us assume that if a pore has $r_{max} > 500$ Å and $r_{min} < 500$ Å, then it is not the entire pore but a combination of two (or more) pores. For example, a pore having $r_{max} = 1000$ Å and $r_{min} = 5$ Å can be divided into

Some macropores with $r_{max} = 1000$ Å and $r_{min} = 500$ Å
Some mesopores with $r_{max} = 500$ Å and $r_{min} = 20$ Å
Some supermicropores with $r_{max} = 20$ Å and $r_{min} = 7$ Å
Some ultramicropores with $r_{max} = 7$ Å and $r_{min} = 5$ Å

The above division allows us to consider separately all parts of the pore, applying the appropriate model to each of the parts.

Since the properties of macropores are determined by the internal surface, we define the characteristic size of macropores by the following equation:

$$r_c = \frac{\int_{r_{min}}^{r_{max}} r\rho(r)\, dr}{\int_{r_{min}}^{r_{max}} \rho(r)\, dr} \tag{1.1}$$

where the integral is taken over all the spectrum of countable values of r (from r_{min} to r_{max}) and $\rho(r)$ is the size distribution of the internal surface (surface area per volume unit for pore having the local size r).

Since the properties of mesopores are determined by both internal surface and internal volume, we define the characteristic size of mesopores by the following equation:

$$r_c = \sqrt{(r'r'')} \tag{1.2a}$$

$$r' = \frac{\int_{r_{\min}}^{r_{\max}} r v(r)\, dr}{\int_{r_{\min}}^{r_{\max}} v(r)\, dr} \tag{1.2b}$$

$$r'' = \frac{\int_{r_{\min}}^{r_{\max}} r \rho(r)\, dr}{\int_{r_{\min}}^{r_{\max}} \rho(r)\, dr} \tag{1.2c}$$

where $v(r)$ is the size distribution of the internal space (volume) in the mesopore (volume of pore having the local size r).

Since the properties of micropores are determined by their internal volume, we define the characteristic size of micropores by the following equation:

$$r_c = \frac{\int_{r_{\min}}^{r_{\max}} r v(r)\, dr}{\int_{r_{\min}}^{r_{\max}} v(r)\, dr} \tag{1.3}$$

Let us note that, according to the above definitions, the characteristic size of a mesopore is geometrically averaged between those of micropore and macropore.

Thus, the above definitions allow us to describe each pore with its characteristic size (or by the characteristic sizes of various parts of the pore).

C. Classification by the Connection to the Surface

According to this classification, a pore can be open, semiopen, or closed, depending on the connection to the surface of the porous sample (or different sides of the same sample).

An example of an open pore is given on Fig. 1.1.

As we can see from Fig. 1.1, the open pore is connected to both opposite sides of the solid sample. That means that this pore (this *open* pore)

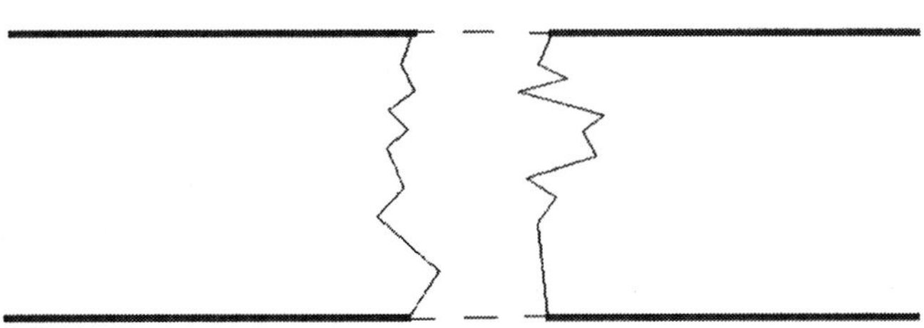

FIG. 1.1 Open pore in a solid sample.

can become a percolation path by which fluid fluxes can penetrate from the head side of the sample to its bottom side. If the open pore is ultra-micropore, it can realize the process of the selective separation.

Let us note that selective separation of components of fluids not only allows separation processes but sometimes makes possible carrying out of some reactions impossible in the same conditions without the separation. An example of such a reaction is the thermal decomposition of hydrogen sulfide with removal of hydrogen through microporous Vycor glass [7].

An example of a semiopen pore is given on Fig. 1.2.

As we can see from Fig. 1.2, the semiopen pore is connected to only one side of the solid sample. That means that this pore (this *semiopen* pore) cannot become a percolation path, but can participate in adsorptive processes. However, on the stage of desorption we may get a problem because the removal of the adsorbed substance from the semiopen pore needs more energy than in the case of an open pore. In such case, the pressure of the removal of the adsorbed substance from semiopen pore is over that of adsorption, and we obtain *hysteresis of adsorption–desorption loop*.

The problem of open and semiopen pores is related to the connectedness (also called *connectivity* or *interconnection*) of the porous structure. *Connectedness* means that the porous structure is continuous from one side of the porous solid sample to its opposite side. If the structure exhibits more tendency to connectedness, open pores predominate. On the contrary, in a porous structure with low connectedness open pores are not largely presented.

Closed pores have no connection to the pore surface. An example of a closed pore is given on Fig. 1.3.

As we can see from Fig. 1.3, the pore is isolated from both opposite sides of the solid sample. That means that this pore (this *closed* pore) cannot participate in mass transfer processes, comprising adsorption–desorption and percolation. However, the presence of closed pores reduces the density

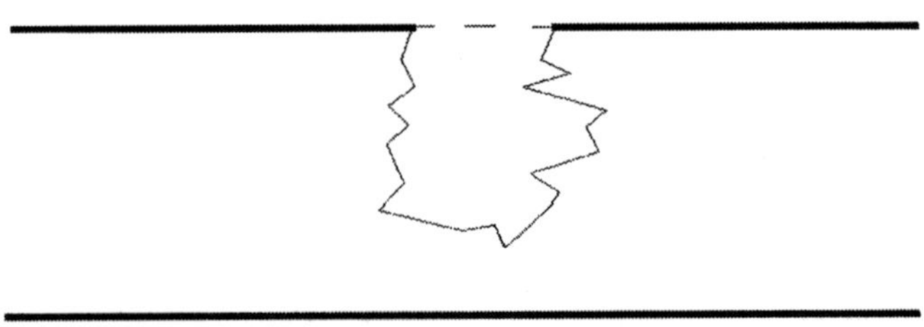

FIG. 1.2 Semiopen pore in a solid sample.

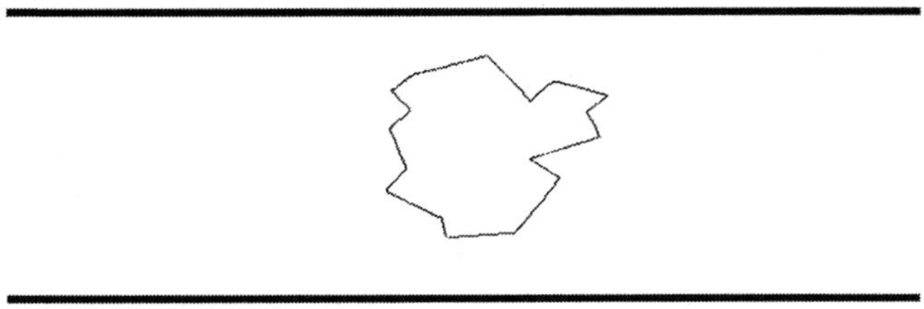

FIG. 1.3 Example of a closed pore inside a solid sample.

of the solid sample and increases its thermal, electrical, and acoustic insulation properties. Let us notice that closed pores (as all pores, of course) influence mechanical properties of the solid material, in most cases reducing the mechanical resistance of the solid sample.

As follows from the above, the classification by the connection to the surface is relative. At the same porous structure, the same porous fragments may provide open, semiopen, or closed pores depending on the manner of cutting the solid material into samples.

Let us note that we need pay more attention to the classification by the connection to the surface in the case (as we have considered above) if the same pore can be divided into macro-, meso-, and micropores. Such division, of course, cannot transform open pores into a combination of semiopen and closed pores. If a micropore is not connected to the surface but is connected to a semiopen or open meso- or macropore, that micropore can be, respectively, semiopen or open. Such a situation is largely spread in heterogeneous catalysts: macropores are connected to the surface (they are, respectively, open or semiopen) and allow the transportation of the reagent-containing fluids into micropores, while the micropores transport the reagents to the catalytic nuclei. When the initial reagents are consumed and the products formed, they are transported through micropores to macropores, through which the products are removed to the exterior surface of the catalyst.

The problem of the relativity of choosing the manner of cutting the porous solid sample will be discussed below when we will consider the problem of percolation and permeability (Chapter 4).

D. Classification by Regularity of Randomness

This classification is tightly related to the process of the synthesis of the porous material. A porous structure is called *regular* if it can be divided into a number of systematically repeating fragments. Two examples of regular porous structures are presented on Figs. 1.4 and 1.5.

FIG. 1.4 Porous structure with repeating fragment A: the location changes regularly, the shape does not change, and the size changes monotonically.

FIG. 1.5 Regular porous structure: repeating fragment "Aa" by location, shape, and size.

Based on Figs. 1.4 and 1.5, we can conclude that not only simple copying but also monotonic change in one of parameters (shape, location, and/or characteristic size) leads to the regularity in the porous structure.

Regular structures are not widespread among microporous materials. Most of these are randomly structured. A porous structure is called *random* if it is impossible to find a little (in comparison with the sample entire) fragment of the structure repeating systematically throughout the sample. An example of random porous structure is given on Fig. 1.6.

It is not a great fault to assume that only zeolites have regular porous structure. The problem of the formation of the regular structure of zeolite will be considered in Chapter 3.

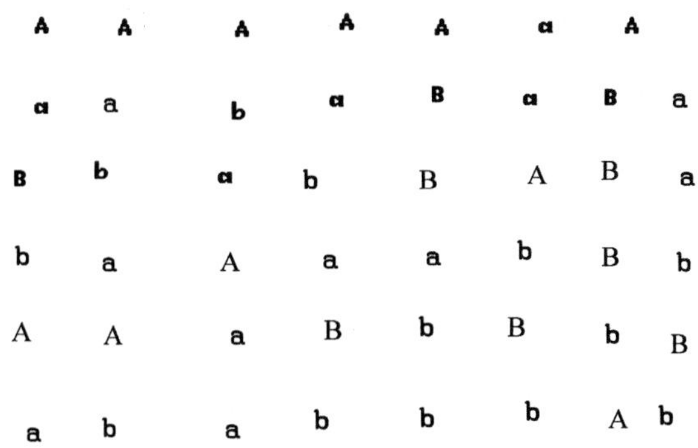

FIG. 1.6 Random porous structure: impossible to find any fragment repeating systematically throughout the structure.

E. Classification by Homogeneity or Heterogeneity

This classification is similar to that by regularity or randomness. We call a porous structure *homogeneous* if the averaged properties of all its fragments, sizes of which are much larger than those of pores, do not significantly differ. On the contrary, if such fragments differ significantly in properties (entropy per volume unit, internal energy per volume unit, chemical composition of the continuous phase, local porosity, size distribution of pores, etc.), the porous structure is called *inhomogeneous* or *heterogeneous*. We notice that the sense of homogeneity and heterogeneity is very relative, because pore itself is heterogeneous, since it contains interface continuous phase void.

Not every regular porous structure is homogeneous. For example, the structure presented on Fig. 1.5 is homogeneous, while that on Fig. 1.4 is heterogeneous. The random structure on Fig. 1.6 can be homogeneous because there is no obvious difference between various fragments, sizes of which are much larger than those of the pores.

II. MICROPORES AND DEFECTS IN CRYSTALS

Now, let us compare micropore to a very similar phenomenon, defect in crystalline structure [3,8–10]. Defect in crystal (dislocation) is understood as a void, eventual vanishing of a particle belonging to a crystalline structure from the spot at which the particle "should be found." Two examples of defects in crystals are given in Figs. 1.7 and 1.8.

Let us compare Fig. 1.7 to Fig. 1.2 and Fig. 1.8 to Fig. 1.3. The similarity between Figs. 1.7 and 1.8 and Figs. 1.2 and 1.3, respectively, is

```
A  X  A     A  X
X  A  X  A  X  A
A  X  A  X  A  X
X  A  X  A  X  A
A  X  A  X  A  X
X  A  X  A  X  A
```

FIG. 1.7 Defect on the surface of a crystalline structure.

```
A  X  A  X  A  X
X  A  X  A  X  A
A  X  A     A  X
X  A  X  A  X  A
A  X  A  X  A  X
X  A  X  A  X  A
```

FIG. 1.8 Defect in the volume of a crystalline structure.

obvious. In Figs. 1.2 and 1.7 the void is connected to the surface, whereas in Figs. 1.3 and 1.8 the void is located inside the solid sample. Is it so important whether the sample is crystalline or amorphous? Is it so important whether the void is big or small? Moreover, there are some specific crystals (zeolites) in which pores can give formation. The difference between micropore and crystalline defect may seem less important than the divergence between micropore and macropore.

The similarity between micropores and defects was discussed in the literature [11], and the physical difference between these phenomena needs serious analysis.

The comparative analysis (given below) aims to emphasize the physical difference between these two phenomena—micropores (comprising micropores in crystals) and defects in crystals.

1. The driving force for micropore formation and defect formation in crystal is the same, that is, the increase of entropy, whereas the total internal energy gets rising too. Hence, the thermodynamics of micropore and defect formation is very similar (but not identical—see below).
2. The mechanism of micropore and crystal defect formation is absolutely different. Micropores are formed due to removal of gas or vapor because of chemical decomposition of the material in which micropores are formed, and this process of gas–vapor particle removal is definitely irreversible. On the contrary, defects in crystals are formed due to statistical fluctuations, and the particles dislocated from their positions inside crystals do not get removed away but remain inside the crystalline structure or on the surface, and the process of defect formation is reversible. Thus, micropore formation is described by thermodynamics of non-equilibrium, while defect formation is described by thermodynamics of equilibrium.
3. Both micropores and defects can be formed in zeolites (regular crystalline structures formed by cocrystallization of some metal-oxide hydrates with further careful removal of water). However,
   a. Micropores in zeolites are formed due to removal of water (irreversible process, as analyzed above in item 2).
   b. Defects in the same zeolites are formed due to eventual removal or dislocations of ions belonging to the regular structure, and this process is reversible in principle (as above given in the general case by item 2).
4. Motion inside the solid structure is possible for particles in crystals but forbidden for particles around micropores. Micropores cannot move inside the solid (continuous) structure, unless this is destroyed. Defects can easily move, that is one of sequences of the reversibility of defect formation. Inside a crystalline structure,

TABLE 1.1 Comparison of Micropores with Crystal Defects

Factor of comparison	Micropore	Defect in crystal
Driving force for formation	Increase of entropy	Increase of entropy
Change of internal energy	Increase	Increase
Possible in nonzeolite crystals?	Negative	Positive
Possible in amorphous structures?	Positive	Negative
Possible in zeolites?	Positive	Positive
Formation is reversible?	Negative	Positive
Motion is possible?	Negative	Positive
Continuous growth is possible?	Positive	Negative

an ion can come into the vacant spot of the defect just due to heat vibrations; then the defect disappears at its previous position but appears at a new position.

5. Continuous growth of micropores is not only possible but very probable if the conditions necessary for the micropore formation take place. On the contrary, defects in crystals cannot get continuous growth (unless the crystal is destroyed), because the eventual neighborhood of two or more defects is the low-probability option.

The comparison of micropores to crystal defects is illustrated by Table 1.1.

III. RELATIONSHIP BETWEEN THE SYNTHESIS OF POROUS MATERIALS AND CHARACTERISTICS OF THEIR STRUCTURES

The following analysis regarding the relationship between the conditions of pore formation and the properties of the resulting porous structure is just qualitative. The numerical analysis is not given in this chapter but will be presented later.

Macropores are formed mostly because of mechanical treatment of solid materials, whereas micropores are formed due to chemical reactions in certain conditions. However, macropore formation due to mechanical effects is always accompanied by the eventual appearance of nanopores—mesopores and micropores in the zones where the mechanical destruction is minor. Micropore formation due to chemical reactions cannot, except few situations, be realized without eventual formation of macropores (see Fig. 1.9).

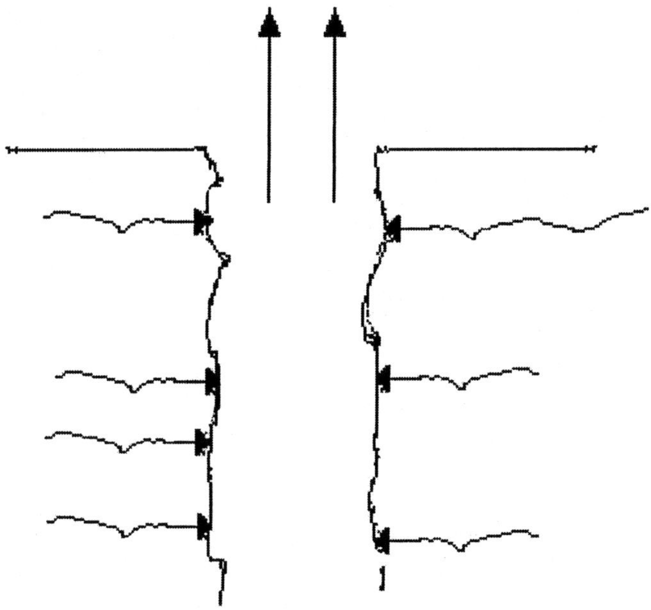

FIG. 1.9 Simultaneous formation of micropores and mesopores or macropores due to chemical reactions inside the solid phase.

Of course, both chemical reactions and mechanical effects cause also formation of mesopores.

All factors related to preparation (synthesis) of porous materials can be divided into two groups:

Characteristics of the raw materials
Parameters of the process of the treatment of the raw materials

We can divide all processes of pore formation in solids into two groups:

1. Pore formation in a solid material previously prepared (which means the solid is prepared *before* being treated for getting pores)
2. Pore formation simultaneously with the preparation of the solid material (which means the processes of synthesis of the solid and its treatment for pore formation are carried out in the same time, in the same volume)

For pore structure formation, one can use all existing amorphous solids. For pore formation simultaneously with the solid formation (synthesis), one can use all known physicochemical processes providing solid phase, such as separation of hydrocarbon mixtures, partial oxidation of hydrocarbons (in liquid or gaseous phase), polymerization, sedimentation, etc.

Among the properties of solid raw materials influencing pore formation we note the following:

1. Polymeric or nonpolymeric structure of the raw material (if polymeric, its structural characteristics comprising branching, cross-linking, etc.)
2. Mechanical properties of the raw material (its mechanical resistance, elasticity, etc.)
3. The content of volatile and unstable components (means first of all, organic groups)
4. The surface tension of the solid phase contacting with air

Polymeric structure of the raw (solid) material is favorable for the formation of micropores more than macropores under the same exterior conditions because natural interactions between neighbor momomeric units stimulate the formation of micropores. For the same reason, among polymeric materials, cross-linked structure with intense branching (first of all rubbers) is very favorable for micropores, but the total porosity decreases.

The high mechanical resistance and high elasticity allow very large freedom of choice for the researcher in varying exterior effects for getting a porous structure. Even if the micropores are formed by reactions inside the solid phase, the large elasticity allows the removal of volatile components with minimal formation of macropores. On the contrary, low elasticity and low mechanical resistance are factors very favorable for the local destruction of the solid phase and, as sequence, formation of macropores. In principle, this process may result even in breaking the solid phase.

The moderate content of volatile components is especially important for the intense micropore formation. If their concentration is low, micropores are rare, isolated, short, and semi open or closed. If volatile components are very abundant, the experiment needs to be very carefully controlled, in order to prevent too deep a decomposition of the solid (otherwise, the materials may be locally broken).

Let us notice that in processes of micropore preparation by decomposition of solid materials, the chemical composition and properties of the solid products may significantly differ from those of the raw materials. In such cases,

1. The kind of the resulting solid structure (polymeric or nonpolymeric) remains important.

2. Mechanical properties of the resulting solid are not so important, unless that is related to its further use.
3. The rest content of the volatile components is not important, unless needed for the further use.
4. That is very important to evaluate the surface tension of the resulting solid phase in contact with the gas or the vapor that is expected inside the micropores at the moment of their formation (σ). We note that *the measurements of* σ *do not need intervention inside micropores.* Instead, it is sufficient to measure the same parameter in standard laboratory conditions for the plate solid surface (having zero curvature) in the atmosphere of the same gas or vapor.

A. Parameters of the Treatment Process

Pore formation is influenced by the following characteristics of the treatment process:

All intensive parameters: pressure P (Pa), temperature T (K), mechanical tension P_m (Pa), concentrations of reagents C_i (mole parts), etc.
Gradients of the intensive parameters and changes of these parameters with time t (s)
Various initiating rays
The total duration of the treatment t_t

The pressure and/or tension loading, especially by vibration, stimulates not only formation of macropores but also transformation of micropores to macropores. If the vibration duration is too long, the solid sample is just destroyed.

The pressure unloading stimulates delivering of volatile components, particularly water vapor. The same effect is obtained if not the total pressure but the partial pressure of water vapor is reduced. In both cases, the obtained structure is mostly microporous, whereas the micropores are open or semiopen. The resulting structure cannot be regular unless it was initially, i.e., before the treatment (e.g., dehydration of zeolites). The porous structure is homogeneous if the treatment was careful and homogeneous: all fragments of the solid phase are under the same intensive parameters (of course, this condition does not work at the border of the condensed phase).

The temperature can influence pore formation in two directions:

1. Temperature rising causes thermal enlargement of the solid phase; if this is heterogeneous, its different parts get different

enlargements, that is, equivalent to mechanical tension, and, respectively, favors the formation of macropores.

2. Temperature rising may stimulate chemical reactions, first of all endothermic decomposition and oxidation of organic components, with delivery of volatile products (water vapor, carbon dioxide, nitrogen, etc.), that is favorable for micropore formation.

Thus, the sequences of thermal treatment of solid are ambivalent and depend on the structure and the chemical composition of the treated solid material.

As it will be shown in the further analysis, the dynamics of temperature changing with time influences significantly the resulting microporous structure, but that is a very specific phenomenon, analysis of which stands out of this chapter.

Initiating rays [ultraviolet (UV), x-rays, acoustic waves like ultrasound, etc.] may cause radical formation. The free radicals attack the unstable organic components in the solid phase, and the result is similar to that in the case of the thermal oxidation or decomposition. However, the ray-initiated chemical reactions have a very serious advantage: in the case of the ray treatment of a heterogeneous solid, macropore formation in which is not desired (e.g., oxidation of pollutants adsorbed on active carbon), macropores almost do not get be formed.

The total duration of the treatment, even without changing the regime of the treatment (e.g., long time heating of the raw material), leads to deep changes in the solid structures. In many cases, this can cause

1. Increase of the total porosity
2. Transformation of micropores and mesopores to macropores
3. Breaking of the solid

The principal sequences of the treatment of organic-containing solids with the above-considered technique are given in Tables 1.2 to 1.5.

Comparing Tables 1.2 and 1.3 to Tables 1.4 and 1.5, respectively, we find that the results of the treatment of similar (in the chemical composition) materials are very similar but not identical, and the initial structure of the raw solid material is extremely important.

B. Effect of Exterior Treatment on the Pore Formation in Processes of Synthesis of Solid

We mentioned above the following processes most important for the synthesis of porous solid from gaseous or liquid raw materials: separation of hydrocarbon mixtures, partial oxidation of hydrocarbons (in liquid or

TABLE 1.2 Structure Formation in Process of the Treatment of Carbon

Characteristics of the resulting structure	Pressure, mechanical tension	Temperature increase	Destroying agents plus initiating rays
Micropore formation	Negligible[e]	Yes[a]	Yes
Mesopore formation	Minor[d]	Yes	Yes
Macropore formation	Very possible[b]	Yes	Minor
Cracking or breaking of the solid	Very possible	Possible[c]	No
Open pores	Yes	Possible	Yes
Semiopen pores	Yes	Yes	Yes
Closed pores	No[f]	Possible	No
Random porous structure	Yes	Yes	Very possible
Regular porous structure	No	No	Minor
Heterogeneous structure	Possible	Very possible	Very possible
Homogeneous structure	Possible	Minor	Minor

[a]Yes means the effect is doubtless.
[b]Very possible means the effect is very probable.
[c]Possible means the probability of the effect is moderate.
[d]Minor means the probability of the effect is low.
[e]Negligible means the probability of the effect is very low.
[f]No means the effect should not appear.

gaseous phase), polymerization and, sedimentation. Let us give some examples of these processes.

1. Separation of Hydrocarbons

Such a known and widespread process as petroleum refining comprises, as one of its sequences, delivering a very hard fraction of hydrocarbons—bitum used as the raw material for asphalt. This consists of a mixture of hard hydrocarbons forming a porous solid phase. The usual synthesis of such a porous product comprises the following stages: condensation, separation from light and intermediary fractions, cooling and keeping as solid, then (for use as asphalt) heated, put onto the chosen surface, and naturally cooled. Asphalt contains all kinds of pores. Many pores are open; this fact makes the removal of water during rains easier.

2. Oxidation of Hydrocarbons and Organics

This technique is used for the synthesis of active carbon. Primary materials (waste organics) are carefully oxidized under high temperature, which causes formation of all kinds of pores, while micropores dominate.

TABLE 1.3 Structure Formation in Process of the Treatment of Polymeric Solid

Characteristics of the resulting structure	Pressure, mechanical tension	Temperature increase	Destroying agents plus initiating rays
Micropore formation	Minor[d]	Yes[a]	Yes
Mesopore formation	Minor	Yes	Yes
Macropore formation	Very possible[b]	Possible[c]	No[e]
Cracking or breaking of the solid	Very possible	No	No
Open pores	Yes	Very possible	Yes
Semiopen pores	Yes	Yes	Yes
Closed pores	No	Possible	No
Random porous structure	Yes	Yes	Very possible
Regular porous structure	No	No	Minor
Heterogeneous structure	Possible	Very possible	Very possible
Homogeneous structure	Possible	Minor	Minor

[a]Yes means the effect is doubtless.
[b]Very possible means the effect is very probable.
[c]Possible means the probability of the effect is moderate.
[d]Minor means the probability of the effect is low.
[e]No means the effect should not appear.

In the general case, the obtained structure is random. However, several studies show the principal opportunity of the synthesis of active carbon, the structure of which is characterized by several regularities (see, e.g., Ref. 12). Such molecular sieving carbons with uniform micropore structure are prepared by (1) carbonization of metal-cation-exchanged resin and (2) pyrolytic carbon deposition from benzene over activated carbon fibers. Both micro- and mesoporosity are under control. High-performance molecular sieving carbon membranes for gas separation are synthesized from the carbonization of several types of organic polymer films. Mesopores are controlled by such novel methods as catalytic activation, polymer blend carbonization, organic gel carbonization, and template carbonization [12].

3. Polymerization

Most of reactions of polymerization (radical or ionic) lead to the formation of microporous solids. Macropores appear mostly in specific processes—e.g., foam synthesis with gas injection into the polymerizing phase—and are always accompanied by micropore formation. Macropores and mesopores take much less volume fraction. The simplest linear polymerization is

TABLE 1.4 Structure Formation in Process of the Treatment of Zeolites

Characteristics of the resulting structure	Pressure, mechanical tension	Temperature increase	Reduction of partial pressure of water vapor
Micropore formation	Negligible[e]	Yes[a]	Yes
Mesopore formation	Minor[d]	Yes	Yes
Macropore formation	Minor	Minor	Minor
Cracking or breaking of the solid	Yes	Yes	No
Open pores	No[f]	Minor	Possible[c]
Semi-open pores	Minor	Yes	Yes
Closed pores	No	Possible	No
Random porous structure	Yes	Yes	Minor
Regular porous structure	No	No	Yes
Heterogeneous structure	Possible	Very possible[b]	Minor
Homogeneous structure	Possible	Minor	Yes

[a]Yes means the effect is doubtless.
[b]Very possible means the effect is very probable.
[c]Possible means the probability of the effect is moderate.
[d]Minor means the probability of the effect is low.
[e]Negligible means the probability of the effect is very low.
[f]No means the effect should not appear.

favorable for the formation of open and semiopen micropores and (much less) open mesopores. Branched polymerization increases the fraction of micropores, while the tendency of the formation of semiopen pores dominates. Cross-linking in macromolecules reduces the porosity and stimulates closing pores. In any case, the product has random porous structure. In a normal situation, the polymer is homogeneous, but if necessary, the polymerization can provide heterogeneous products (e.g., due to using block polymers or composites).

4. Sedimentation

Processes of sedimentation can be carried out in very different conditions. If solid particles are allowed to move down due to simple gravitation, the obtained solid structure is microporous. In many cases, the products of sedimentation (e.g., silica gel or alumina gel) get secondary treatment, mainly thermal. The changes of the porous structure during the secondary treatment do not differ from those described in Tables 1.2 to 1.5. If the rate of sedimentation is increased, e.g., by centrifugation, the porosity decreases, the characteristic size of micropores decreases, and the obtained pores are semiopen or closed. The resulting porous structure is random and is not

TABLE 1.5 Structure Formation in Process of the Treatment of Silica–Alumina Gel

Characteristics of the resulting structure	Pressure, mechanical tension	Temperature increase	Reduction of partial pressure of water vapor
Micropore formation	Negligible[e]	Yes[a]	Yes
Mesopore formation	Minor[d]	Yes	Yes
Macropore formation	Minor	Possible[c]	Minor
Cracking or breaking of the solid	Yes	Minor	No[f]
Open pores	No	Very possible[b]	Possible
Semiopen pores	Minor	Yes	Yes
Closed pores	No	Yes	No
Random porous structure	Yes	Yes	Minor
Regular porous structure	No	No	Yes
Heterogeneous structure	Possible	Very possible	Minor
Homogeneous structure	Possible	Minor	Yes

[a]Yes means the effect is doubtless.
[b]Very possible means the effect is very probable.
[c]Possible means the probability of the effect is moderate.
[d]Minor means the probability of the effect is low.
[e]Negligible means the probability of the effect is very low.
[f]No means the effect should not appear.

homogeneous but has the cone symmetry: heterogeneity is found by the vertical (because of the gravitation), whereas by the horizontal the structure is homogeneous (see Fig. 1.10).

A specific situation takes place when sedimentation is not carried out by simple gravitation but is realized in the regime of crystallization. In most of cases, such situation does not regard pore formation. However, if the obtained crystals contain water, this can be sometimes carefully removed by increasing temperature and/or reducing the partial pressure of water vapor, obtaining molecular sieves, zeolites.

We note that since micropores are formed mainly due to chemical reactions, the notion of *synthesis* seems well applicable to micropore formation, though that is not synthesis in the traditional sense (as is understood in inorganic and organic chemistry). This term allows us to emphasize the physical difference between micropore formation (due to chemical reactions) and macropore formation (mostly due to mechanical effects). Moreover, in the situation when micropores are formed simultaneously with the solid phase, the notion of *synthesis of micropores* becomes logically valid even in the traditional sense.

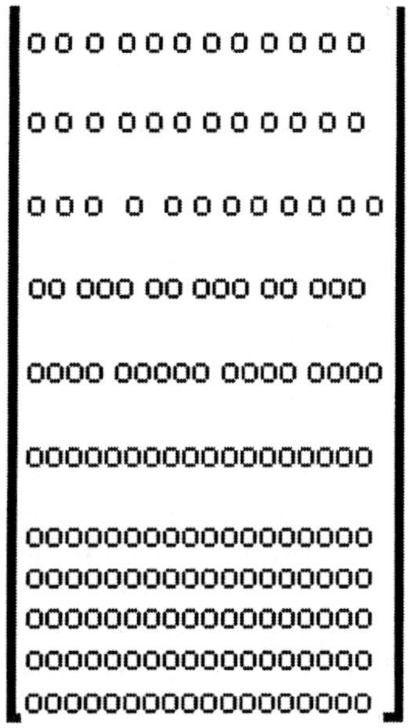

FIG. 1.10 Horizontal homogeneity and vertical heterogeneity in porous structure after simple sedimentation.

C. Influence of the Treatment on the Homogeneity and Heterogeneity of Porous Structure

Since the porous structure is determined by both raw material and the regime of its treatment, homogeneous porous structure can be obtained in one of the following situations:

1. The raw material is initially homogeneous and gets the homogeneous treatment. The homogeneity of the treatment means that all fragments of the raw material get mechanical, chemical, and/or thermal treatment and/or initiating irradiation at the same intensive parameters (pressure, temperature, concentrations of chemical reagents, wavelength) at the same duration. An example of such a process is transformation of alumosilicate hydrates to zeolites.
2. The raw material is initially heterogeneous, but its treatment is homogeneous and so intense that the initial heterogeneity is neutralized. An example of such a process is the recovery of active carbon from the products of the adsorption of organic pollutants with active carbon (the organic pollutants are oxidized and not only deliver the active carbon but, moreover, form some additional amount of active carbon). Though such raw material

was initially heterogeneous, the careful treatment allows its transformation to homogeneous, pure active carbon.

D. Open, Semiopen, and Closed Pores

The pore opening and closing are very specific characteristics that do not significantly depend on the raw material. Their dependence on the parameters of the treatment process is significant but not decisive. Much more important for pore opening and closing is the manner of the sample preparation. If the ready porous material is cut to thin films (membranes), these contain mostly open pores. Such a procedure can be especially important in the preparation of microporous membranes of zeolites, several polymers, and composites or silica gel: most of their micropores get opening. That transforms these films, due to selective properties of ultramicropores, to a powerful technique for precious separation processes (of course, in the case of zeolites the micropores remain regular, while in other mentioned materials they remain random). However, since opening pores is very significant for heat or mass transfer through porous structures, it is very commendable to pay attention to some technologies allowing preparation of microporous structures with high tendency to opening. Let us give some examples of such processes.

(a) Example 1.1. Microporous ceramic materials having an open microporous cell structure with an internal surface area $>50\,m^2/g$ (microporous structure comprises a volume fraction above $0.015\,cm^3/g$ of the ceramic) can be synthesized on the base of nonsilicon containing ceramics, carbon, and inorganic compounds having a decomposition temperature over 400°C [13].

(b) Example 1.2. Microporous ceramic materials having an open microporous cell structure with an internal surface area $>70\,m^2/g$ (microporous structure comprises a volume fraction above $0.03\,cm^3/g$ of the ceramic) can be synthesized on the base of silicon carbide, silicon nitride, and silicon–carbide–nitride ceramics [13].

(c) Example 1.3. A porous material having open pores can be prepared by casting a slurry from a mixture containing a bisphenol-type epoxy resin, an amide compound as a hardener, a filler, and water in a water-impermeable mold, hardening the slurry while it contains the water, further dehydrating the hardened article. The mean pore size estimated 4.0 nm, and the apparent porosity can be up to 41% [14].

(d) Example 1.4. An aluminum-oxide-based ceramic catalyst support with open interconnecting pores can be prepared by mixing fine-grained

aluminum oxide with bonding clay, organic binding agents, and water to form a suspension. This is added to a polystyrene framework formed by treating polystyrene balls with aqueous acetone. The aluminum oxide and the supporting polystyrene framework are dried, the polystyrene is removed by heating, and the aluminum oxide support containing open interconnecting pores is sintered at 1600°C for 2 h. The aluminum oxide support is used with a nickel-based catalyst [15].

(e) Example 1.5. Chen et al. [16] noticed that, in many applications, usefulness of conventional hydrogels was limited by their slow swelling. To improve the swelling property of the conventional hydrogels, Chen et al. [16] have synthesized superporous hydrogels (SPHs) that swell to equilibrium size in minutes due to water uptake by capillary wetting through numerous interconnected open pores. The swelling ratio was also large in the range of hundreds. The mechanical strength of the highly swollen SPHs was increased by adding a composite material during the synthesis. The composite material used in the synthesis of SPH composites was Ac-Di-Sol(R) (croscarmellose sodium). Their study indicated that SPH composites possessed three required properties: fast swelling, superswelling, and high mechanical strength [16].

The results of the analysis of the relationship between the characteristics of the simultaneous synthesis of microporous structure with the solid phase itself and the characteristics of the porous structure are given in Table 1.6.

TABLE 1.6 Influence of the Parameters of the Process of Synthesis of Porous Solid onto the Characteristics of the Porous Phase[a]

Technique of treatment	Main pore size	Opening and closing pores	Homogeneous or heterogeneous[b]
Polymerization:			
Linear	Micropores, mesopores	Open	Homogeneous
Branched	Micropores	Open, semiopen	Homogeneous
Cross-linked	Micropores	Semiopen, closed	Homogeneous
Oxidation of organics	Micropores, mesopores	Open, semiopen	Homogeneous
Sedimentation (by gravitation)	Micropores	Open, semiopen	Heterogeneous (cone symmetry)

[a]Note: the column regularity or randomness is not included into the table because *all* known structures prepared simultaneously (solid + pores) are random only.
[b]Assume the condition that the synthesis process is homogeneous in the sense claimed above.

IV. RELATIONSHIP BETWEEN THE POROUS STRUCTURE OF A SOLID MATERIAL AND ITS PROPERTIES HAVING TECHNICAL MERIT

As in the previous article, the below analysis of the relationship between the structure of a porous material and its measurable properties is just qualitative. The mathematical models and the numerical studies of microporosity will be presented in Chapters 3 to 5.

Among numerous interesting properties of porous materials, in the further analysis we will deal with

1. Specific properties of porous materials that are not found for nonporous materials
2. Specific properties of porous materials finding technical and/or domestic applications

In the Preface to this book, we gave a list of technically important specific properties of porous materials: heat, electrical, or acoustic isolation, adsorption of various substances (and, as sequences, catalytic properties and such uses as fluid component separation and poison removal), percolation and permeability (as their sequence, fine selective separation of components of fluids), and regulated mechanical properties.

Now, let us consider the influence of various qualitative characteristics of porous structure on the above measurable properties.

A. Insulation Properties

The large size of a pore seems very favorable for the insulation properties of a porous material. However, the most important is the total porosity (of course, many little pores over the volume of the solid sample provide much better insulation than one big pore, the volume of which is less than the total volume of the mentioned little pores). At the same value of porosity, better insulation is found for microporous materials, because micropores constrain mass transfer and heat transfer, respectively, by convection. Open pores do not allow enough good insulation, because allow convective heat transfer [17]. Open pores allow also percolation [18] that harms the acoustic and electric insulation. Semiopen pores cancel percolation and seriously reduce convective heat transfer. However, only closed pores make convective heat transfer negligible (which can exist just because of the motion of the interior gas inside the closed pores). Moreover, in a closed micropore such motion is practically impossible, and the closed microporous structure is the best insulator in comparison with alternative structures. Regular and random porous structures do not significantly differ in the aspect of insulation properties, and their divergence can be estimated only on the stage of mathematical models. Heterogeneous structures provide weaker insulation because the limiting rate of heat or mass transfer is determined by the maximal available local rate, which is larger on the zones of low local porosity, whereas the homogeneous structures have equal local thermal, mass transfer, and electrical resistance.

B. Adsorption and Heterogeneous Catalysis

Adsorption is understood as "capturing" several substances (component of gas or liquid phase) by solid surface (exterior or interior), mostly micropores. Let us compare adsorption to absorption—capturing of substance by liquid or (very rarely) solid volume. We notify that since adsorption in micropores is due to their active volume, microporous adsorption has some features of absorption (see Table 1.7).

We can deduce that adsorption by micropores can be considered as a phenomenon intermediary between the traditional adsorption and absorption.

Let us note that *desorption* is usually understood as the process reverse to adsorption; however, to avoid misunderstanding, one needs to remember that *desorption* means also the process reverse to absorption. Thus, *desorption* can be defined as the process of delivering of captured substance from the condensed phase, but the exact sense of desorption becomes clear only after determining the mechanism of the process precedent to desorption—absorption or desorption.

Now, let us consider some examples of adsorption and heterogeneous catalysis on porous adsorbents.

(a) Example 1.6. The adsorption of mainly butyl and heptyl alcohols from water solutions by means of six different charcoals ranging widely in pore size was investigated by Dzhigit et al. [19] to show the effect of the pore structure onto the limited adsorption of alcohols of limited solubility in water. It was shown that the charcoal structure was having a strong effect on the shape of the adsorption isotherms and on the values of the limit adsorption. The rule of constant adsorbed limit volumes was found approximately true for each of the carbons adsorbing one of four alcohols (from butyl to heptyl). It was notified that the micropores of the charcoals are equally accessible to the different alcohols and densely filled by the alcohol molecules at limit adsorption [19].

(b) Example 1.7 (water purification). Active carbons for water treatment can be obtained by steam activation of olive-waste cakes. This raw material is an abundant and cheap waste by-product of oil production that

TABLE 1.7 Comparison of Adsorption in Micropores to Adsorption in Mesopores and Macropores and Absorption

Factor of comparison	Adsorption on plate surface or macropore	Adsorption in mesopores	Adsorption in micropores	Absorption in liquid	Absorption in solid[a]
Driving force of the process	Decrease of internal energy	Decrease of internal energy	Decrease of internal energy	Decrease of internal energy	Decrease of internal energy
Sorption by surface	Yes	Yes	No	No	No
Sorption by volume	No	Yes	Yes	Yes	Yes
Capturing by solid phase	Not relevant	Not relevant	Not relevant	No	Yes
Capturing by liquid phase	Not relevant	Not relevant	Not relevant	Yes	No
Capturing by continuous phase	No	No	No	Yes	Yes
Transport of captured substance through all the continuous phase	No	No	No	Yes	Yes

[a]An example of absorption by solid: absorption of hydrogen by platinum.

makes the obtained activated carbons economically feasible. These activated carbons from olive-waste cakes showed a high capacity to adsorb herbicides (2,4-dichlorophenoxyacetic acid, 2,4-D- and 2-methyl-4-chlorophenoxyacetic acid, MCPA) from water, with adsorption capacity values higher than those corresponding to a commercial activated carbon used for treating drinking water [20].

(c) Example 1.8 (technology of separation of organics). Separation of paraxylene from metaxylene and/or ethylbenzene by adsorption [21], in the technical aspect, is preceded by the process of zeolite preparation. Adsorbents like X and Y zeolites are exchanged with potassium and barium cations and associated with an adequate solvent. After this procedure, the zeolites are efficient for paraxylene separation by adsorption. Paraxylene selectivity can be observed only at high loading and for bulky and weakly charged cations. Entropy effects allow paraxylene to be more efficiently packed in the zeolite micropores. Due to this fact, the value of paraxylene–metaxylene selectivity is over 3, while paraxylene–ethylbenzene over 2, which is enough for the effective separation [21].

(d) Example 1.9 (technology of catalytic synthesis). This example demonstrates the synthesis of fine and intermediate chemicals on immobilized catalysts [22]. Homogeneous catalysts (e.g., homogeneous transition metal complexes supported on zeolites with mesopores completely surrounded by micropores, homogeneous transition metal complexes on mesoporous MCM-41, and ionic liquids on various carriers) can be transformed to recyclable heterogeneous catalysts. The catalysts are very effective in reactions such as oxidation, hydrogenation, and carbon–carbon bond formation [22].

(e) Example 1.10 (technology of catalytic synthesis). Multiuse selective catalysts–amorphous microporous mixed oxides (AMM) can be described as catalysts prepared to contain isolated catalytically active centers in the shape selective environment of micropores. The lack of inherent limitations with respect to chemical composition and pore size makes AMM materials versatile catalysts for many uses. AMM can be effectively used in such processes as hydrocracking, redox catalysis with organic hydroperoxides, redox catalysis with hydrogen peroxide, acid-catalyzed etherification and esterification reactions, oxidative dimerization of propene with air, selective oxidation of toluene to benzaldehyde with air, and selective aromatic alkylation. AMM can also work in microporous catalyst membranes [23].

(f) Example 1.11 (technology of organic synthesis). Catalytic versions of the Baeyer–Villiger oxidation are particularly attractive for practical

applications because catalytic transformations simplify processing conditions while minimizing reactant use as well as waste production. Corma et al. [24] evolved the Baeyer–Villiger oxidation into a versatile reaction widely used to convert ketones—readily available building blocks in organic chemistry—into more complex and valuable esters and lactones. These authors [24] expected some benefits from replacing peracids, the traditionally used oxidant, by cheaper and less polluting hydrogen peroxide. Dissolved platinum complexes and solid acids, such as zeolites or sulphonated resins, efficiently activated ketone oxidation by hydrogen peroxide. However, it was noted in Ref. 24 that these catalysts lack sufficient selectivity for the desired product if the starting material contains functional groups other than the ketone group; they perform especially poorly in the presence of carbon–carbon double bonds. It was shown in Ref. 24 that upon incorporation of 1.6 wt% tin into its framework, zeolite beta acts as an efficient and stable heterogeneous catalyst for the Baeyer–Villiger oxidation of saturated as well as unsaturated ketones by hydrogen peroxide, with the desired lactones forming more than 98% of the reaction products. This high selectivity was ascribed to the direct activation of the ketone group, whereas other catalysts first activate hydrogen peroxide, which can then interact with the ketone group as well as other functional groups [24].

(g) Example 1.12 (technology of inorganic synthesis). Classic three-phase catalytic cocurrent down-flow trickle bed reactor was compared by Medeiros et al. [25], under the same operating conditions, with a three-phase cocurrent down-flow vector named *Verlifix*. This reactor is an association of a Venturi device set above a fixed bed. The Venturi device allows a good gas–liquid mass transfer and a homogeneous distribution of the gas and liquid flow. The chemical reaction studied, in both reactors, is the catalytic oxidation of sulfur dioxide on active carbon particles. All experiments have been carried out at 25°C. It was found in Ref. 25 that the presence of a Venturi distributor increased the gas–liquid mass transfer, the wetting efficiency of the catalyst and improve the conversion of sulfur dioxide. The Verlifix reactor is particularly adapted for heterogeneous reactions limited by gas–liquid and solid–liquid mass transfer [25].

(h) Example 1.13 (specific reactions impossible in ordinary conditions and allowed due to adsorption in micropores). Some effects such as phase change and phase growth in porous media [26] occur in micropores only due to their excess energy, such as the confinement in phase transitions in micropores, the growth and dissolution of gas and liquid phases. Applications of these effects range from capillary condensation to drying to gas evolution to condensation.

Due to higher internal energy, micropores allow the best characteristics of adsorptive processes. However, if the adsorbent is regenerated by desorption, this process is seriously hindered in micropores, and the resulting energetic efficiency of the cycle adsorption–desorption, to be compared to other separation techniques, needs numerical evaluations. On the other hand, micropores have the definite advantage in the amount of the adsorbed fluid per mass unit of the adsorbent. Microporous adsorbents are absolutely preferable if the adsorbed substance is not desorbed but decomposed or oxidized (as was considered above, the oxidation of the adsorbed organic pollutants on active carbon not only regenerates the active carbon but even increases its amount, due to the products of the oxidation).

Adsorption is realized mostly due to open pores, much less to semiopen. Closed pores do no participates in the adsorption of all. The regular structure may have a significant advantage in comparison with the random structure: pores are optimally organized and do not hinder one other in the process of adsorption. That is one of the reasons why the selective separation by zeolites is preferable.

Heterogeneous structures have several disadvantages in comparison with homogeneous structures, because of different local partial pressure (concentration) of the adsorbed substance against various parts of the heterogeneous adsorbent [27,28]. For the same reason, desorption processes are better organized on homogeneous adsorbents.

C. Percolation and Permeability

Percolation and permeability are found only in open pores. The value of permeability is larger in the case of macropores, just for geometrical reasons. However, only micropores allow the selective separation of components in fluid flows [29,30].

The fact of percolation does not depend on the regularity or the randomness of a porous structure, however, if the percolation is found, the regular structure has more permeability—again, because its pores are "optimally organized."

Heterogeneity is favorable for increasing permeability—for the same reason that heterogeneity is not favorable for the insulation.

(a) Example 1.14 (selectivity of ultramicropores and high-selectivity membranes). The selectivity of micropores and ion channels was examined for simple pore topologies within the framework of density function theory of highly confined fluids [29]. It was shown that in an infinite cylindrical pore purely steric (excluded volume) effects lead to strong, nontrivial size selectivity, which is highly sensitive to the pore radius [29].

(b) Example 1.15. In the estimation of Canan [31], vapor-phase sorption is the most influential process governing the transport and the fate of volatile organic compounds in soil. The single-pellet moment technique was used to investigate sorption and diffusion of trichloromethane (TCM) and carbon tetrachloride (CTC) at varying relative humidities (0–80%) of synthetic humic–clay complex pellets consisting of clay (montmorillonite) and different amounts of organic matter (humic acid). The effective diffusivities of TCM and CTC did not show a noticeable change with moisture and humic acid content. On the other hand, with increasing humic acid content of clay at zero relative humidity, an appreciable decrease of the equilibrium sorption constants of the tracers (TCM, CTC) was found because of the blockage of some sites of the mineral surfaces and especially micropores by the humic acid. The presence of water also reduced dramatically the sorption of TCM and CTC on synthetic humic–clay complexes. Above 20% relative humidity, the sorption coefficient of TCM and CTC varied only slightly with humic acid content. It was concluded that the sorption of TCM and CTC in synthetic humic–clay complexes was strongly effected by the moisture and humic acid content [31].

D. Mechanical Properties

For most known materials, microporosity and homogeneity are favorable for better mechanical resistance and elasticity. Opening and closing pores does not influence the mechanical characteristics. Regular materials, as in many similar situations, have better mechanical properties than randomized ones—for the reason of "optimal organization" of the structure.

The results of the analysis of the relationship between the structural characteristics and the measurable properties of porous materials are given in Table 1.8.

The structure of pores is very difficult to control; therefore, we are much more interested in the relationship between the conditions of preparation of a porous material and its measurable properties. An example of such a relationship is given by Table 1.9.

V. ENERGY DISTRIBUTION OF PORES

Above, presenting the classification of pores, we mentioned the classification by the characteristic size and noticed that micropores are characterized by the energy much better than by the size. The main reason for the micropore pore classification by size is the tradition of IUPAC but not physical considerations. In Chapters 3 through 5, presenting mathematical

TABLE 1.8 Relationship Between the Structural Characteristics and the Measurable Properties of Porous Materials

Structural factor	Insulation	Adsorption	Percolation or permeability	Selective separation	Mechanical resistance or elasticity
Size of pores	Micropores are preferable	Micropores are preferable	Macropores are preferable	Micropores required	Micropores are preferable
Opening and closing of pores	Closing is preferable	Opening required	Opening required	Opening required	Indifferent
Randomness or regularity	Randomness is preferable	Indifferent	Regularity is preferable	Regularity is desired	Regularity is preferable
Homogeneity or heterogeneity	Homogeneity is preferable	Homogeneity is preferable	Heterogeneity is preferable	Homogeneity is preferable	Homogeneity is preferable

TABLE 1.9 Conditions of Preparation of Porous Materials, Allowing the Obtainment of Several Needed Properties

Property desired	Needed regime of treatment of solid raw material[a]	Needed regime for the simultaneous synthesis of pores with the solid[a]
Insulation	Careful oxidation of organic solid	Polymerization with intense cross-linking and moderate oxidation
Adsorption	Moderate oxidation of organic solid	Thermal polymerization without cross-linking
Percolation or permeability	Deep thermal oxidation of organic solid	Moderate-degree polymerization, no cross-linking
Selective separation	Dehydration of zeolites	Special careful oxidation of organics
Mechanical resistance and elasticity	Moderate chemical oxidation of organic solid	Polymerization with careful cross-linking

[a]The table presents not all available solutions but some examples only.

models related to various aspects of microporosity, we will use not characteristic size but energy.

Energy of macropores (per internal volume unit) is very low; the energy of mesopores (per internal surface or volume unit) is moderate; the energy of micropores (per volume unit) is high or very high. In principle, the characteristic energy per volume unit could be the criterion for the classification of pores, instead the characteristic size.

However, since micropores work in assemblages, their characterization should be collective. For this aim, we use the function of *energy distribution of micropores* or *micropore distribution in energy* $F(\varepsilon)$ (more exactly, free energy by Gibbs). Here, F has the statistical sense and presents the volume fraction of micropores, energy of which is from ε to $\varepsilon + d\varepsilon$, where $d\varepsilon$ is a very small value. For statistical reasons,

$$\int F(\varepsilon) = 1 \tag{1.4}$$

where the integration is taken on all the spectrum of values of ε. The energy distribution is related to the differential enthalpy, differential entropy, and immersion enthalpy via traditional thermodynamic relationships. For macropores and mesopores, one can find analytical relationship between energy and size distributions.

For some systems, one can evaluate $F(\varepsilon)$ from quantum considerations [32].

The energy distribution of micropores in a material can be estimated on one of the following ways:

Assumption of energy distribution
Derivation of the energy distribution from certain model considerations
Indirect estimation of the energy distribution from experimental data

Experimental methods for studies of micropores, particularly for the estimation of their energy distribution, will be considered in Chapter 2.

VI. FRACTALS IN POROUS STRUCTURES

Fractal is defined as a structure in which each fragment repeats the same configuration, and the entire structure has the same configuration (see Fig. 1.11).

Obviously, all regular fractals are regular structures. However, not all regular structures are fractals. All crystals are regular fractals.

Fractals on Fig. 1.11 have simple configurations. Of course, if the configuration is more complex, that does not change the sense of the phenomenon of fractal. One may ask whether the configuration can be random? The answer is positive. Such fractals, fragments of which randomly repeat on different scales, are defined as *random fractals*.

The first observation of random fractal was related to maps of the British coast line: its configurations in various scales were found very similar.

A	BBBBBBB
AA	BB BB
AAA	BB BB
AAAA	BBBBBBB
AA AA	BB BB
AA AA	BB BB
AA AA	BBBBBBBB

FIG. 1.11 Regular fractals: configurations A and B, respectively.

Random fractals correspond only to homogeneous random structures. Random fractals are observed for many known macro- and mesoporous structures. What about microporous structures? As it was mentioned above, their direct observations are impossible. Moreover, the scale of below 1 nm makes the existence of fractals theoretically impossible. However—it is much more important—*properties of homogeneous random micropororosity are well described by the random fractal approach.*

The theoretical and numerical aspects of random fractals will be analyzed in Chapters 3 and 5.

The priority of porous structures in the fractal concept is absolutely due to Benoit Mandelbrot.

In the theory of porosity, the fractal approach is largely used for modeling formation and structure of microporous media, comprising such complicated processes as microporous catalyst synthesis. One of methodological advantages of the fractal approach consists in its compatibility with most of existing theoretical methods. For example, Tsekouras and Provata [33] considered spontaneous local clustering on lattice and homogeneous initial distributions turn into clustered structures, at the surfaces of which the reactions take place, using mean-field analysis and Monte Carlo simulations [33].

The aspect of the compatibility of the fractal approach with other theoretical methods will be considered in more detail in Chapter 5.

The usefulness of the fractal approach in modeling of macromolecule separation processes in packed columns was illustrated by Ref. 34. In these columns filled with porous particles the macromolecule retention mechanism is determined by size-exclusion chromatography (SEC), hydrodynamic chromatography (HDC), or their combination [34]. It was proved that the degree to which SEC and HDC are mixed depends on the particle diameter, the relative size of the pores, and the macromolecule size. Guillaume et al. [34] showed the fractal character of the apparent selectivity between two adjacent peaks on the chromatogram.

The fractal analysis seems also effective in accounting nonhomogeneous distribution of porosity and quantifying the drag correction for flow through effects, e.g., by treating an individual aggregate as a self-similar object made up of a settling swarm of permeable spheres that are themselves made up of a settling swarm of subspheres and so on to the level of primary particles [35].

One can notify some other applications of the random fractal concept (not related to porosity):

1. *Biochemistry of proteins*. Fractal structural tendencies are found for distributions of small-angle neutron scattering (SANS) used

to study the structure of protein–sodium dodecylsulfate complexes [36]. Analyses of SANS distributions clearly showed that the arrangement of micellelike clusters resembles a fractal packing of spheres. Guo et al. [36] showed that a protein–SDS complex can be characterized by four parameters related to the fractal model and extracted from the scattering experiment (the average micelle size and its aggregation number, the fractal dimension characterizing the conformation of the micellar chains, the correlation length giving the extent of the unfolded polypeptide chains, and the numbers of micellelike clusters in the complex) [36].

2. *Biochemistry of proteins.* Liebovitch and Todorov [37] presented an analysis of opportunities related to the use of fractal concept and nonlinear dynamics in studies of ion channel proteins:
   a. Existence of long-term correlations in channel function and self-similar properties of the currents recorded through individual ion channels, characterized by a scaling function.
   b. Possibility of fractal form for the whole-cell membrane voltage recorded across many channels.
   c. The appearance of the same distributions of open and closed times for random events and the deterministic dynamics [37].
   d. Liebovitch and Todorov [37] debate whether random, fractal, or deterministic models best represent the functioning of ion channels.
3. *Theoretical biochemistry.* Aiming to describe cellular cytoplasm, Aon and Cortassa [38] suggested its analogy with a percolation cluster, a sort of "random fractal," with a quantification of the fractal dimension D in micrographs of cytoskeleton components or microtrabecular lattice. Their hypothesis dealt with
   a. The existence of the percolation threshold.
   b. The reactivity increase when enzymes (or targets) and substrates (or effectors) coexist in the same topological dimension [38].
4. *Anatomy.* Aiming to investigate the resting heart rate (HRV), Yamamoto et al. [39,40] assumed the basic fractal nature of its variability, relative to that in breathing frequency (BFV) and tidal volume (TVV), and tested the hypothesis that fractal HRV is due to the fractal BFV and/or TVV in humans. In addition, the possibility of the fractal nature of respiratory volume curves (RVC) and HRV was observed [39,40].
5. *Theoretical biophysics.* According to Ref. 41, recent mathematical analyses had shown fractal properties for some seemingly irregular

biological signals. A fractal time series was characterized by the property of self-similarity (self-affinity) and had long-range time correlation. The aim of the study [41] was to investigate the question of whether the fluctuation of H-reflex was fractal with strong time correlation. The results obtained in [41] indicated that

a. The fractal correlation found in the H-wave sequences was caused neither by the conduction through nerve fibers nor by the transmission at the neuromuscular junction because the M-wave sequence had a significantly weaker time correlation.
b. Special impulses in a motor nerve induced by the stimulation made a minor contribution to the generation of fractal correlation in the H-wave sequences because it was preserved when the stimulation intensity was below MT. It was suggested that the fractal correlation in human H-reflex was generated at the synaptic connections to alpha-motoneurons in the spinal cord [41].

6. *Studies of human brain.* Mather [42] studied images of three-dimensional scenes that inevitably contain regions that are spatially blurred by differing amounts, owing to depth-of-focus limitations in the imaging apparatus. Computational modeling had shown that the effect of blur on single-step edges was very similar to its effect on random fractal patterns, because the two stimuli had similar Fourier amplitude spectra [42].
7. *General biology and pharmacology.* It was found in [43] that fractal analysis is suitable for analyzing liver fibrosis and has excellent reproducibility. Moal et al. [43] estimated fractal analysis as the only quantitative morphometric method able to discriminate among the models of fibrosis and sensitive enough to detect pharmacologically induced changes in liver fibrosis [43].
8. *Environment engineering.* Fractal approach is used also in modeling processes in multiphase (soil–atmosphere–oil–water) flow systems [44]. That allows the application of light transmission method (LTM), based on the use of the hue and intensity of light transmitted through a slab chamber to measure fluid content, since total liquid content is a function of both hue and light intensity, when comparing experiments with LTM and synchrotron x-rays. As result, the contaminant concentrations are measured, and the risk to human health and the environment because of contaminating groundwater resources is minimized [44].

9. *Geophysics* (certainly, because that was the beginning of the random fractal concept). Numerous natural network systems have fractal structures that are optimal for minimizing energy expenditure during material transport. It was shown in Ref. 45 that a fractal magma "tree" provides a network in which magma rapidly loses diffusive chemical "contact" with its host matrix.

VII. CLASSIFICATIONS OF MODELS OF POROSITY

As it was noted above, interactions in micropores pores and around them have quantum origin. Therefore, for the correct description of microporous structures, one should solve the Schroedinger equation for the entire solid system— which is absolutely impossible. In the real practice, one uses several assumptions based on the general physical comprehension of porosity. According to the kind of such assumptions, the existing models of porosity can be classified according to their

- Applicability to such or such aspects of porosity (pore formation, characteristics of structure, measurable properties, or their combinations)
- Determinism or randomness
- Continuousness or discreteness
- Limited or unlimited number of counted elements of the porous structure
- Analytical or numerical method of simulations
- Level of assumptions (fundamental physical, just for this model, empirical treatment of experimental data)

Let us note that some models are intermediate and partly satisfy both criteria in several classification.

Of course, in the methodological aspect the fundamentally based models have a preference, but we know some situations when models were formulated initially based on experimental results or intuitive assumptions and later received fundamental support. For example, Dubinin's assumption of Gauss energy distribution got explanation in Ref. 11 as sequence of casual (Monte Carlo) approach. Fractal model suggested by B. Mandelbrot was due to observations and empirical suggestions presented in Ref. 11, a thermodynamic explanation.

In Table 1.10 we present some examples of existing models of porosity based on fundamental physical assumptions in the light of the above classifications.

TABLE 1.10 Classification of Some Existing Models of Porosity

Model	Application	Determinist or random	Continuous or discrete	Limited or unlimited counting	Analytical or numerical
Monte Carlo	All aspects	Random	Discrete	Limited	Numerical
Gauss energy distribution	Structure only	Random	Continuous	Unlimited	Analytical
Gauss size distribution	Structure only	Random	Continuous	Unlimited	Analytical
Gibbs energy distribution	Structure only	Random	Continuous	Unlimited	Analytical
Chemical model	Preparation and structure	Randomness with elements of determinism	Discrete	Unlimited	Analytical
Statistical polymer	All aspects	Randomness with elements of determinism	Discrete	Limited	Numerical
Monte Carlo with thermodynamic limitations	All aspects	Randomness with elements of determinism	Discrete	Limited	Numerical

The detailed analysis of existing models of microporous structures, with discussing numerical results, will be presented in Chapter 5.

VIII. CONCLUSIONS

1. Porous structures can be classified in accordance to the following features:
   a. Pore size or energy distribution
   b. Fractions of open, semiopen, and closed pores
   c. Regularity or randomness
   d. Homogeneity or heterogeneity
   Micropores are better characterized by energy than size.
2. All characteristics of porous structures are determined by the technology of their preparation and determine all measurable properties of porous materials. Processes of formation of micropores, mesopores, and macropores compete, and their rates depend on both the kind of raw material and the technology of its treatment. Microporous substructures are formed due to chemical reactions in the condensed phase and determine such specific properties of porous materials as heat or mass transfer, adsorption, percolation or permeability, and mechanical resistance.
3. Models of porosity can be classified in accordance to the following characteristics:
   a. Applicability to aspects of porosity (pore formation, characteristics of structure, measurable properties, or their combinations)
   b. Determinism or randomness
   c. Continuousness or discreteness
   d. Limited or unlimited number of counted elements of the porous structure
   e. Analytical or numerical method of simulations
   f. Level of assumptions (fundamental physical, just for this model, empirical treatment of experimental data)
   Theoretical modeling of microporosity is especially important for research of porous materials because the experimental information about them is indirect and characterized by significant error of measurement.
4. Though the fractal concept of porosity exists due to observations of large pores, the fractal approach is especially important for the description of micropores, because of technical difficulties related to the measurement of their properties.

REFERENCES

1. Romm, F. Microporous materials: modeling of internal structure. In *Encyclopedia of Interface Science*; Hubbard, A., Ed.; Marcel Dekker Inc., 2002.
2. Sing, K.S.W. Characterization of porous solids: An introductory survey. In *Characterization of Porous Solids II*, Proceedings IUPAC Symposium (COPS II), Alicante, Spain, May 6–9, 1990, pp. 1–9.
3. Dubinin, M.M.; Kadlec, O. New ways in determination of the parameters of porous structure of microporous carbonaceous adsorbents. Extended Abstracts and Program—Biennial Conference on Carbon 1975, *12* (no issue), 59–60.
4. Dubinin, M.M. Physical adsorption of gases and vapors in micropores. Prog. Surface Membrane Sci. **1975**, *9*, 1–70.
5. Dubinin, M.M. Physical feasibility of Brunauer's micropore analysis method. J. Coll. Interface Sci. **1974**, *46* (3), 351–356.
6. Schroeder, J. Detection and importance of micropores. GIT Fachzeitschrift fuer das Laboratorium **1986**, *30* (10), 978–980, 982–984.
7. Gavalas, G.R.; Jiang, S. Method for forming improved H.sub.2 –permselective oxide membranes using temporary carbon barriers 1996, U.S. Patent 5,503,873.
8. Suri, V.; Yu, K.-O. Defects formation [in castings]. Mater. Eng. (New York, NY, United States) **2002**, *19* (Modeling for Casting and Solidification Processing), 95–122.
9. Watts, R.K. *Point Defects in Crystals*; Wiley-Interscience: New York, 1977; 312 pp.
10. Phillips, R. *Crystals, Defects, and Microstructures: Modeling Across Scales*; Cambridge University Press: Cambridge**, 2001**; 780 pp.
11. Romm, F. Thermodynamics of microporous material formation. In: *Surfactant Science Series "Interfacial forces and fields: Theory and applications"*; Monographic series, Jyh-Ping Hsu, J.-P., Ed.; Marcel Dekker: New York, Chapter 2, 1999; 35–80.
12. Kyotani, T. Control of pore structure in carbon. Carbon **2000**, *38* (2), 269–286.
13. Dismukes, J.P.; Johnson, J.W.; Corcoran, Edward William, Jr.; Vallone, Joseph; Pizzulli, J.J.; Anderson, M.P. Synthesis of microporous ceramics. U.S. Patent 5,902,759, 1999.
14. Kishima, N.; Ueda, Y.; Sato, T. Method of producing porous material having open pores. US Patent 4,464,485, 1984.
15. Heide, H.; Hoffmann, U.; Broetz, G.; Poeschel, E. Process for producing catalyst supports or catalyst systems having open pores, U.S. Pat. 3,957,685, 1976.
16. Chen, J.; Blevins, W.E.; Park, H.; Park, K. Gastric retention properties of superporous hydrogel composites. J. Controlled Release **2000**, *64* (1–3), 39–51.
17. Nicholson, D.; Travis, K. Molecular simulation of transport in a single micropore. Membrane Sci. Technol. Ser. **2000**, *6* (no issue), 257–296.
18. Stubos, A.K.; Mitropoulos, A.; Steriotis, T.; Katsaros, F.K.; Romanos, G.E.; Kanellopoulos, N.K. Ceramic membranes: industrial applications. NATO ASI

Ser., Series C: Mathematical and Physical Sciences **1997**, *491* (Physical Adsorption: Experiment, Theory and Applications), 485–510.

19. Dzhigit, O. M.; Dubinin, M. M.; Kiselev, A. V.; Shcherbakova, K. D. Active charcoals and adsorption from solutions. Compt. Rend. Acad. Sci. U.R.S.S. **1946**, *54*, 141–144.
20. Bacaoui, A.; Dahbi, A.; Yaacoubi, A.; Bennouna, C.; Maldonado-Hodar, F.J.; Rivera-Utrilla, J.; Carrasco-Marin, F.; Moreno-Castilla, C. Experimental design to optimize preparation of activated carbons for use in water treatment. Environ. Sci. Technol. **2002**, *36* (17), 3844–3849.
21. Methivier, A. Separation of π-xylene by adsorption. Catal. Sci. Ser. **2002**, *3*(Zeolites for Cleaner Technologies), 209–221.
22. Holderich, W.F.; Wagner, H.H.; Valkenberg, M.H. Immobilized catalysts and their use in the synthesis of fine and intermediate chemicals. Special Publication. Royal Soc. Chem. **2001**, *266* (Supported Catalysts and Their Applications), 76–93.
23. Maier, W.F. Amorphous microporous mixed oxides, new selective catalysts with chemo- and shape-selective properties. Preprints of Symposia. Amer. Chem. Soc. Division of Fuel Chemistry, **1998**, *43* (3), 534–537.
24. Corma, A; Nemeth, L.T; Renz, M; Valencia, S. Sn-zeolite beta as a heterogeneous chemoselective catalyst for Baeyer-Villiger oxidations. Nature **2001**, *412* (6845), 423–425.
25. Medeiros, E.B.M.; Petrissans, M.; Wehrer, A.; Zoulalian, A. Comparative study of two cocurrent down-flow three phase catalytic fixed bed reactors: application to the sulphur dioxide catalytic oxidation on active carbon particles. Chem. Eng. Proc. **2001**, *40* (2), 153–158.
26. Yortsos, Y.C.; Stubos, A.K. Phase change in porous media. Curr. Opin. Coll. Interface Sci. **2001**, *6* (3), 208–216.
27. Cerofolini, G.F.; Rudzinski, W. Theoretical principles of single- and mixed-gas adsorption equilibria on heterogeneous solid surfaces. Stud. Surf. Sci. Cat. **1997**, *104* (Equilibria and Dynamics of Gas Adsorption on Heterogeneous Solid Surfaces), 1–103.
28. Jaroniec, M.; Rudzinski, W. Adsorption of gas mixtures on heterogeneous solid surfaces. J. Res. Inst. Cat., Hokkaido Univ. **1977**, *25* (3), 197–210.
29. Hahn, R.; Jungbauer, A. Peak broadening in protein chromatography with monoliths at very fast separations. Anal. Chem. **2000**, *72* (20), 4853–4858.
30. Aminabhavi, T.M.; Rudzinski, W.E.; Kulkarni, P.V.; Antich, P.; Soppimath, K.S.; Kulkarni, A.R. Polymeric membranes. Polym. News **2000**, *25* (11), 382–384.
31. Canan C.H. Effects of humidity and soil organic matter on the sorption of chlorinated methanes in synthetic humic–clay complexes. J. Hazardous Mat. **1999**, *68* (3), 217–226.
32. Jaroniec, M.; Choma, J. Characterization of geometrical and energetic heterogeneities of active carbons by using sorption measurements. Stud. Surf. Sci. Cat. **1997**, *104* (Equilibria and Dynamics of Gas Adsorption on Heterogeneous Solid Surfaces), 715–744.

33. Tsekouras, G.A.; Provata, A. Fractal properties of the lattice Lotka-Volterra model. Phys. Rev. E: Statistical, Nonlinear, and Soft Matter Physics **2002,** *65* (1–2), 16204/1–16204/8.
34. Guillaume, Y.-C.; Robert, J.-F.; Guinchard, C. A mathematical model for hydrodynamic and size exclusion chromatography of polymers on porous particles. Anal. Chem. **2001**, *73* (13), 3059–3064.
35. Woodfield, D.; Bickert, G. An improved permeability model for fractal aggregates settling in creeping flow. Water Res. **2001**, *35* (16), 3801–3806.
36. Guo, X.H.; Zhao, N.M.; Chen, S.H.; Teixeira, J. Small-angle neutron scattering study of the structure of protein/detergent complexes. Biopolymers **1990**, *29* (2), 335–346.
37. Liebovitch, L.S.; Todorov, A.T. Using fractals and nonlinear dynamics to determine the physical properties of ion channel proteins. Crit. Rev. Neurobiol. **1996**, *10* (2), 169–187.
38. Aon, M.A.; Cortassa, S. On the fractal nature of cytoplasm. FEBS Letters **1994**, *344* (1), 1–4.
39. Yamamoto, Y.; Fortrat, J.-O.; Hughson, R.L. On the fractal nature of heart rate variability in humans: effects of respiratory sinus arrhythmia. Am. J. Physiol. **1995**, *269* (2, Pt. 2), H480–H484.
40. Yamamoto, Y; Hughson, R.L. On the fractal nature of heart rate variability in humans: effects of data length and beta-adrenergic blockade. Am. J. Physiol. **1994**, *266* (1 Pt 2), R40–R49.
41. Nozaki, D.; Nakazawa, K.; Yamamoto, Y. Fractal correlation in human H-reflex. Exper. Brain Res. **1995**, *105* (3), 402–10.
42. Mather, G. The use of image blur as a depth cue. Perception **1997**, *26* (9), 1147–58.
43. Moal, F.; Chappard, D.; Wang, J.; Vuillemin, E.; Michalak-Provost, S.; Rousselet, M.C.; Oberti, F.; Cales, P. Fractal dimension can distinguish models and pharmacologic changes in liver fibrosis in rats. Hepatology **2002,** *36* (4 Pt 1), 840–849.
44. Darnault, C.J.G.; Dicarlo, D.A.; Bauters, T.W.J.; Steenhuis, T.S.; Parlange, J.-Y.; Montemagno, C.D.; Baveye, P. Visualization and measurement of multiphase flow in porous media using light transmission and synchroton x-rays. Ann. NY Acad. Sci. **2002**, *972* (Visualization and Imaging in Transport Phenomena), 103–110.
45. Hart, S.R. Equilibration during mantle melting: a fractal tree model. Proc. Nat. Acad. Sci. **1993**, *90* (24), 11914–11918.

2

Experimental Methods for Study of Microporous Media

I. PRINCIPLES OF EXPERIMENTAL STUDIES OF POROSITY

A. Goals and Classification of Experimental Studies of Porosity

Experimental measurements and tests of properties of porous materials are the main criteria for the estimation of their technical merit, their choice for various needs, estimation of the correctness and the usefulness of the relevant theoretical models, and economic forecasting regarding the perspective of the industrial and/or domestic applications of the tested materials. For studying porous materials, in principle, one may use all existing methods of study of condensed phase and multiphase systems. The main idea of such experiments is the comparison of measured characteristics of a porous solid sample with those of the similar nonporous sample having the same chemical composition and structure as the continuous phase of the porous sample. The difference between those results of measurements is attributed to the contribution of porosity. In the case of a homogeneous porous sample, this phenomenological comparison can be presented as the following equation:

$$X(P, \xi) = X_0(P, \xi = 0) + X_\xi(\xi) \tag{2.1}$$

where P is the measured parameter, X is the result of the measurement, ξ is the porosity, X_0 is the result of the measurement for the nonporous sample, and X_ξ is the contribution of the homogeneous porosity. The form of the analogous equation for the measurements of heterogeneous porosity is very similar to Eq. (2.1) but comprises a sum over all factors of heterogeneity:

$$X(P, \Sigma\,\xi) = X_0(P, \xi = 0) + \Sigma X_\xi(\xi) \tag{2.2}$$

In principle, for the homogeneous porosity Eq. (2.1) should provide

$$X(P, \xi) = X_0(P) + \xi\left(\frac{\partial X}{\partial \xi}\right)(\mathrm{P}, \xi = 0) + \Sigma\left[\left(\frac{1}{\mathrm{k}!}\right)\xi^{\mathrm{k}}\left(\frac{\partial^{\mathrm{k}} X}{\partial \xi^{\mathrm{k}}}\right)(\mathrm{P}, \xi = 0)\right] \tag{2.3}$$

where the sum is taken over all accounted powers of derivatives of X. Equation (2.2) means that two or more experiments of the same kind accomplished on the same sample should provide the same results. However, this is not always right (for micropores one can claim that it is never right!), and in real conditions $X(P, \xi)$ depends also on time and the number of measurements. This well-known experimental fact is caused by two phenomena:

1. "Aging" of pores
2. Hysteresis of properties of porous structure

B. Aging

Aging of a pore can be defined as a spontaneous change of its structure. The driving force for aging is the excess internal energy of the pore. As it was discussed in Chapter 1, micropores are characterized by the upper level of the internal energy. Therefore, the effect of aging is especially important for micropores. The increase of the characteristic size of a pore reduces its energy and, respectively, the influence of aging. For macropores, errors related to aging are so minor they can be neglected in comparison with the level of the error of the measuring equipment.

Let us notify that the practical effect of the phenomenon aging can be not always negative but sometimes positive. Let us consider some examples of technical solutions, in which authors used the phenomenon of aging for obtaining positive results.

(a) Example 2.1. Experimental studies accomplished in 1998–2001 in AMSIL Ltd. (Migdal HaEmek, Israel) showed that some adhesive coatings based on tetra-buthyl ammonium silicate [1] just after preparation, absolutely fresh, were very sensitive to water and could be easily destroyed by intense washing. However, coatings based on the same composition and exposed to air for 2 y in room conditions got high stability not only against

water but also against diluted acid and alkali solutions; moreover, even high-content acids and alkalis were unable to destroy these coatings effectively. This fact was due to aging of micropores that allowed the stabilization of the structure of silica oxide inside the coatings.

The effect of the increase of protective properties of some coatings is a well-known fact that can be used in practice. For example, it was shown that when the micropores and various defects of coating are impregnated with inhibitors, they seriously reduce the rate of the penetration of corrosive agents [2]. The effectiveness of the inhibition cannot be easily related to the chemical composition of inhibitors [2].

(b) Example 2.2. Motojima and Kawamura [3] used the process of aging for the synthesis of heterogeneous catalysis. Zeolite was contacted with an aqueous solution containing at least one of copper, nickel, cobalt, manganese, and zinc salts (preferably copper and nickel salts, particularly preferably copper salt) in such forms as sulfate, nitrate, or chloride, thereby adsorbing the metal on the zeolite in its pores by ion exchange, then the zeolite was treated with a water-soluble ferrocyanide compound, e.g., potassium ferrocyanide, thereby forming metal ferrocyanide on the zeolite in its pores. Then, the zeolite was subjected to aging treatment, thereby producing a zeolite adsorbent impregnated with metal ferrocyanide in the pores of zeolite. The obtained adsorbent was able to selectively recover cesium with a high percent cesium removal from a radioactive liquid waste containing at least radioactive cesium, for example, a radioactive liquid waste containing cesium and coexisting ions such as sodium, magnesium, calcium, and carbonate ions at the same time at a high concentration. The zeolite adsorbent had a stable ability to adsorption for a long time [3].

(c) Example 2.3. Takumi et al. [4] used aging for the preparation of heterogeneous catalyst carrier comprising gamma-alumina, the total pore volume being in the range from 0.55 cm^3/g to 0.65 cm^3/g, while the content of pores having a diameter in the range of from 60 Å to 110 Å occupied at least 75% of the total pore volume, the surface area being in the range of from 210 m.sup.2/g to 250 m.sup.2/g and the attrition loss was less than 0.5 wt%. This sort of catalyst carrier can be manufactured through a series of steps: (1) preparing a first alumina hydrosol, (2) preparing a second alumina hydrosol by adding hydrochloric acid to said first hydrosol, (3) preparing an alumina precursor by commingling said second hydrosol with a suitable gelling agent necessary to neutralize the chloride contained in the second hydrosol, and (4) dispersing this precursor as droplets in a suspending medium thereby forming hydrogel particles, aging the thus obtained hydrogel particles in said suspending medium and then in an aqueous ammonia, washing the aged particles in water, drying, and calcining thereafter [4].

Though several technical solutions get profit from micropore aging, in most situations this process is undesirable and should be neutralized if possible.

(d) Example 2.4. Willkens et al. [5] prepare aging-resistant carbonized silica ceramics comprising (1) coarse SiC particles having a particle size of at least 30 μm, (2) a coating of recrystallized alpha-SiC, which coats and connects the coarse SiC particles, and (3) porosity of 8–10 vol%. The content of coarse SiC particles was 40–60 wt%, free Si, 2 wt%, and the recrystallized coating, 40–60 wt% of the total SiC content. A process for making an oxidation-resistant SiC article includes (1) forming a green body containing fine SiC particles having a particle size of maximum 10 μm and coarse SiC particles, (2) firing the green body to form a recrystallized first-fired SiC body, (3) infiltrating the first-fired body with a slurry comprising SiC particles to obtain an impregnated body, and (4) refiring at 2000–2200°C the infiltrated body in a nonoxidizing atmosphere. The slurry further comprises 0.01–3 wt% solids of alumina particles. It is interesting to notify that the authors perform the infiltration by sonication [5].

Uses of acoustics allows very interesting perspectives for experimental studies of porosity, and we will analyze this aspect in this chapter.

(e) Example 2.5. This example of preventing aging is related to zeolites. Kim et al. [6] describe mesoporous hexagonal, cubic, lamellar, wormhole, or cellular foam aluminosilicates, gallosilicates, and titanosilicates derived from protozeolitic seeds using an ionic structure directing agent. Kim et al. [6] believe that the Si and the Al, Ga, or Ti centers in the described structures are stable, so that the framework of the structure does not collapse when heated in the presence of water or water vapor (steam). The process for forming a porous aluminosilicate compound comprises (1) providing protozeolitic aluminosilicate seeds selected from an aqueous solution, gel, suspension wetted powder, and mixtures thereof, (2) reacting in a mixture the seeds in an aqueous medium with an organic surfactant, (3) aging the mixture of step 2 at 250–2000°C, and (4) separating the components from the mixture of step 3. The steam-stable components can be used as catalysts for hydrocarbon conversions, including the fluidized bed catalytic cracking and the hydrocracking of petroleum oils and other reactions of organic compounds [6].

(f) Example 2.6. This example is related to aging-resistant micropores. Taking into account the extreme complexity of this problem, the attempt of Wu et al. [7] is very brave.

Wu et al. [7] improve nanoporous (mesoporous and microporous) dielectric film useful for the production of semiconductor devices, integrated circuits, and the like. Such improved films are produced by a process that includes (1) preparation of a Si-based precursor component including a porogen (component stimulating pore formation); (2) coating a substrate with the Si-based precursor to form a film; (3) aging or condensing the film in the presence of water; (4) heating the gelled film at a temperature and for a duration effective to remove substantially all of the porogen, and in which the applied precursor component is substantially aged or condensed in the presence of water in liquid or vapor form, without the application of external heat or exposure to external catalyst [7].

We can also mention the invention [8] in which aging-resistant composite SiC igniter, consisting of second interlayers of recrystallized SiC, is prepared by infiltration of SiC coarse grains into the open porosity of silicon carbide ceramics. Willkens et al. [8] believe that the resistivity of such heating element SiC igniters was increased by approximately 8% over 6000 cycles [8].

C. Hysteresis

The second principal factor of errors in experimental studies of pores, *hysteresis* (also written frequently as *histeresis*, while the first version seems more appropriate because of the Greek origin of this word), can be defined as the change of the structure of pore because of the exterior effects related to the measurement. If an exterior force is loaded and then unloaded up to the initial values of all exterior intensive parameters (pressure, temperature, concentrations, etc.), the porous structure may get so significant changes that its measurable properties differ more or less significantly from the initial state. In the concept of micropores, the notion of *hysteresis* is related mostly to adsorption–desorption processes: if the cycle adsorption–desorption is carried out two and more times, the obtained isotherms of adsorption are different though the identical conditions of all cycles. One of the causes for the appearance of the hysteresis loop adsorption–desorption is related to closing of semiopen pores and the most narrow among open micropores by the molecules of the adsorbate. This kind of hysteresis is considered in more details in Chapter 4.

The fundamental cause for the phenomenon of the hysteresis loop consists in the well-known quantum effect: every exterior intervention into quantum system (while microporous systems are obviously quantum) causes irreversible uncountable changes in this system. In the thermodynamic aspect, both loading and unloading of the exterior force (in the case of adsorption–desorption: loading is involving the substance to adsorb, unloading is its removal) are irreversible processes increasing the entropy of the system.

Among various physical causes for hysteresis (meaning not only adsorption–desorption but all kinds of exterior action loaded–unloaded), experimental researchers must first of all remember the danger of the destruction of pores. Pores, especially micropores, can be destroyed by all kinds of exterior forces, comprising tension loading–unloading, fluid intrusion, aggressive chemicals and/or high-energy radiation attacking the continuous phase, etc. Pore destruction changes the structure of the studied material so much that all the obtained results of measurements may become nonsense. On the other hand, pore destruction on purpose can sometimes provide a very important information unavailable for alternative experimental methods.

While pore aging can sometimes be useful, it is very difficult to find in the literature any example of the positive use of hysteresis.

Since hysteresis of adsorption–desorption is very widely discussed, we will consider this kind of hysteresis later, and the following examples present other cases of hysteresis.

(a) Example 2.7. Mechanical hysteresis, could it be reduced? Lawson et al. [9] present the title elastomers having improved raw polymer viscosity and good compound viscosity state and reduced hysteresis and low rolling resistance in the cured state, prepared by initially coupling a portion of ionically polymerized living diene polymer using sulfur and polyhalide coupling agent and terminating the remaining diene polymer chains using fused ring polynuclear aromatic compounds or aromatic nitrile compounds [9].

Another example of reducing mechanical hysteresis (for rubbers) is given by Lawson et al. [10], the same authors as Ref. 9; we do not consider its details.

An example of intrusion–extrusion hysteresis was analyzed in Ref. 11.

Positive or negative, effects related to aging and hysteresis should be always taken into account by experimental researchers of micropores.

The principal scheme of the experimental installation for studies of porosity is presented on Fig. 2.1.

An exterior factor (exterior force) comes from the source of the exterior force and acts onto the porous sample in the vessel. Because of interactions with the porous material, the parameters of the loaded force get changes, and the resulting force is registered. Then, the result of measurement is normally compared to that found for a standard sample, e.g., nonporous material.

In several situations, it is not easy to distinguish aging from hysteresis. For example, scattering by x-rays causes secondary reactions in micropores, which causes their irreversible changes (obviously, that is hysteresis!); however, after the irradiation is removed, the secondary reactions continue and cause spontaneous changes without exterior action (of course, that is aging!). Thus, the same exterior action may result in both hysteresis and aging.

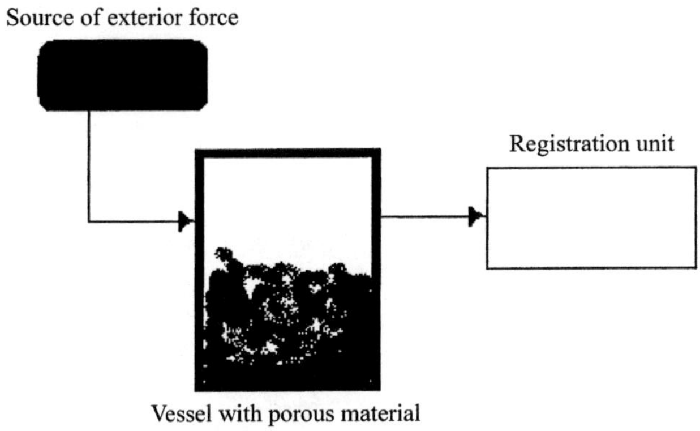

FIG. 2.1 Principal scheme of measurements of characteristics of porous materials.

II. CLASSIFICATION OF EXPERIMENTAL METHODS FOR STUDYING PORES

Existing methods of experimental study of porosity can be classified by:

1. Sphere of the applicability of each method: preparation or synthesis of porous materials, their structure and/or properties.
2. Physical base of the method: fundamental physical phenomena, on which the method is based, and the type of interactions used for the treatment of the porous sample.
3. Whether the method is destructive, what are the concrete sources of experimental error (the principal factors, of course – aging and hysteresis), what is the level of the experimental error, and how can this be taken into account?
4. Provided information: What are the parameters measured?

Talking about destructive methods, one usually means the situation when the sample is mechanically destroyed before or during the tests. Such criterion does not seem good. Whether or not cutting the primary material to some samples is destructive test? What about tests in which the samples are destroyed chemically, or not destroyed but get too deep structural and chemical changes? In the author's opinion, also such tests should be considered as destructive. The following analysis will consider all measurements related to very deep irreversible changes of the porous structure as destructive tests.

In Table 2.1 we present some principal experimental methods used in studies of porosity, in the light of the proposed classification. A more detailed analysis of the related experimental technique will be given below.

TABLE 2.1 Classification of Principal Experimental Methods for Studying Pores

Group of methods	Sphere of applicability	Physical base for the method	Destructive or not, source(s) of error(s)	Information obtained
Control and monitoring of reaction	Synthesis of microporous material	Reproducibility of synthesis	Nondestructive; errors related to equipment, nonpreviewed factors, aging	Exterior intensive parameters of synthesis
Periodical tests of samples	Synthesis; studies of aging	Continuous change of parameters of the treated material with time	Destructive; errors because of the dilution and relaxation processes	Change of measurable parameters with time
Electrical conductivity and capacity tests	Studies of structure	Different electrical properties of continuous phase and fluid in pores	Nondestructive	Interactions caused by electrical fields
Treatment by electromagnetic field	Studies of structure	Interactions of electromagnetic waves with continuous phase and voids	Destructive; errors because of changes in micropores due to the electromagnetic treatment	Size and shape of pores; their energy distribution; regularity, randomness; opening, semiopening, closing

Scattering with high-energy elementary particles	Studies of structure	Interactions of the particles with continuous phase and voids	Nondestructive. Errors because of changes in micropores due to the interactions with the particles	Size and shape of pores; their regularity, randomness; opening, semiopening, closing
Acoustic treatment	Studies of structure; percolation, permeability, and mechanical properties	Interactions of the sound with the interface "continuous phase voids"	Destruction, possible for high frequencies of sound; errors because of the interactions of micropores with the sound	Opening of pores, form of pores, connectedness, percolation paths
Fluid intrusion	Studies of structure; percolation, permeability	Pass-through of the fluid in open pores	Destructive; errors because of destruction of micropores having thin walls	Energy distribution, percolation paths
Adsorption isotherms	Studies of structure and adsorption–desorption	Adsorption–desorption of the fluid in open and semiopen pores	Nondestructive; errors because of irreversible events	Energy distribution, internal surface, internal volume
Mechanical tests	Studies of mechanical properties	Interactions of particles of the continuous phase under tension loaded	Destructive in most cases; errors because of irreversible events.	Mechanical parameters

Let us note that Table 2.1 contains the information only about the principal methods and the results obtained directly by these methods. Opportunities of their combinations, information obtained due to the use of theoretical models, etc., are discussed in the further analysis.

Now, let us consider the above methods in more details.

III. CONTROL AND MONITORING OF SYNTHESIS OF MICROPOROUS MATERIAL

The goal of monitoring of the synthesis of microporous material consists in assuring the reproducibility of the same material prepared in different laboratories. The process of monitoring comprises the control of all exterior intensive parameters influencing the synthesis: temperature, pressure, concentrations of the reagents, electromagnetic fields, etc. A very important part of monitoring is the measurement of the surface tension of the product. Though this parameter cannot be varied by experimental researchers, it is one of important "symptoms" if the reproduction of the needed material does not succeed.

Now, let us consider some technique for monitoring synthesis of porous materials.

Some aggregates and methods for synthesis in solid phase are described in Refs. 12 and 13. We do not analyze these examples because they are just technical information.

Technique for measuring surface tension of solids using tertiary systems gas–liquid–solid with angle measurements is described in Refs. 14 and 15.

Errors in the monitoring of synthesis of microporous materials are related to

Errors of measurement by the available equipment
Various factors that cannot be previewed (especially when dealt with novel materials)
Retarding of studies of the structure and properties of the synthesized material after the synthesis is finished (aging)

Now, let us consider some typical examples of control of synthesis of porous materials.

(a) Example 2.8. Synthesis of active carbon by carbonization and activation with water vapor of the microporous structure of coals (anthracite and lean coal) at 900°C. It was found that the obtained coals were characterized by a great polydispersity of micropores. At low

degrees of activation, fine micropores predominanted in the active carbons obtained [16].

(b) Example 2.9. Active carbon impregnated with metal. Bekyarova and Mehandjiev [17] suggest the following way for the preparation of active carbon impregnated with metal cobalt.

Active carbon from apricot shells with known characteristics is impregnated with an aqueous 9.88% cobalt nitrate solution. The samples were destroyed in air at 200, 300, 400, and 550°C. The processes accompanying the thermal treatment are studied by differential thermal analysis (DTA). Two processes are established during calcination of Co-impregnated active carbon: (1) destruction of the support as a result of oxidation catalyzed by the impregnated cobalt and (2) interaction of the active phase (Co_3O_4) with the support (active carbon), during which Co_3O_4 is reduced to CoO and Co. The presence of Co_3O_4 and CoO phases is proved by x-ray measurements, while that of metal Co is established by magnetic measurements. The porous structure changes are studied by adsorption studies. The characterization of the samples is performed by physical adsorption of nitrogen (77.4 K) and carbon dioxide (273 K). The micropore volume is determined by two independent methods: t/F method and D-R plot [17].

The controlled parameters are

On the stage of impregnation with the water solution of cobalt nitrate: temperature, pressure, composition of the liquid phase, and weight of both phases

On the stage of the thermal treatment: temperature

(c) Example 2.10. Preparation of silica-zeolite from aluminosilicate. The aluminium oxide is removed from a crystalline aluminosilicate compound having a SiO_2-Al_2O_3 mole ratio of (3–12) : 1 at 50–100°C in the presence of a cationic form of Cr(III) in an aqueous solution of >0.01 N Cr in a mineral acid, whereby the $pH < 3.5$. The aluminosilicate compound retains its substantial crystallinity. The dealuminized materials are useful as catalysts in organic compound conversions. Laboratory tests showed that zeolites that were dealuminized by $CrCl_3$ in the above procedure compared favorably with untreated zeolites as catalysts in the cracking of gas oils. The dealuminized zeolites have the advantage of having high oxidation activity, which is useful in promoting the conversion of CO to CO_2 during catalyst regeneration [18].

The controlled parameters are temperature, pressure, and the composition of the phases.

Let us note that, according to Ref. 18, the structure does not get changes in the process of the selective removal of a component (in this example, aluminium oxide). This fact is found not only for zeolites but all microporous materials. This is a base for some theoretical studies in Chapter 3.

(d) Example 2.11. Preparation of microporous polymeric surfaces by photodecomposition. Nakayama and Matsuda [19] presented a method to prepare microporous polymer surfaces, using an excimer laser microprocessing. The irradiation of a KrF excimer laser (248 nm) was applied to several polymer films by passing a laser pulse through an optical microscope, resulting in ablative photodecomposition. The control unit was designed to control the fluence of the laser, pulse number, size of irradiated areas, and irradiation micropositioning. The ablation depth linearly increased with an increase in the accumulation of pulses. The chemical composition of the ablated surfaces did not vary with the accumulation of pulses, confirmed by x-ray photoelectron spectroscopy measurements. Excellent structuring quality of ablated micropores was obtained for polyurethane, polyimide, and polycarbonate films. As an application of the ablation technique, polyurethane films were micropored by the excimer laser ablation technique in conjunction with open-cell structured, small diameter grafts under development. In vitro cell growth and compliance on the micropored polyurethane films were examined. Rapid growth was observed on pore sizes of a few microns to several tenths of a micron in diameter. Higher density micropores provided enhanced elastomeric properties [19].

The controlled parameters are parameters of the laser (given above) and the chemical composition of the treated material.

(e) Example 2.12. Synthesis of microporous alumina ceramics. Heide et al. [20] described the preparation of an aluminium oxide ceramic catalyst support with open interconnecting pores. Fine-grained aluminium oxide was mixed with bonding clay, organic binding agents, and water to form a suspension. This was added to a polystyrene framework formed by treating polystyrene balls with aqueous acetone. The aluminum oxide and the supporting polystyrene framework were dried, the polystyrene was removed by heat, and the aluminum oxide support containing open interconnecting pores was sintered at 1600°C for 2 h. The ceramic support was used with a Ni catalyst [20].

The controlled parameters are

On the stage of mixing: temperature, pressure, mixing rate, and chemical composition of the suspension

On the stage of adding the suspension to polystyrene framework: temperature, pressure, and chemical composition of the product

On the stage of drying and heating: temperature and pressure (maybe also the composition of the gas phase)

(f) Example 2.13. Synthesis of metal foams. Haack et al. [21] prepare metallic foams in three stages:

a. Coating a polymeric foam with metal powder
b. Placing the powder-coated polymeric foam in contact with the metal component
c. Heat treating the assembly to volatilize the polymeric foam and to sinter the metal powder coating

The process is especially suitable for manufacturing of sintered metal tube assembly with the internal core of the sintered metal foam from the powder-coated polymer foam or of laminated plates having the foam core. The metal or alloy powders are typically selected from Fe, steel, Al, Cu, brass, bronze, Ni, Co, Pt, Pd, Ag, Pb, Sn, Zr, or the associated alloys. The polymer foam is reticulated organic foam, especially polyurethane [21]. In this process, the controlled parameters are

Stage a: temperature, pressure, parameters of the polymeric foam (comprising its weight, volume, pore distribution) and the completeness of the coating
Stage b: temperature, pressure, amount of the metal applied
Stage c: temperature, pressure, composition of the gas phase

IV. PERIODIC TESTS OF SAMPLES

This experimental technique is applicable mostly to the process of the synthesis of porous media, but also obtaining information about aging of pores is possible. The main problem related to measurements of aging consists in long-term measurements.

In principle, the process of aging can be considered as a specific form of the preparation of a new porous material from another, already existing one.

The periodic tests obtaining experimental measurements on the stage of the preparation (synthesis) or aging of porous material have the following goals:

Monitoring of the preparation conditions, aiming to accomplish the technological process in accordance with the plan or assuring some specified conditions for aging
Avoiding eventual declinations of technological parameters of the preparation–synthesis–aging process (pressure, temperature,

concentrations, their gradients and changes with time) from the required values

Avoiding heterogeneity in the conditions of the treatment of the raw material, unless such heterogeneity is desirable

Monitoring of the preparation–synthesis process is the most complicated problem in experimental studies of porosity, because that means the dynamic measurement of characteristics of the porous structure. In the perfect case, the treated raw material should be continuously treated by an exterior factor, e.g., some electromagnetic or acoustic waves, the changes in the spectrum of the waves being continuously registered. However, any intervention to the process of synthesis leads to the change of the conditions of preparation of the porous material. In the case of micropores such change can be fatal. Therefore, in the real conditions of research or industrial laboratory, the only available solution consists in the periodical tests of little amounts of the material. Without breaking the process of synthesis, one removes a little part of the product and treats it according to the scheme on Fig. 2.1. The fundamental principle allowing such procedure consists in the assumption that the properties of the solid phase change monotonically with time. For such technical solution, one finds the following sources of errors:

1. Error because of the change of the amount of the treated material: This error can be neglected in condition if the total amount of the taken samples is negligible in comparison with the initial amount of the raw material. In the methodological aspect, this kind of error is analogous to hysteresis (because the process of removal of solid samples for tests is irreversible). The level of the error caused by the change of the amount of the treated material is estimated from the following equation:

$$\mathrm{Er}_a \sim \exp\left(\frac{W_s}{W_a} - 1\right) \tag{2.4}$$

   where W_a is the initial amount of the treated material, and W_s is the maximal amount of the samples taken away.
2. Relaxation processes: the fresh samples are characterized by very intense processes of aging because of nonequilibrium conditions of taking the sample. The process of taking a sample and measuring its characteristics is not instant. From the moment of taking the sample until the moment of achieving the measurement, the properties of the microporous structure may show very significant changes. If the period of measurements is τ_m and the averaged

relaxation period τ_r, the error caused by aging during the measurement is proportional to

$$\mathrm{Er}_m \sim \exp\left(\frac{\tau_m}{\tau_r} - 1\right) \tag{2.5}$$

As follows from Eq. (2.5), such measurement brings physically acceptable results only in condition that $\tau_m \ll \tau_r$.

Let us note that this technique of measurements is applicable only to processes in which the exterior intense parameters are constant. If they are changed periodically, the period of changing the exterior intense parameters should be much larger than the period of relaxation:

$$\tau_f \gg \tau_r \tag{2.6}$$

where τ_f is the period of changing of the exterior intense parameters, e.g., temperature. Respectively, if the exterior intense parameters are changed monotonically, the condition of the applicability of the method of testing samples is given by

$$\partial Y/(Y\partial t) \ll 1/\tau_r \tag{2.7}$$

where t is time and Y is the changed exterior parameter, e.g., the temperature. Equations (2.6) and (2.7) mean that the process of changing the exterior parameters is slow enough to allow the relaxation processes to get accomplishment.

The control of exterior parameters is performed by the ordinary equipment for measuring temperature, pressure, etc.

The method of periodic tests of little samples allows measurements of all measurable parameters of the samples, comprising their temperature at surface, chemical composition, porosity, etc.

Now, let us consider an example of measurement of parameters of a porous structure with aging.

(a) Example 2.14. Mellema et al. [22] noted that during aging of casein or skim milk gels, structural changes take place that affect gel parameters such as pore size and storage modulus. These changes could be explained in terms of rearrangements of the gel network at various length scales. Mellema et al. [22] presented results of rheological experiments on rennet-induced casein gels and a general model on rearrangements. The results of experiments (e.g., microscopy, permeametry) and computer simulations, the model, and recent literature on casein gels and other types of particle gels

were compared to each other. Experiments presented included measurements of storage and loss moduli and maximum linear strain of the casein gels. Parameters varied were pH (5.3 and 6.65) and temperature (25 and 30°C). Also the casein volume fraction (5–9 vol%) was varied, which enabled, in the authors' estimation, the application of fractal scaling models. For rennet-induced casein gels, it was demonstrated that at the lower pH, all types of rearrangements proceeded significantly faster. The rearrangements included an increase in the size of compact building blocks, partial disappearance of fractal structure, and the formation of straightened strands, some of which eventually break. All of these rearrangements were attributed to a consequence of particle fusion. There were indications of universality of the relation between particle fusion and gel syneresis for gels composed of viscoelastic particles [22].

In the experimental study of preparation of porous materials, one can use also various flash pyrolysis methods (PyGC, PyMS, PyGC-MS) for the qualitative and quantitative analysis of additives in polymeric materials. Additive analysis may be carried out by the examination of polymeric materials, by hydrolysis, by nondestructive spectroscopic (in-polymer) testing of solid or melt, or by destructive testing using thermal methods (thermal analysis, pyrolysis, thermal desorption, and laser ablation) mainly through the examination of volatiles released. Thermoanalytical methods are especially useful in characterizing cross-linked materials [23].

V. STUDY OF POROUS STRUCTURE AFTER PREPARATION

The main assumption in all methods of after-preparation study of porous materials consists in neglecting aging processes during the measurements. In most of existing technical solutions, such an assumption is valid and can be wrong only in the case of very slow processes of measurements. On the other hand, fast measurements need the serious declination of the treated samples from equilibrium. Thus, we obtain again two factors of errors: aging because of finite rate of measurements and hysteresis because of irreversibility of effects influencing the treated samples.

All methods of study of porous structure are based on some interactions of the structure with exterior factors. These factors can be

Electromagnetic fields
Acoustic fields
High-energy elementary particles
Fluids

A. Electromagnetic Fields and Waves

All measurements of porous media using electromagnetic (EM) treatment are based on specific interactions of EM field with the continuous phase around the pores, the gaseous phase inside the pores, and the interface presented by the pore walls. The effects caused by interactions of EM field with the continuous phase do not significantly differ from those found for nonporous solids, whereas the analogous interactions with the gas inside pores are very similar, with several exceptions, to the same interactions with the same gas in ordinary laboratory conditions. The exceptions are related to the curvature of the pores and their energy. Both effects get significance mostly for micropores. Interactions of EM field and waves to the pore walls is analogous to the interactions with solid surface, and the related effects are very similar, e.g., the reflection of waves having several frequencies.

Let us give the principal effects related to interactions of EM fields and waves to pores. In all cases, the same phenomena for micropores are especially important:

Absorption of EM waves by the substance of both continuous phase and pore gas, excitation of molecules, secondary radiation, acceleration of aging, and/or initiation of spontaneous reactions of excited molecules
Diffraction of EM waves by the transparent phase (in most cases, that is pore gas)
Reflection by the pore walls

Following the scheme given by Fig. 2.1, we note that the experimental researcher can register

1. Secondary radiation from the excited molecules, the wavelength of the secondary rays being larger than that for the primary rays (This red shift is because of the dissipation of a part of primary ray energy)
2. Spectrum of the products of the spontaneous reactions
3. The changes in the behavior of the primary rays (their declination from the initial trajectory, their energy distribution on different directions)

One of possible sequences of EM–pore interactions can sometimes consist in the appearance of photoluminescence. Such photoluminescence is found, for instance, at room temperature in porous silicon (PS) [24]. PS may have porosity over 70%. PS is prepared by anodic dissolution technique and finds various optical electronic applications such as LED, laser, solar cells, and optical interconnections in the integrated circuits. The most widely studied property of PS is photoluminescence. The photoluminescence emission in the red region is understood due to quantum confinement and opening of the band gap. In the estimation of Mehra et al. [24], the broad photoluminescence spectrum is probably due to a wide distribution of the energy band gap caused by different pore sizes. However, as Mehra et al. [24] believe, the red photoluminescence decay in microsecond time scale may be related to deeper states produced by Si–H bonds on the surface of PS. Though an explanation for blue photoluminescence could consist in the direct recombination through shallow surface states arising due to SiO_2, the required correlation between the nanocrystallites size and the photoluminescence peak is not being observed [24].

The measurement of all above factors of information allows the estimation of the chemical composition of both phases, the distribution of voids in the volume, the fraction of open and semiopen pores, and the energy distribution of pores.

However, not all kinds of EM waves can be used for experimental studies of pores. Vibrations in EM field are perpendicular to the direction of the wave propagation, therefore, according to principles of quantum physics, a structural element having the characteristic size r_c can be "measured" only by the waves, the length of which satisfies the following condition:

$$\lambda_{EM} \ll r_c \tag{2.8}$$

where λ_{EM} is the wavelength.

Table 2.2 illustrates the opportunity of using various EM waves for measurements of porous structures.

As follows from Table 2.2, only macropores are available for visual studies by microscopes. Gamma-rays are applicable to all kinds of pores, but experimental researchers get difficulties with the technique of safety and the registration of γ. Therefore, electromagnetic methods are not largely used for studies of the structure of micropores. However, EM treatment of porous materials can be effectively combined with other methods of experimental study (analyzed below). For macro- and mesopores, x-ray scattering is largely applicable [25].

Of course, all EM rays more or less intensively interact with the substance of porous materials. Table 2.3 presents possible sequences of such interactions for the waves that can, according to Table 2.2, in principle be used for studies of porosity.

Now, let us consider some examples of scattering of porous structures by x-rays.

TABLE 2.2 Usefulness of Various EM Waves for Measurements of Porous Structures

Type of waves	Wavelength	Pores available to measurements by these waves
ULF, LF, RF, HF, VHF, UHF, MW, MMW	>0.1 mm	None
IR	10^{-4}–10^{-2} cm	Very large macropores
Visible light	400–750 nm	Most of macropores
UV	200–400 nm	All macropores
Vacuum UM	50–200 nm	All macropores
Soft X	10–500 Å	Macropores, some mesopores
X	0.1–10 Å	Macropore, mesopores, some micropores
γ	10^{-3}–1 Å	All

(a) Example 2.15. Kruk et al. [26] investigated properties of ordered porous silicas with unprecedented loadings of pendant vinyl groups have been synthesized via co-condensation of tetraethyl orthosilicate (TEOS) and triethoxyvinylsilane (TEVS) under basic conditions in the presence of cetyltrimethylammonium surfactant. The resulting organosilicate-surfactant composites exhibited at least one low-angle x-ray diffraction (XRD) peak up to the TEVS:TEOS molar ratio of 7:3 (70% TEVS loading) in the synthesis gel. The surfactant was removed from these composites without any structural collapse. Nitrogen adsorption provided strong evidence of the presence of uniformly sized pores and the lack of phase separation up to TEVS:TEOS ratios as high as 13:7 (65% TEVS loading), whereas (29) Si MAS NMR and high-resolution thermogravimetry showed essentially quantitative incorporation of the organosilane. Thus, a hitherto unachieved loading level for pendant groups, considered by many to be impossible to achieve for stable organosilicas because of the expected framework connectivity constraints, has been obtained.

The resulting vinyl-functionalized silicas exhibited gradually decreasing pore diameter (from 2.8 to 1.7 nm for TEVS loadings of 25–65%) and pore volume as the loading of pendant groups increased, but the specific surface area was relatively constant. Because of the reactivity of vinyl groups, ordered silicas with very high loadings of these groups are expected to be robust starting materials for the synthesis of other organic-functionalized ordered microporous materials. Kruk et al. [26] demonstrated that these starting materials could also be transformed via

TABLE 2.3 Possible Sequences of Interactions of Electromagnetic Waves with the Substance of Continuous Phase and Pore

Type of EM rays	Interaction with the continuous phase	Interaction with gas in pores	Interactions with pore walls
IR	Weak secondary IR	Secondary IR	Slow aging
Visible	Secondary IR	Secondary IR, visible rays with red shift	Aging
UV	Secondary IR and visible rays, decomposition of organics, formation of new micropores	Reactions in gas phase, formation of ozone, nitrogen oxide, free radicals, ionization, various luminescence	Very fast aging
Vacuum UV	Various secondary rays, fast decomposition of organics, radical formation, ionization, intensive formation of new micropores	Fast reactions in gas phase, intense formation of ozone, nitrogen oxide, free radicals, ionization, various luminescence	Destruction of micropores
Soft X	Various secondary rays, radical formation, ionization, fast decomposition of organics, intensive formation of new micropores	Very fast reactions in gas phase, very intense formation of ozone, nitrogen oxide, ionization, free radicals, various luminescence	Destruction of micropores
X, gamma	Various secondary rays, very fast decomposition of organics, intensive formation of new micropores	Very fast reactions in gas phase, very intense formation of ozone, ionization, nitrogen oxide, free radicals, various luminescence	Destruction of micropores

calcination into ordered microporous silicas with pore diameters from 2.5 to as little as 1.4 nm just by using an appropriate loading of the vinyl-functionalized precursor. The novel ordered microporous materials were found promising as adsorbents and catalyst supports [26].

B. Electrical Measurements of Pores

In addition to various methods of scattering, EM tests of pores can be carried out in some cases just by the traditional electrical measurements.

That is possible in the situations when the conductivity of the continuous phase differs enough from that of voids (that is very close to the conductivity of air). On the other hand, one may also fill pores with a conductive liquid and measure the electrical parameters of such system. For example, Lira-Olivares et al. [27] suggest a technique for measuring characteristics of porous structures of rocks by electrical tests, comprising picnometry, conductance, and capacitance measurements. The results of these measurements were compared to traditional methods of study of porosity, and Lira-Olivares et al. [27] confirmed the effectiveness of their method. Electrical studies of porosity are, in most situations, non destructive. The errors of such tests come from interactions and irreversible structural changes caused by the electrical field.

VI. SCATTERING BY ELEMENTARY PARTICLES

As follows from the above analysis, the main disadvantage of all EM rays, except gamma, consists in too large a wavelength, while measurements of micropores need waves as short as possible according to Eq. (2.8). This disadvantage can be neutralized if the solid-phase scattering is performed not by particles of EM field (photons) but other particles having much shorter wavelength.

As follows from quantum mechanics, all elementary particles can be characterized by several wavelengths. This is evaluated from the Louis de Broglie equation:

$$\lambda_{EM} = \frac{h}{mc} \tag{2.8a}$$

or

$$mc^2 = hf \tag{2.8b}$$

where h is Planck's constant ($h = 6.626 \times 10^{-34}$ J·s), m is the mass of the particle in the state of immobility, and c is the velocity of light. As follows from Eq. (2.8b), if we replace photons with enough massive particles, the wavelength is enough for the treatment of micropores. (About photons, there is debate in the physical literature whether their mass in immobility is zero or just very low, but, in any case, the mass in immobility is negligible in comparison with electron.)

In the real practice, experimental researchers usually employ electrons or neutrons. Their wavelengths by Eq. (2.8a) are very short: for electron, about 0.01 Å and for neutron about 10^{-5} Å.

However, use of electrons and neutrons for scattering porous structure is related to some additional problems:

1. Problem related to electrons: Electrons are not neutral particles; they have negative electrical charge, and their interactions with substances cause a very intensive ionization and/or radical formation.
2. Problems related to neutrons:
   a. Neutron is not a stable particle; it is able to decompose to electron–proton pair,
   b. Neutron is so hard a particle that its collision with protons (atoms of hydrogen contained in many of substances, especially hydrates and organics) may cause beat out of protons,
   c. Large mass of the neutron makes difficult its reflection by solid structure, and neutrons penetrate into the structure and destroy it.

In principle, most of interactions between porous structures and photons as mentioned in Table 2.3 are found also for electrons and neutrons.

Because of difficulties in getting reflected neutrons, these are not used in microscopes, and narrow pores are observed in electron microscopes.

We notify that electrons and neutrons are easily registered; even the decomposition of neutrons allows the registration of electrons and/or protons, as the researcher prefers.

In principle, as the tool for measurements of microporosity, electrons can be considered as intermediate between gamma and neutrons.

The results of the comparison of neutrons and electrons to short-wavelength photons as particles for scattering of porous solids are given in Table 2.4.

Let us note that all types of scattering destroy micropores: x-rays and gamma-rays accelerate aging, electrons stimulate chemical reactions, while neutrons destroy the continuous phase. Therefore, the results obtained by these methods should be accepted with criticism.

Now, let us consider some examples related to scattering of porous structures by electrons and neutrons.

TABLE 2.4 Comparison of Neutrons and Electrons to Short-Wavelength Photons as Particles for Scattering of Porous Solid

Parameter for comparison	Electron	Neutron	X-rays	Gamma
Mass in the state of immobility	Moderate	Massive	"Zero"	"Zero"
Quantum wavelength	Short	Very short	Long	Moderate
Applicability to measurements of micropores	Applicable	Applicable	Problematic	Applicable
Reflection by solid structure	Moderate	Problematic	Good	Moderate
Destruction of solid phase	Intensive	Very intensive	Moderate	Intensive
Destruction of micropores	Very intensive	Very intensive	Very intensive	Very intensive
Penetration through solid films	Moderate	Low	Moderate	High
Absorption by solid phase	Strong	Very strong	Moderate	Low
Beat out of protons	None	Very strong	None	None
Stability of particle	Stable	Unstable	Stable	Stable
Ionization, radical formation in condensed phase	Very intensive	Very intensive	Intensive	Very intensive
Registration of particles	Easy	Easy	Easy	Problematic

A. Scattering by Electrons

(a) Example 2.16. Equipment for electron and x-ray scattering. Energy-dispersive x-ray analyzers are described by in Ref. 28 as provided with nonmagnetic collimators. Modifications include the use of different materials and a corrugated structure on the inner surface of the collimator while shaping the collimator so that its input aperture is narrower than its exit. The collimators may be used in electron microscopes employing x-ray spectrometers, where the collimator is provided in the head portion of the spectrometer and a part of the collimator is arranged in a leakage magnetic field of an objective lens included in the electron microscope, whereby the paths of scattering electrons are curved and thus prevented from colliding

with the x-ray spectrometer, thereby removing background noise from the x-ray spectrum [28].

(b) Example 2.17. Use of electron scattering for studies of biochemical processes. Flanagan et al. [29] used scanning transmission electron microscopy (STEM) for direct mass determination of individual particles in the system "*Escherichia coli* ATP-dependent caseinolytic protease (Clp), protease, ClpP, and ATPase, ClpA." Active ClpP was overexpressed to approximately 50% of soluble protein in *E. coli* and purified to homogeneity. Direct mass determination of individual particles using scanning transmission electron microscopy (STEM) yielded a mean native molecular mass of 305 ± 9 kDa for the ClpP oligomer, suggesting that it was having a tetradecameric structure. Small-angle x-ray scattering (SAXS) curves were determined for ClpP in solution at concentrations of 1–10 mg/mL. A combination of STEM and SAXS data was used to derive a model for ClpP, comprising a cylindrical oligomer about 100 Å in diameter and about 75 Å in height with an axial pore about 32–36 Å in diameter. The volume of the pore was estimated to be approximately 70,000 $Å^3$, similar in size to those found in chaperone proteins, and was found large enough to accommodate unfolded polypeptide chains, although most globular folded proteins would be excluded [29].

B. Scattering by Neutrons

(a) Example 2.18. Use of neutron scattering for studies of molecular sieves. Mansour et al. [30] described high-resolution inelastic neutron scattering measurements of the molecular dynamics of water confined to a porous host, the molecular sieve known as MCM-41 having a hexagonal array of parallel pores with average pore diameter of 27 Å. Previous neutron measurements probing higher energy transfers, and thus shorter time scales, have been analyzed with both a rotation–translation diffusion model and a stretched exponential intermediate scattering function. The dynamics on longer time scales presented here are modeled well with a stretched exponential relaxation in a confining geometry. The observed molecular dynamics of water are three orders of magnitude slower than has been previously reported for water confined in MCM-41 [30].

(b) Example 2.19. Use of neutron scattering for water in porous media. Fratini et al. [31] proposed a method of analysis of high-resolution quasielastic neutron-scattering (QENS) spectra of water in porous media, applied to the case of water in hydrated tricalcium and dicalcium silicates. They plotted the normalized frequency-dependent susceptibility as a

function of a scaling variable [omega]/omega(p), where omega(p) was the peak position of the susceptibility function. QENS data were scaled into a single master curve and fitted with an empirical formula proposed by Bergman to obtain three independent parameters describing the relaxation dynamics of hydration water in calcium silicates [31].

(c) Example 2.20. Use of neutron scattering for studies of silica. Webber et al. [32] applied a combination of neutron scattering with NMR cryoporometry to studies of sol–gel silicas with nominal pore diameters ranging from 25 to 500 Å were studied by NMR cryoporometry and by neutron diffraction and small-angle scattering from dry silicas over the Q range 8, 10(−4) Å (−1) ≤ Q ≤ 17 Å (−1). Density and imbibition experiments were also performed. Geometric models of porous systems were constructed and were studied by both analytic techniques and Monte Carlo integration. These models, combined with the information from the above measurements, enabled the calculation of the fully density corrected solid–solid density correlation functions $G(r)$ for the sol–gel silicas, deduction of the (voidless) silica matrix density, measurement of the silica fraction in the grain and of the packing fraction of the silica grains, and an estimation of the water equivalent residual hydrogen on the dried silica surface. In addition, the pore diameter D, pore-diameter-to-lattice-spacing ratio D/a, and pore and lattice variance sigma could also be measured. While the NMR cryoporometry pore diameter measurements for the sol–gel silicas showed excellent colinearity with the nominal pore diameters as measured by gas adsorption and the calculated pore diameters from the measured neutron scattering show surprisingly good agreement with these measurements at large pore diameters, Webber et al. [32] notified a divergence between the calibrations for pore diameters below about 100 Å [32].

(d) Example 2.21. Combination of x-ray and neutron scattering for characterizing the surface roughness and porosity of a natural rock. Broseta et al. [33] used small-angle x-ray and neutron scattering to characterize the surface roughness and porosity of a natural rock that were described over three decades in length scales and over nine decades in scattered intensities by a surface fractal dimension $D = 2.68 \pm 0.03$. When this porous medium was exposed to a vapor of a contrast-matched water, neutron scattering reveals that surface roughness disappears at small scales, where a Porod behavior typical of smooth interfaces was observed instead. In the estimation of Broseta et al. [33], water-sorption measurements confirmed that such interface smoothing was mostly due to the water condensing in the most strongly curved asperities rather than covering the surface with a wetting film of uniform thickness [33].

VII. ACOUSTIC STUDIES OF POROSITY

(a) Physical processes related to sound acting solid surface. As follows from the above analysis, many of difficulties related to the use of EM waves for studies of porosity result from the nature of EM, namely, the perpendicularity of the vibration of EM field to the direction of the propagation of EM waves. This problem can be neutralized if one uses waves of another nature, e.g., acoustic waves, in which the vibration is parallel to the direction of the propagation of waves. Let us note such an obvious fact that sound waves, independently of their length, could penetrate through very narrow pores having the right cylinder form. Since most pores do not have cylinder form, of course, they are not permeable for long acoustic waves, but combinations of different effects related to internal reflections inside the porous structure allow sometimes the obtainment of such information about pores as their opening, form, connectedness of different fragments of porous substructures, minimal/maximal size, etc. Thus, one could obtain the important information about a porous materials just exploring the intensity of waves penetrating through the sample, as the function of acoustic wavelength. Acoustic waves are easily registered. Varying the interior fluid inside the pores (e.g., filling them with water perfectly transporting acoustic vibrations), one can obtain some additional information.

However, as it was in the case of neutron and electron scattering, one should analyze also the factors related to interactions of sound with interface.

The problem of the interaction of sound with various kinds of multiphase condensed systems was largely studied in the physical literature [34–43]. One finds numerous very special and interesting effects, such as

Decomposition of liquids and formation of free radicals (sonolysis)
Degassing of liquid phase
Chain reactions in the liquid phase
Luminescence (sonoluminescence)
Destruction of solid phase

Among all these effects, the destruction of the solid phase contacting with vibrating liquid can cause us most serious problems. On the other hand, various effects caused by vibration allow a variety of options in the indirect registration of acoustic waves.

Let us compare sound to neutron, electron, and x-rays as tools for experimental study of porosity (see Table 2.5).

Now, let us consider some examples of acoustic studies of pores.

TABLE 2.5 Comparison of Acoustics to Neutron, Electron, and X-ray Scattering of Pores

Parameter for comparison	Acoustics (long waves)	Acoustics (short waves)	Neutron scattering	Electron scattering	X-ray scattering
Applicability to micropores	Problematic	Possible	Possible	Possible	Problematic
Destructive or nondestructive	Nondestructive	Destructive	Destructive	Sometimes destructive	Nondestructive in most cases
Stimulation of chemical reactions	Rare	Yes	Yes	Yes	Yes
Availability for registering	Yes	Yes	Yes	Yes	Yes
Stimulation of aging	Negligible	Yes	Yes	Yes	Yes

(b) Example 2.22. Sound absorption measurements (acoustic insulation). Lu et al. [44] presented a combined experimental and theoretical study for the feasibility of using such perspective porous material as aluminum foams with semiopen cells for sound-absorption applications. The foams were processed via negative-pressure infiltration, using a preform consisting of water-soluble spherical particles. The dependence of pore connectivity on processing parameters, including infiltration pressure, particle size, wetting angle, and surface tension of molten alloy was analyzed theoretically. Normal sound-absorption coefficient and static flow resistance were measured for samples having different porosity, pore size, and pore opening. The predicted sound-absorption coefficients are compared with those measured. To help select processing parameters for producing semiopen metallic foams with desirable sound-absorbing properties, emphasis was placed on revealing the correlation between sound absorption and morphological parameters such as pore size, pore opening, and porosity [44].

(c) Example 2.23. Acoustic study of microporous structure. Miller et al. [45] explored the ultrasonic activation of free microbubbles, encapsulated microbubbles, and gas-filled micropores using available linear theory. Encapsulated microbubbles, used in contrast agents for diagnostic ultrasound, were found to have relatively high resonance frequencies and damping. At 2 MHz the resonance radii were found about 1.75 μm for free microbubbles, 4.0 μm for encapsulated microbubbles, and 1.84 μm for gas-filled micropores. Higher-pressure amplitudes were needed to elicit equivalent subharmonic, fundamental, or second-harmonic responses from the encapsulated microbubbles, and this behavior increases for higher frequencies. If an encapsulated microbubble became destabilized during exposure, the resulting liberated microbubble would be about twice the linear resonance size, which would be likely to produce subharmonic signals. The authors suggested that scattered signals used for medical imaging purposes might be indicative of bioeffects potential because the second-harmonic signal was proportional to local shear stress for a microbubble on a boundary and a strong subharmonic signal might imply destabilization and nucleation of free-microbubble cavitation activity. The potential for bioeffects from contrast agent gas bodies decreased rapidly with increasing frequency. This information was found valuable for understanding of the etiology of bioeffects related to contrast agents and for developing exposure indices and risk management strategies for their use in diagnostic ultrasound [45].

(d) Example 2.24. Use of ultrasound in studies of porous structure. Langton et al. [46] analyzed the effectiveness of studies of porous structure with ultrasound. They reported on the measurement of trabecular perimeter

and fractal dimension on the two-dimensional images used to create the stereolithography models. Adjusted coefficients of determination (R2) with *n*BUA were 94.4% ($p < 0.0001$) and 98.4% ($p < 0.0001$) for trabecular perimeter and fractal dimension, respectively [46].

(e) Example 2.25. Technique for acoustic studies of porous solids. The technique for determining a borehole fluid property based on finding the acoustic impedance of the drilling fluid using reflections from a precise metal disk is described in Ref. 47. Because the reverberation characteristics of an acoustic wave depend in part on the acoustic wave shape, the first reflection from the metal disk may be used to calibrate the measurement. A method for determining a borehole fluid property includes (1) generating an acoustic signal within a borehole fluid, (2) receiving reflections of the acoustic signal from the fluid, and (3) analyzing a reverberation portion of the acoustic signal to determine the property. The analyzing of the reverberation portion may include obtaining a theoretical reverberation signal and relating the measured reverberation signal with the theoretical reverberation signal to determine the borehole fluid property [47].

(f) Example 2.26. Technique for studies of thin films and solid surfaces. Maznev and Mazurenko [48] present an optoacoustic measuring device for thin films and solid surfaces, in which the probe beam is split into a first probe beam portion and a second reference beam portion. The splitting of the probe beam is achieved using a phase mask that also splits the excitation beam. The probe beam is aligned using a retroreflector on a motorized stage to control the beam angle. Excitation and probe or reference beams are overlapped at the sample surface. The first probe beam portion gets diffracted by material disturbances generated by excitation beams. The diffracted part of the first probe beam portion is colinear with the second reference beam portion, resulting in heterodyning. The heterodyne signal measured by the detector is analyzed in order to determine thickness and/or other properties of a thin film or solid surface. The invention improves magnitude and reproducibility of the optoacoustic signal which results in enhanced precision of measurements [48].

(g) Example 2.27. Technique for measuring or monitoring the noise and porosity of road asphalts. Osele [49] proposed a device for measuring or monitoring the noise and porosity of road asphalts, consisting of at least two microphones associated, respectively, with at least two wheels of a vehicle, so as to be able to measure the acoustic pressure caused by contact of the tires of the wheels on the asphalt on which the vehicle is traveling, a position transducer able to measure and/or monitor the advancement of the vehicle on the asphalt, and a multichannel spectrum analyzer connected to

the output of the microphones and the position transducer to analyze signals coming from the microphones and from the position transducer and output data concerning the noise level and porosity of the portion of asphalt covered [49].

(h) Example 2.28. Technique for drying of saturated porous solids. Meyer [50] proposed a method for drying a porous solid saturated with a fluid including the steps of subjecting the saturate porous image to an ultrasonic signal to release the fluid from the saturated porous image layer and removing the fluid from said saturated porous solid, said subjecting step including applying a predefined acoustic slow wave frequency based upon the particle sizes in the said porous image layer thereby causing the fluid to move from an interior of said porous particulate image to an outer surface of said porous image, where fluid could be removed by conventional image [50].

(i) Example 2.29. Study of acoustic waves on liquid in pores. Poesio et al. [51] investigated experimentally and theoretically the influence of high-frequency acoustic waves on the flow of a liquid through a porous material. The experiments were performed on Berea sandstone cores. Two acoustic horns were used with frequencies of 20 and 40 kHz and with maximum power output of 2 and 0.7 kW, respectively. Also, a temperature measurement of the flowing liquid inside the core was made. A high external pressure was applied in order to avoid cavitation. The acoustic waves were found to produce a significant effect on the pressure gradient at constant liquid flow rate through the core samples. During the application of acoustic waves the pressure gradient inside the core decreases. This effect turned out to be due to the decrease of the liquid viscosity caused by an increase in liquid temperature as a result of the acoustic energy dissipation inside the porous material [51].

(j) Example 2.30. Estimation of tortuosity (characteristics of the complexity of the form of percolation paths) and characteristic length of pores. Moussatov et al. [52] suggested a new ultrasonic method of acoustic parameter evaluation for porous materials saturated by air (or any other gas). The method is based on the evolution of speed of sound and the attenuation inside the material when the static pressure of the gas saturating the material is changed. Asymptotic development of the equivalent fluid model of Johnson–Allard is used for analytical description. The method allows an estimation of three essential parameters of the model: the tortuosity and the viscous and thermal characteristic lengths. Both characteristic lengths are estimated individually by assuming a given ratio between them. Tests are performed with industrial plastic foams and granular substances (glass beads, sea sand) over a gas pressure range from

0.2 to 6 bar at the frequencies 30–600 kHz. The present technique has a number of distinct advantages over the conventional ultrasonic approach: operation at a single frequency, improved signal-to-noise ratio, possibility of saturation the porous media by different gases. Moussatov et al. [52] concluded that in the case when scattering phenomena occur, the present method permits a separate analysis of scattering losses and viscothermal losses [52].

VIII. STUDIES OF STRUCTURE AND PROPERTIES BASED ON PORE–FLUID INTERACTIONS

Above we have considered existing methods of study of microporous structures, using various kinds of scattering, by electromagnetic waves, acoustic waves, and elementary particles. The main conclusion we can do from this above analysis is that there is no direct method for precious experimental study of micropores. Hence, one may try indirect methods for estimations of properties of microporous structures. The most obvious solution is indirect study of microporous structure, using measurements of its various properties related to interactions with fluids. Such indirect method for study of micropores is, in principle, more spread in the science of porosity than direct measurements. Two most important experimental methods using pore–fluid interactions are

Adsorption–desorption measurements
Percolation–permeability (intrusion) measurements

One can mention also other techniques of such kind, e.g., thermoporometry, but their use is much more rare.

A. Adsorption–Desorption

Traditional adsorption–desorption studies are based on different adsorptive properties of different micropores. The general principles of adsorption were considered in Chapter 1; theoretical aspects of adsorption and desorption will be considered in Chapter 4; in this chapter we consider only experimental aspects of adsorption and desorption.

A typical experimental device for adsorption measurements is presented on Fig. 2.2.

A gaseous component is injected from a gas source through manometer measuring the pressure to a vessel containing a porous material. The process of gas injecting should be as slow as possible (on purpose) to minimize nonequilibrium effects. The change in the weight of the vessel

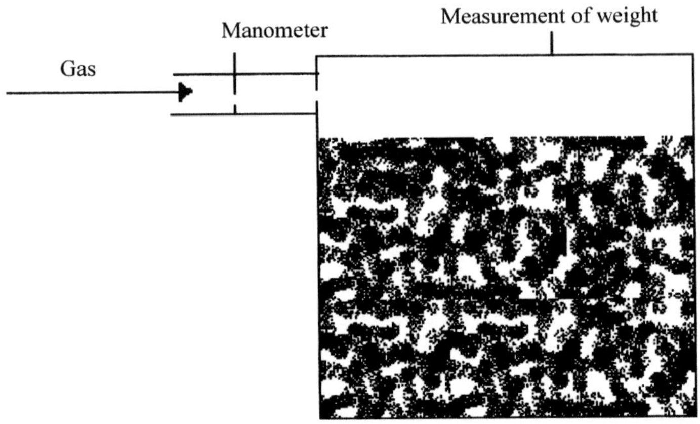

FIG. 2.2 Typical experimental device for adsorption isotherm measurements.

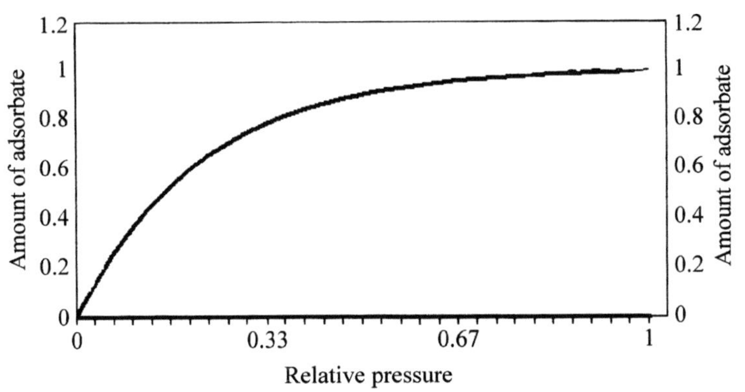

FIG. 2.3 Typical isotherm of adsorption.

is preciously measured. The temperature is controlled. As a result, one obtains the function $\theta(P)$ at temperature T, where θ is the amount of the adsorbed substance. Of course, the function $\theta(P)$ is always characterized by monotonic rising. However, over several values of pressure P_0 the value of θ remains without change. This pressure is called *saturation pressure*. The typical form of the adsorption isotherm curve is given on Fig. 2.3.

Of course, for $P=0$, $\theta(P=0)=0$, and for $P/P_0=1$, $\theta(P/P_0=1)=1$. The curve has negative second derivative; physically that means that free fragments in the porous material, able to capture the adsorbate, dilute.

The above considerations are valid for a material having the simple structure, e.g., macropores without nanopores—mesopores or micropores. However, for a real porous materials, all these groups of pores are more or less found, and the curve of the adsorption isotherm has much more complicated form (see Fig. 2.4).

Why is the adsorption isotherm so complicated this time? As in the case presented by Fig. 2.3, the process of adsorption starts when the gas appears. The initial part of the curve is very similar to Fig. 2.3: the adsorption is almost only on micropores, due to their very high energy, and gets to saturation when micropores are diluted. However, contrary to Fig. 2.3, the further increase of the pressure makes mesopores take part in the adsorption process. The second upstair of the adsorption curve is related to the adsorption on mesopores—of course, the pressure needed for this process is much more than for adsorption on micropores. When mesopores are diluted, the curve gets the second saturation plate. If the pressure is always increased, macropores start adsorbing the injected gas, which provides the third up-stair and, respectively, after the dilution of macropores, the third saturation plate. In principle, very sensitive equipment can register the fourth upstair caused by the adsorption on the exterior surface, but its energy is almost the same as that for macropores. The fourth upstair is very low and short.

What happens if we start reducing the pressure of the gas? Theoretically, the system should get back all the trajectory of the adsorption isotherm curve. Although, in real practice this is absolutely impossible

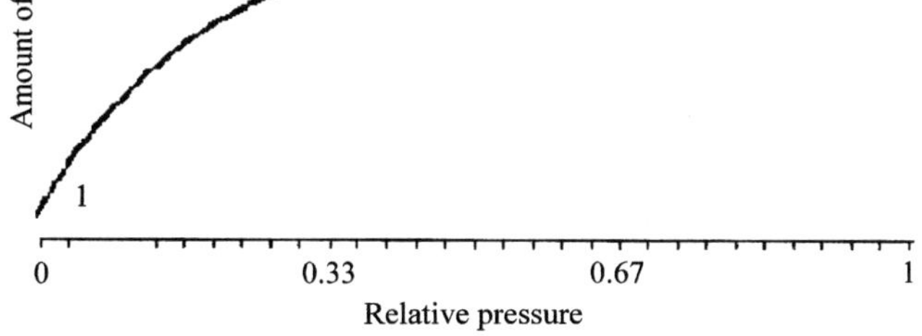

FIG. 2.4 Typical isotherm of adsorption on a porous material having micropores, mesopores, and macropores.

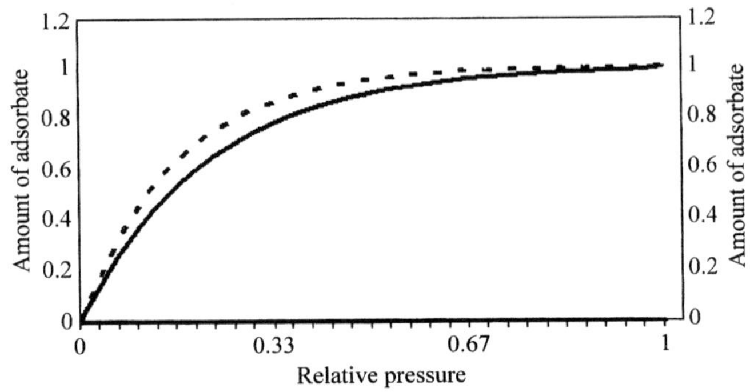

FIG. 2.5 Typical hysteresis loop for micropores: (——) adsorption isotherm, (- - - -) desorption isotherm.

because of hysteresis related to irreversible effects. For the plate surface and macropores the hysteresis loop is negligible, and it is very difficult to measure it against the error of the equipment. For mesopores, this is much more observable, and for micropores the hysteresis loop is comparable to the isotherm curve itself (see Fig. 2.5).

Let us consider some examples of using adsorption isotherms for the characterization of microporous structures.

(a) Example 2.31. Studies of heterogeneity of microporous structures by adsorption isotherms. Madey and Jaroniec [53] analyzed opportunities in the characterization of heterogeneity of microporous structures by adsorption isotherms.

Adsorption isotherms were used to evaluate the adsorption-energy and micropore-size distributions. In addition to the internal surface area and the mesopore- and macropore-size distributions, the adsorption-energy and micropore-size distributions were recommended for characterizing the energetic and structural heterogeneities of microporous adsorbents [53].

B. Thermoporometry

Though the technique of desorption measurements is very similar to that for adsorption, there are some technical solutions providing additional information about micropores, namely, due to desorption. A very interesting example of such technique is thermoporometry.

The main idea of thermoporometry is based on the different behavior of adsorbed substances in different micropores. For example, if we adsorb water in micropores and then heat the product, the adsorbed water does not leave the adsorbent simultaneously: first is water adsorbed in macropores (maybe also exterior surface), then water vapor from mesopores, later from supermicropores, and after all from ultramicropores. Of course, measurements of the amount of vapor removed should be very careful, as always in adsorption studies.

Of course, instead heating, one may use freezing (cryoporometry).

The principal form of thermoporometric graph is given on Fig. 2.6.

Let us consider some examples of thermo- and cryoporometry technique.

(a) Example 2.32. Cryoporometric study of microporous silica. Webber et al. [32] studied sol–gel silicas with nominal pore diameters ranging from 25 to 500 Å by NMR cryoporometry and by neutron diffraction and small-angle scattering from dry silicas. Density and imbibition experiments were also performed. Geometric models of porous systems were constructed and studied by both analytic techniques and Monte Carlo integration. These models, combined with the information from the above measurements, enabled the calculation of the fully density-corrected solid–solid density correlation functions $G(r)$ for the sol–gel silicas, deduction of the (voidless) silica matrix density, measurement of the silica fraction in the grain and of the packing fraction of the silica grains, and an estimation of the water equivalent residual hydrogen on the dried silica surface. In addition, the

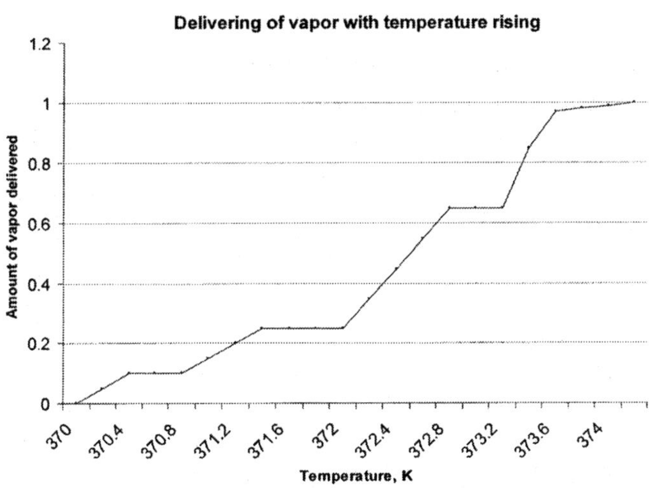

FIG. 2.6 Typical thermoporometric curve for a material containing macropores, mesopores, and micropores.

pore diameter D, pore-diameter-to-lattice-spacing ratio D/a, and pore and lattice variance sigma could also be measured. While the NMR cryoporometry pore diameter measurements for the sol–gel silicas show excellent colinearity with the nominal pore diameters as measured by gas adsorption and the calculated pore diameters from the measured neutron scattering show surprisingly good agreement with these measurements at large pore diameters, there was found a divergence between the calibrations for pore diameters below about 100 Å [32].

(b) Example 2.33. Thermoporometric study of microporous silica. Van Bommel et al. [54] studied the porous texture of fumed silica–aerogel composite materials by thermoporometry, investigating the effect of fumed silica (Aerosil) powder in aerogels, added during the sol–gel processing, and the textural change due to the autoclave drying process as a function of the fumed silica-to-aerogel ratio of the systems. The meso- and macroporosity of the fumed silica–aerogel composites was found mainly influenced by the ratio of fumed silica to aerogel and only slightly by the autoclave drying process. Addition of fumed silica powder resulted in an increase of the mean pore radius of the system and in a decrease of the meso- and macropore volume. By contrast, the microporosity was found hardly affected by the addition of the fumed silica powder; it was only influenced by the autoclave drying process [54].

C. Measurements of Kinetically Limited Adsorption

As it was notified above, the main technique of adsorption isotherm measuring is based on the assumption of the equilibrium in the studied system. However, also nonequilibrium of adsorption can be studied. In addition to the technique described above, the following methods are used for this purpose:

Maximum-bubble-pressure method [55]
Ellipsometry [55]
Surface-light scattering [55]
Laser Doppler velocimetry [55]
Neutron reflection [55,56]

D. Combined Methods

In many cases, indirect and direct (scattering) methods of study of micropores are combined. That allows one to obtain the maximal available information. In most cases, such combinations comprise adsorption isotherm measurement. For example, in the above-considered process of the synthesis of metal-impregnated active carbon (Example 2.9) the presence of Co_3O_4 and CoO phases was proved by x-ray measurements, while that of metal Co was established by magnetic measurements. The porous structure changes were studied by adsorption studies. The characterization of the samples was performed by physical adsorption of N_2 (77.4 K) and CO_2 (273 K) [17].

Let us notice that since adsorption–desorption studies do not provide any direct information about properties of microporous structures, their results need some theoretical treatment before being combined to the direct results obtained by scattering (that concerns not only Ref. 17 but all combined explorations). An appropriate theory should provide some knowledge about the relationship between the structural characteristics and the measurable data on adsorption–desorption.

E. Percolation, Permeability, and Porosimetry

Traditional percolation–permeability studies are based on different penetration of different molecules and ions through different micropores.

Notions of percolation and permeability are very important for the characterization of solid structures and related processes of solid–fluid interactions. However, the use of these notions is not always correct; moreover, they are frequently mixed. As a result, misunderstanding appears in existing concepts of percolation and permeability, in addition to physical and mathematical difficulties in their studies.

Methods developed in theories of percolation and permeability, are largely used in other fields, even having no relation to porosity, e.g., modeling of colloid systems [57], electric conduction and superconductivity [58–62], optics [63,64], studies of mechanical stability [65], and even urban problems of communications and transport [66]. In the methodological aspect, this is possible due to the analogy between the mentioned phenomena. For example, electrical properties of a structure can be modeled as a result of interactions of its conductive and nonconductive fragments; in such case, the total electrical conduction of the structure is attributed to the connection of the conductive fragments. Such urban problems as transport and radio-net optimization in cities can also be formulated in terms of percolation and permeability, respectively, and then solved by the same technique that derived for solid–liquid interactions.

In the terminological aspect, the notions of percolation and permeability are very close, almost synonyms. However, their use in the physical literature seriously differs. At first look, one may easily find that keyword *percolation* provides references mostly in theoretical physics, while *permeability* is used in applied physical chemistry.

In most of references, *percolation* means the ability of a structure to allow the penetration of fluids, whereas *permeability* means the amount of the fluid penetrating through the same structure. In electrical terms, percolation describes whether or not the material is conductive, while permeability (if this was applied to electrical phenomena) might show the level of the conductivity. The same comparison can be presented also for mechanical stability, urban processes, etc. Thus, the practical use of both terms is mostly determined by traditions existing in a given discipline: just *percolation* is widely accepted term, while *permeability* is mostly replaced with *conductivity*, etc.

If we analyze the terminological literature, we find that the *percolation problem* is defined as "the problem of determining the critical threshold concentration of conducting links in a percolation network at which an infinite cluster of conducting links is formed and the lattice transforms from an insulator to a conductor" [67]. According to Ref. 67, the notion of permeability may have both senses: qualitative (whether or not percolation exists) and quantitative (if the percolation exists, what is the amount of the fluid or current through the structure). There are two principally different definitions for permeability proposed by Ref. 67:

The ability of a membrane or other material to permit a substance to pass through it [67]

The amount of substance which passes through the material under given conditions [67]

A similar definition for permeability was proposed in Ref. 68: permeability is defined as the rate of diffusion of gas or liquid under a pressure gradient through a porous material. Let us note that this definition is also problematic: what happens if the mass transfer is not due to pressure gradient but due to thermodiffusion or electrical potential? Does Ref. 68 mean that the physical sense of the phenomenon changes? In this aspect, definitions given in Ref. 67, though contradictive, are preferable—of course, if their contradiction is taken into account.

In the scientific literature, the main divergence between these notions is that *percolation* means the qualitative effect ("yes–no" to the penetration through the solid structure), while *permeability* describes this effect (if yes to the penetration, while this yes may have exceptions, as it is shown below) quantitatively: how much fluid penetrates?

In addition to the main divergence between the notions of percolation and permeability, we may note the following factors of their difference:

Percolation characterizes the solid structure only, its porosity, complexity, opening–semiopening–closing of substructures, while permeability describes the entire system "solid + fluid", and in this case the solid structure is additionally characterized by connectedness and tortuosity (see below).

Percolation tests, if these are correctly organized, cannot cause any destruction of the tested solid structure, whereas permeability tests can easily be destructive, e.g., Hg intrusion, because of destroying thin pore walls (especially for closed and semiopen pores) under the exterior pressure. Moreover, as it will be shown below, the structure destruction during the permeability tests may cause hysteresis: the curves permeability vs. the pressure of the fluid may get changes because of the accumulation of destructions.

The obligatory condition of percolation is the existence of percolation path: a group of voids forming a trajectory inside the solid structure from one its side to its opposite side.

If we return to Chapter 1, we find that Fig. 1.1 presents not only an open pore but also a percolation path in a very thin film. All open pores and only open pores are percolation paths. All closed and semiopen pores are lost for permeability, unless they are destroyed and transformed to open pores (that may happen, for example, because of the pressure of a fluid flux).

Of course, the length, the shape, and the number of percolation paths influence the measurable value of the permeability. In principle, that is very similar to electrical conductivity determined by the geometry and the number of conductive fragments. As in electricity, the resistance to passing through of fluids can be characterized by a special notion, *tortuosity*. Tortuosity is completely analogous to electrical resistance. As electrical resistance, tortuosity can be measured only when passing through of a fluid is over zero. As electrical resistance, the value of the tortuosity increases with the length of percolation paths and decreases with their thickness.

In the structural aspect, the tendency of open pores to form low-tortuosity fragments can be characterized by *connectedness* (this notion was introduced in Chapter 1): a parameter showing the tendency of voids to form a neighborhood. Numerically, connectedness can be characterized by the following parameter:

$$\phi = [\Theta(m = n_0)/\Theta(m = 0)]^{1/(n_0+1)}\xi \tag{2.9}$$

where n_0 is the coordination number of cells belonging to the structure (also, n_0 is the maximal number of neighbors around the minimal void corresponding to one empty cell only), m is the number of voids around a chosen void, and $\Theta(m=0)$ is the number of voids having (each) m empty neighbors. As in Chapter 1, ξ is the porosity of the considered structure.

At the same porosity, the connectedness parameter in Eq. (2.9) increases when voids show the tendency to get together [then $\Theta(m = n_0)$ is maximal] but decreases when the voids show the tendency to be isolated. Of course, even a very high value of the connectedness does not assure that the percolation exists: if we compare Figs. 1.1–1.3, we see that the connectedness is approximately the same, but only the structure on Fig. 1.1 provides a percolation path. However, high connectedness means that if the percolation exists, the value of permeability can be very high. High-connectedness structures may give sharp risings of the permeability, e.g., if the high-connectedness substructure is semiopen but suddenly gets open (for example, due to removal of the closing fragment under the exterior pressure), the resulting permeability immediately gets a high value.

In the numerical aspect, percolation is characterized by percolation threshold: this can be defined as the combination of a material's characteristics under which percolation just appears. If no high-connectedness substructure gets an opening, at the moment of the appearance of the percolation, the permeability is zero.

Below, to avoid misunderstanding, we will use the notion of permeability only in its quantitative sense, while the qualitative aspect ("yes–no" to the passing through) is characterized by percolation.

As mentioned above, in most of real situations the value of the permeability is zero when the percolation is absent (below the percolation threshold), whereas the appearance of percolation (over the percolation threshold) should mean the positive value of the permeability. However, we can give two examples when this rule does not work.

1. Zero Permeability over the Percolation Threshold (Percolation Retard)

Such effect can be caused by some specific properties of the fluid (especially if this one is a viscous liquid), particles of which are too big and cannot penetrate into the percolation paths up to their end or just the fluid particles are "repulsed" by the solid medium. For example, if we try to inject a melted polymer into very narrow pores, this can fail just because of geometry. Of course, if we replace such fluid with another, more compatible with the geometry of the porous structure, the permeability will not be zero.

Let us notice that percolation retard allows us the selective separation of liquid and gaseous mixtures; of course, since some components in a fluid can pass through a microporous membranes while others do not, the separation process does not need additional efforts.

The effect of "repulsion" of the liquid by the medium is not typical for micropores (as we noted in Chapter 1, micropores in hydrophobic carbon accept even water) but is possible in principle.

2. Positive Permeability Below the Percolation Threshold (False Percolation)

This effect is extremely specific, and the majority of experimental researchers of porosity have no chance of finding it. That can happen when the solid structure dissolves (absorbs) the fluid, as in the case of the absorption of hydrogen by platinum. Such absorption allows the dissipation of the fluid inside the solid structure even without pores. The absorbed molecules penetrate through the solid structure due to the simple diffusion and leave the solid from the opposite side. Of course, if this fluid is replaced with another substance, the effect of permeability below the percolation threshold disappears. Let us note that the substances typically used in intrusion studies—nitrogen and mercury—do not specifically interact with traditional porous materials listed in Chapter 1.

F. Technique of Percolation–Permeability Studies

Let us note that the technique for percolation and permeability (intrusion) tests is principally different.

The principal scheme of the percolation tests is given by Fig. 2.7.

The tested solid sample (film) is positioned horizontally. Above it, one places a source of a fluid providing no specific interactions with the material of the tested sample; also, the particles of the fluid must be little enough to avoid geometrical difficulties—hence, both percolation retard and false percolation are avoided. Also, the experiment needs to avoid chemical interactions between the tested sample and the fluid (e.g., forbidden to test a sodium chloride structure by water), in order to avoid irreversible changes in

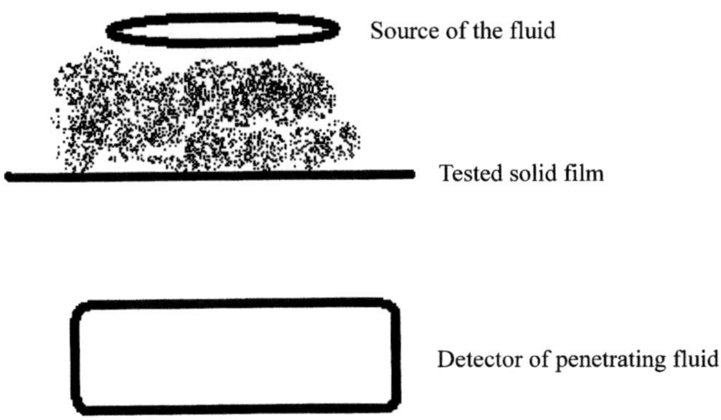

FIG. 2.7 Principal scheme for percolation tests.

the solid structures. The fluid is dropped onto the tested sample, then a detector under the sample checks whether a drop of the fluid is found on the bottom surface of the tested sample. Such a scheme makes impossible the destruction of the solid structure; although, the disadvantage of such technical solution results in waiting too long for the result. Can the tested sample considered as nonpermeable if no percolation is found in 1 d of measurements? Maybe just the percolation path is too long? For this reason, in real laboratory conditions, researchers use for the same purpose the technique designed for measurements of permeability (see Fig. 2.8).

The fluid is closed in a volume limited by the tested sample (film), the walls of the vessel, and the piston. When the pressure on the piston gets increased the fluid presses onto the film. As in the previous case, the fluid penetration through the film is checked by a detector.

The permeability tests accelerate much the rate of the penetration, but the pressure causes destructions in the tested sample. In some situations (e.g., when percolation tests are combined with mechanical tests) such destructions are even desired, because they allow the estimation of the mechanical stability of the sample, but the information about the initial structure of the sample, obtained on such a way, is obviously wrong. If the pressure is unloaded, the permeability curve measured by the detector differs from that obtained on the stage of pressure loading, and one obtains a permeability hysteresis loop.

A typical form for the permeability hysteresis loop is given by Fig. 2.9.

We see from Fig. 2.9 that the measured permeability on the stage of the pressure loading is negligible until the exterior pressure is increased up to several value. Could one assume that this value is the percolation threshold of the initial solid structure? No: if the sample got destructions because of

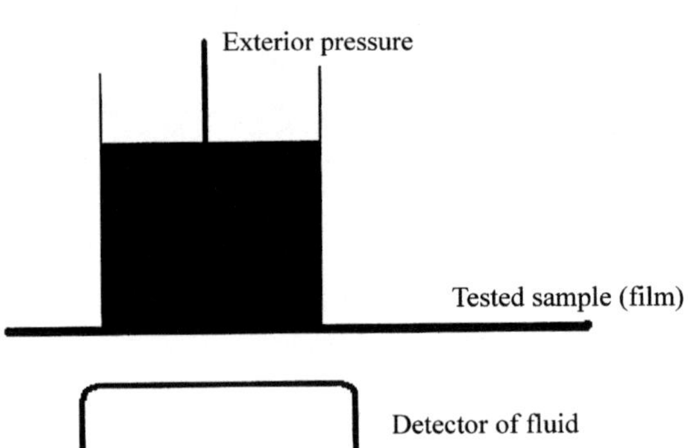

FIG. 2.8 Principal scheme for permeability measurements.

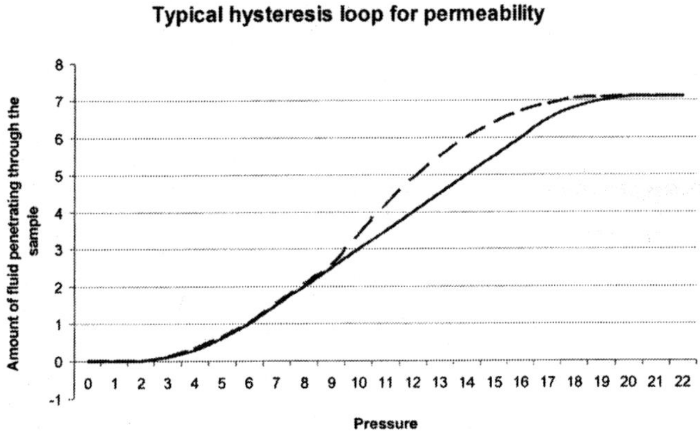

FIG. 2.9 Typical hysteresis loop for permeability.

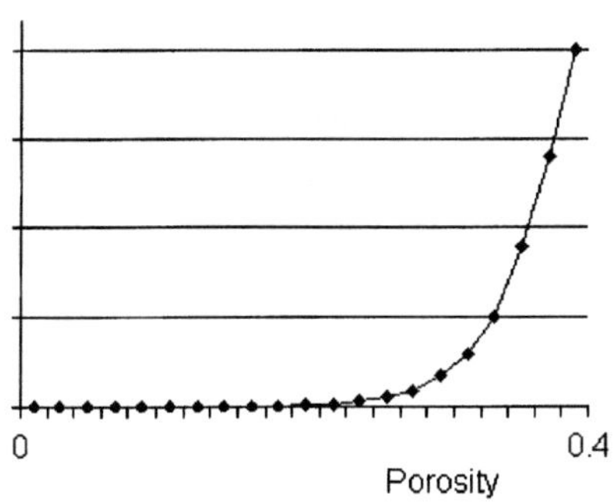

FIG. 2.10 Observation of percolation threshold in tests of porous samples having different porosity.

the exterior pressure, its structure differs significantly from the initial one, whereas the assumption that the destructions were negligible just means that the system was initially over the percolation threshold, but the pressure was not sufficient for enough fast passing through of the fluid.

Let us compare Fig. 2.9 to the graph for percolation threshold appearance (see Fig. 2.10).

Though the curve at Fig. 2.10 may seem well similar to the rising curve at Fig. 2.9, there are some principal factors of difference between them:

Figure 2.10 presents results of measurements on *different* samples, whereas Fig. 2.9 presents results on the same sample.

In Figure 2.9, the system is always over the percolation threshold, while in Fig. 2.10 the system is only over the percolation threshold for several sample.

Hysteresis. After the measured passing through is positive, the function permeability vs. exterior pressure is approximately linear until, because of nonlinear effects (e.g., specific interactions due to high pressure), the rising of the permeability is slowed. If the pressure loading is continued, we just would observe a sudden sharp vertical rising of measured permeability because of the complete destruction of the sample. If we do not intend to destroy it (as it is in the case presented by Fig. 2.9), we slowly unload the pressure. If the permeability tests were nondestructive, we would obtain the curve of depressing exactly as the curve for pressure loading. However, that is not the case: at the same pressure, the unloading curve is over the loading curve, because the film has interior destructions increasing its permeability. In any case, when the pressure is completely unloaded, the fluid penetration is stopped (or, more exactly, becomes too slow for the detection).

Let us consider some examples of permeability measurements (not nitrogen or mercury porosimetry).

(a) Example 2.34. Permeametry with air. Eriksson et al. [69] compared permeametry surface areas with surface areas calculated from analysis of granule dimensions by microscopy and ring gap sizing, measured with a steady-state and a transient permeameter for beds of coarse (0.9–1.0 mm), spherically shaped, porous granules. The permeametry surface areas were calculated with an aspect factor predetermined on beds of monodispersed steel balls and with an effective bed voidage calculated from the effective particle diameter of the granules. In the estimation of Eriksson et al. [69], the permeametry methods were also capable of giving significantly different surface areas between granule masses of similar granule size but of small differences in granule shape. They concluded that the air permeametry principle was suitable for the assessment of the external surface area and the geometrical shape (sphericity) of granules [69].

Montillet et al. [70] used permeametry to determine the dynamic surface area and tortuosity of nickel foams.

G. Mercury Intrusion

Many existing standard permeability tests use mercury intrusion [71–76]. Chemical interactions of mercury with the solid structures are almost always negligible. Mercury flux is easily divided into some fluxes when one finds a number of available paths. The detection of mercury is easy. Specific interactions and capillary effects are easily taken into account. If the pressure is well regulated and the detector has a sufficient sensitivity, the results of the tests do not provide linear curves (as it is in Fig. 2.9) but show the contribution of micro- and mesopores. However, working with mercury, one must take into account such important negative features of this substance as a very high toxicity.

Studies using intrusion porosimetry are very widespread, and we will consider some specific examples.

In many cases, researchers combine intrusion studies with other methods. For example, Weishauptova et al. [72] and Domingo-Garcia et al. [73] studied porosity combining nitrogen and carbon dioxide adsorption to mercury porosimetry. In Ref. 74, mercury intrusion porosimetry was used for the estimation of the structure of microporous films used in ink-jet recording medium. Fathima et al. [75] used mercury intrusion porosimetry for analysis of pores in skin.

H. Combination of Permeability to Thermal Conductivity in Structural Studies

Since experimental methods related to permeability of microporous structures are very selective and sensitive, it is very attractive to apply them to problems not related directly to permeability, first of all by specific combinations of permeability-based measurements to others studies. Let us consider an example of such combination.

(a) Example 2.35. Diffusion permeability. Leofanti et al. [76] studied microporous fibers by using the thermal diffusivity of carbon fiber–polymer composite materials. Thermal diffusion is a very specific phenomenon related to linear effects in nonequilibrium that consist in causing diffusion processes without gradients of concentration, due only to temperature gradient; thermal diffusion is described by Onsager equations. This depends on fiber, matrix, and porosity volume fractions. Since the carbon fiber thermal conductivity is much larger than the same parameter for the epoxy resin matrix, thermal diffusivity is estimated in Ref. 77 as a good indicator of the fiber volume fraction. Through-the-thickness thermal diffusivity images were obtained nondestructively on carbon fiber reinforced composite plates [T300 fiber/934 epoxy resin]. The fiber and porosity volume fractions were measured destructively. The destructive test data were used as a standard for comparison of three commonly used transverse equivalent thermal conductivity models: the stacked plate, the mixed flow cylindrical fiber model, and the composite circular assemblage model. In these models, a porosity correction was also studied. Zalameda

[77] found a good agreement between the composite circular assemblage and cylindrical fiber model when porosity levels were less than 2%. The porosity correction helped to reduce the chi-squared value by 24% in the cylindrical filament model [77].

Let us note that, as it was in the case of adsorption–desorption, percolation–permeability data need some theoretical treatment before being applied to estimations of structural properties. An appropriate theory should provide some knowledge about the relationship between the structural characteristics and the measurable data on percolation and permeability.

IX. MECHANICAL TESTS OF POROUS SOLIDS

Mechanical tests of porous materials are not as important for the characterization of microporous structures as scattering, adsorption, and permeability tests. That is related to the fact that the contribution of micropores to mechanical properties of solids are very difficult for distinguishing from the contribution of other pores. However, they can provide indirect information about the structure. Let us consider an example of such studies.

(a) Example 2.36. Combined mechanical studies of macromolecular systems. Ferrari et al. [78] studied rheological as well as gel-strength properties of nonmedicated gel systems based on polymers of different chemical nature and source: a synthetic product, polyacrylic acid, and a natural product, scleroglucan, both known to produce true gels, and two semisynthetic materials, sodium carboxymethylcellulose and hydroxypropylmethylcellulose, both known to originate polymeric colloidal solutions. The rheological analysis involved the comparison of the different polymers on the basis either of viscosity (constant rate test) or of viscoelastic measurements (oscillation test). The possible relationships between either viscosity or viscoelastic parameters and the gel-strength parameter were investigated in Ref. 78. The mechanical resistance of the pharmaceutical gels examined was found not to depend on the viscosity but rather on the viscoelasticity of the systems [78]. Ferrari et al. [78] concluded that mechanical resistance measurements, similarly to viscoelastic ones, give information on the structural properties of the samples and therefore represent a suitable tool in the optimization of gel formulations.

More than other indirect methods of structural studies, mechanical tests need theoretical treatment, without which they are absolutely useless for estimations of structural properties of microporous media.

X. CONCLUSIONS

1. Methods of experimental study of porosity can be classified by
   a. Sphere of the applicability of each method: preparation or synthesis of porous materials, their structure, and/or properties.
   b. Physical base of the method: fundamental physical phenomenon/phenomena, on which the method is based, and the type of interactions used for the treatment of the porous sample.
   c. Whether the method is destructive, what are the concrete source or experimental error (the principal factors, aging and hysteresis), what is the level of the experimental error, and how can this be taken into account?
   d. Provided information: what are the parameters measured?
2. Existing methods of study of microporous materials on the stage of their synthesis allow the solution of two problems:
   a. Assuring reproducibility of the same materials with the same properties.
   b. Simplification of further experimental studies of the obtained microporous structures.
3. Two principal factors of errors in experimental measurements of properties of microporous structures are
   a. Aging related to spontaneous irreversible changes in the pore structure.
   b. Hysteresis related to irreversibility of interactions of micropores to exterior forces.
4. Applicability of electromagnetic scattering to micropores is limited by wavelength of electromagnetic waves (too long). Only gamma-rays and some x-rays can be used in studies of microporosity.
5. The applicability of neutron scattering to studies of microporosity is limited by the very heavy mass and low stability of neutrons.
6. The applicability of electron scattering to studies of microporous media is limited by the negative electrical charge of electrons.
7. The general problem in studies of microporous structures by scattering consists in significant hysteresis and acceleration of aging of micropores.
8. Because of the above-mentioned difficulties of direct measurements of parameters of microporous structures, indirect methods based on adsorption and permeability measurements become very important. The most detailed information about microporous structures can be obtained due to a combination of results of

direct measurements (scattering) to indirect studies. However, results of indirect studies need several theoretical interpretation before being compared to directly obtained data.

9. Traditional adsorption–desorption studies, based on different adsorptive properties of different micropores, need some theoretical interpretation of the obtained results based on several assumptions about the relationship between structural characteristics and measurable adsorption–desorption properties.
10. Traditional percolation–permeability studies, based on different penetration of different molecules and ions through different micropores, allow good sensitivity of results to various features of studied structures but also need some theoretical interpretation of the obtained results, based on several assumptions about the relationship between structural characteristics and measurable percolation–permeability properties.
11. Nontraditional methods of indirect exploration of microporosity, like thermoporometry and electrical studies, seem to have very interesting perspectives.
12. Existing mechanical tests of porous structures especially need theoretical interpretations before being applied to estimations of properties of microporous structures.

REFERENCES

1. Karchevsky, V.; Figovsky, O.; Romm, F.; Shapovalov, L. Polymeric composition having self-extinguishing properties. U.S. Patent 6329059, 2001.
2. Figovsky, O.L.; Romm, F.A. Improvement of anticorrosion protection properties of polymeric materials. Anticorrosion Meth. Mater. **1998**, *45* (5), 312–320.
3. Motojima, K.; Kawamura, F. Process for producing zeolite adsorbent and process for treating radioactive liquid waste with the zeolite adsorbent. U.S. Patent 4,448,711, 1984.
4. Takumi, S.; Hashimoto, T.; Akimoto, F. High apparent bulk density gamma alumina carrier and method of manufacture of same. U.S. Patent 4301033, 1981.
5. Willkens, C.A.; Arsenault, N.P.; Olson, J.; Lin, R. Aging resistant, porous silicon carbide ceramic igniter. U.S. Patent Appl. Publ. 20020010067, 2002.
6. Kim, S.S.; Liu, Yu.; Pinnavaia, T.J. Ultrastable MSU-G molecular sieve catalysts with a lamellar framework structure and vesiclelike particle texture. Mesopor. Micropor. Mater. **2001**, *489* (no issue), 44–45.
7. Wu, H.-J.; Brungardt, Lisa; Smith, Douglas M.; Drage, James S.; Ramos, Teresa A. Simplified Method to Produce Nanoporous Silicon-based Films for Device Applications. EP 0 881 678, 2001.
8. Willkens, C.A.; Arsenault, N.P.; Olson, J.; Lin, R. Aging-resistant Porous Silicon Carbide Ceramic Igniters Produced by Infiltration of SiC Coarse

Grains in Porous SiC Ceramic and Recrystallization. U.S. Patent, 6297183, 2001.
9. Lawson, D.F.; Stayer, M.L., Jr.; Antkowiak, T.A. Preparation of Diene Elastomers having Reduced Hysteresis and Improved Raw Viscosity Using Partial Coupling and Chain Termination. EP 543305, A1 19930526, 1993.
10. Lawson, D.F.; Stayer, M.L., Jr.; Antkowiak, T.A. Manufacture and Use of Diene Rubbers having Reduced Hysteresis and Improved Raw Viscosity Incorporating Primary Partial Coupling and Terminals Formed form Vinyl Compounds Containing Heterocyclic Nitrogen Groups. EP 538789, A1 19930428, 1993.
11. Shively, M.L. Analysis of mercury porosimetry for the evaluation of pore shape and intrusion-extrusion hysteresis. J. Pharmac. Sci. **1991**, *80* (4), 376–379.
12. Frisina, D.; Rong, F.-G.; Ferriell, M.R.; Saneii, H. Reaction block assembly for chemical synthesis. PCT Int. Appl. **2001**, WO 0192557 A2 20011206.
13. Anderson, J.E.; Tarczynski, F.J.; Walker, D.S. Method and apparatus for monitoring solid phase chemical reactions. PCT Int. Appl. **2001**, WO *0129537*, A2 20010426.
14. Lilburne, M. T. Development of the bubble technique for the measurement of the surface energy of solids. J. Mater. Sci. **1970**, *5* (4), 351–356.
15. Driedger, O.; Neumann, A.W. Surface activity on solid surfaces. Tenside **1964**, *1* (1), 3–7.
16. Dubinin, M.M.; Plavnik, G.M. Microporous structures of carbonaceous adsorbents. Carbon **1968**, 6(2), 183–192.
17. Bekyarova, E.; Mehandjiev, D. Effect of calcination on cobalt-impregnated active carbon. J. Coll. Interface Sci. **1993**, *161* (1), 115–119.
18. Garwood, W.E.; Chen, N.Y.; Lucki, S.J. Alumina Removal from Crystalline Alumino-silicates with Cr(III) Solutions. U.S. Patent 3937791, 1976.
19. Nakayama, Y.; Matsuda, T. Microporous Polymer Surfaces Prepared by an Excimer Laser Ablation Technique. ASAIO J. *40* (3), M590-M593, 1994.
20. Heide, H.; Hoffmann, U.; Broetz, G.; Poeschel, E. Catalyst Supports or Catalyst Systems having Open Pores. U.S. Patent 3957685, 1976.
21. Haack, D.F.; Lin, C.L.; Speckert, M. Method of Coforming Metal Foam Articles and the Articles Formed by the Method Thereof. U.S. Patent Appl. Publ. 2002104405, A1 20020808, 2002.
22. Mellema, M.; Walstra, P.; van Opheusden, J.H.J.; van Vliet, T. Effects of structural rearrangements on the rheology of rennet-induced casein particle gels. Adv. Coll. Interface Sci. **2002**, *98* (1), 25–50.
23. Bart, J.C.J. Polymer/additive analysis by flash pyrolysis techniques. J. Analyt. Appl. Pyrolysis **2001**, *58–59* (no issue), 3–28.
24. Mehra, R.M.; Agarwal, V.; Mathur, P.C. Development and characterization of porous silicon. Diffusion and Defect Data—Solid State Data, Pt. B: Solid-State Phenomena **1997**, *55* (Semiconductor Materials and Technology), 71–76.

25. Frojdh, C. Triggering of solid state x-ray imagers with nondestructive readout capability. PCT Int. Appl. WO 0201925 A1 20020103. 2002.
26. Kruk, M.; Asefa, T.; Jaroniec, M.; Ozin, G.A. Metamorphosis of ordered mesopores to micropores: periodic silica with unprecedented loading of pendant reactive organic groups transforms to periodic microporous silica with tailorable pore size. J. Am. Chem. Soc. **2002**, *124* (22), 6383–6392.
27. Lira-Olivares, J.; Marcano, D.; Lavelle, C.; Garcia S.F. Porous ceramic simulation of reservoir rocks determination of porosity by electric permittivity measurements. Ceram. Trans. **2001**, *125* (Fundamentals of Refractory Technology), 103–134.
28. Koshihara, S.; Sato, M.; Suzuki, N. Energy Dispersive X-ray Analyzer. U.K. Patent Appl. GB 2295454 A1 19960529, Application: GB 95-23856, 19951122, 1996.
29. Flanagan, J.M.; Wall, J.S.; Capel, M.S.; Schneider, D.K.; Shanklin, J. Scanning transmission electron microscopy and small-angle scattering provide evidence that native *Escherichia coli* ClpP is a tetradecamer with an axial pore. Biochemistry **1995**, *34* (34), 10910–10917.
30. Mansour, F.; Dimeo, R.M.; Peemoeller, H.F. High-resolution inelastic neutron scattering from water in mesoporous silica. Phys. Rev. E Stat. Nonlin. Soft. Matter. Phys. **2002**, *66* (4 Pt. 1), 41307.
31. Fratini, E.; Chen, S.H.; Baglioni, P.; Cook, J.C.; Copley, J.R.D. Dynamic scaling of quasielastic neutron scattering spectra from interfacial water. Phys. Rev. E Stat. Nonlin. Soft. Matter. Phys. **2002**, *65* (1 Pt 1), 10201.
32. Webber, J.B.; Strange, J.H.; Dore, J.C. An evaluation of NMR cryoporometry, density measurement, and neutron scattering methods of pore characterization. Magnet. Reson. Imag. 2001, *19* (3–4), 395–399.
33. Broseta, D.; Barre, L.; Vizika, O.; Shahidzadeh, N.; Guilbaud, J.P.; Lyonnard, S. Capillary condensation in a fractal porous medium. Physical Review Lett. **2001**, *86* (23), 5313–5316.
34. Mason, T.J.; Lorimer, J.P. *Sonochemistry*: *Theory, Applications, and Use of Ultrasound in Chemistry*; Ellis Horwood Ltd: Chchester, U.K. 1988.
35. Auld, B.A. *Acoustic Fields and Waves in Solids*, 2nd Ed.; Krieger: Malabar, FL, **1990**; Vol. 2.
36. Didenko, Y.T.; McNamara, W.B., III; Suslick, K.S. Temperature of multibubble sonoluminescence in water. J. Physical Chemistry A **1999**, *103* (50), 10783–10788.
37. Margulis, M.A. *Sonochemical Reactions and Sonoluminescence*. Khimia: Moscow (in English), 1982.
38. Luche, J.L. Sonochemistry from experiment to theoretical considerations. Adv. Sonochem. **1993**, *3* (no issue), 85–124.
39. Margulis, M.A. Fundamental problems of sonochemistry and cavitation. Ultrason. Sonochem. **1994**, *1* (2), S87–S90.
40. Suslick, K.S.; Doktycz, S.J.; Flint, E.B. On the origin of sonoluminescence and sonochemistry. Ultrasonics **1990**, *28* (5), 280–90.

41. Henglein, A. Sonochemistry: historical developments and modern aspects. Ultrasonics **1987**, *25* (1), 6–16.
42. Romm, F. Modeling of interface phenomena in liquid under vibration, using the chemical model. J. Coll. Interface Sci. **2002**, *246* (2), 321–327.
43. Pestman, J.M.; Engberts, J.B.F.N.; de Jong, F. Sonochemistry: Theory and applications. Rec. Travaux Chim. Pays-Bas **1994**, *113* (12), 533–542.
44. Lu, T.J.; Chen, F.; He, D. Sound absorption of cellular metals with semiopen cells. J. Acoust. Soc. Am. **2000**, *108* (4), 1697–1709.
45. Miller, D.L. Frequency relationships for ultrasonic activation of free microbubbles, encapsulated microbubbles, and gas-filled micropores. J. Acoust Soc. Am. **1998**, *104* (4), 2498–2505.
46. Langton, C.M.; Whitehead, M.A.; Haire, T.J.; Hodgskinson, R. Fractal dimension predicts broadband ultrasound attenuation in stereolithography models of cancellous bone. Phys. Med. Biol. **1998**, *43* (2), 467–471.
47. Mandal, B. Self-calibrated Ultrasonic Method of in situ Measurement of Borehole Fluid Acoustic Properties. U.S. Patent, Appl. Publ. U.S. 20030029241 A1 20030213, 2003.
48. Maznev, A.; Li, Z.; Mazurenko, A. Optoacoustic Apparatus with Optical Heterodyning for Measuring Solid Surfaces and Thin Films. PCT Int. Appl. WO 0310518 A2 20030206, 2003.
49. Osele, C. Device and method for measuring the noise and porosity of asphalts. Eur. Patent Appl. EP 1229327 A2 20020807, 2002.
50. Meyer, R.J. Ultrasonic Drying of Saturated Porous Solids via Second Sound. U.S. Patent 6376145, 2002.
51. Poesio, P.; Ooms, G.; Barake, S.; van der Bas, F. An investigation of the influence of acoustic waves on the liquid flow through a porous material. J. Acoust. Soc. Am. **2002**, *111* (5, Pt. 1), 2019–2025.
52. Moussatov, A.; Ayrault, C.; Castagnede, B. Porous material characterization–ultrasonic method for estimation of tortuosity and characteristic length using a barometric chamber. Ultrasonics **2001**, *39* (3), 195–202.
53. Madey, R.; Jaroniec, M. Characterization of microporous carbonaceous materials. Trends Phys. Chem. **1992**, *3* (no issue), 281–297.
54. van Bommel, M. J.; den Engelsen, C. W.; van Miltenburg, J. C. A thermoporometry study of fumed silica–aerogel composites. J. Porous Materials **1997**, *4* (3), 143–150.
55. Eastoe, J.; Rankin, A.; Wat, R.; Bain, C.D. Surfactant adsorption dynamics. Intern. Rev. Phys. Chem. **2001**, *20* (3), 357–386.
56. Eastoe, J.; Dalton, J. S. Dynamic surface tension and adsorption mechanisms of surfactants at the air–water interface. Adv. Coll. Interface Sci. **2000**, *85* (2–3), 103–144.
57. Stell, G. Continuum theory of percolation. J. Phys.: Condensed Matter **1996**, *8* (25A), A1–A17.
58. Bruneel, E.; Ramirez-Cuesta, A. J.; Van Driessche, I.; Hoste, S. Simulation and study of the percolation effect in the magnetic susceptibility of

high-temperature superconducting composites. Phys. Rev. B: Condensed Matter Mater. Phys. **2000**, *61* (13), 9176–9180.
59. Chiteme, C.; Mclachlan, D.S. Measurements of universal and nonuniversal percolation exponents in macroscopically similar systems. Physica B: Condensed Matter (Amsterdam) **2000**, *279* (1–3), 69–71.
60. Haslinger, R.; Joynt, R. Theory of percolative conduction in polycrystalline high-temperature superconductors. Phys. Rev. B: Condensed Matter Mater. Phys. **2000**, *61* (6), 4206–4214.
61. Skal, A. Classical and superconductive percolation. Scientific Israel—Technol. Advant. **2000**, *2* (1), 1–15.
62. Dieterich, W.; Durr, O.; Pendzig, P.; Bunde, A.; Nitzan, A. Percolation concepts in solid-state ionics. Physica A: Stat. Mech. Its Appli. (Amsterdam) **1999**, *266* (1–4), 229–237.
63. Klochikhin, A.; Reznitsky, A.; Permogorov, S.; Breitkopf, T.; Grun, M.; Hetterich, M.; Klingshirn, C.; Lyssenko, V.; Langbein, W.; Hvam, J. M. Luminescence spectra and kinetics of disordered solid solutions. Phys. Rev. B: Condensed Matter Mater. Phys. **1999**, *59* (20), 12947–12972.
64. Sarychev, A.K.; Shalaev, V.M. Giant high-order field moments in metal-dielectric composites. Physica A: Stat. Mech. and Its Appl. (Amsterdam) **1999**, *266* (1–4), 115–122.
65. Menezes-Sobrinho, I.L.; Moreira, J.G.; Bernardes, A.T. Scaling behaviour in the fracture of fibrous materials. Eur. Phys. J. B: Condensed Matter. Phys. **2000**, *13* (2), 313–318.
66. Franceschetti, G.; Marano, S.; Pasquino, N.; Pinto, I.M. Model for urban and indoor cellular propagation using percolation theory. Phys. Rev. E: Stat. Physics, Plasmas, Fluids, Rel. Interdiscipl. Topics **2000**, *61* (3), R2228–R2231.
67. Parker, S.P. McGraw-Hill concise encyclopedia of science and technology. Chem. Eng. Prog. **1994**, *90* (11), 83.
68. Walker, P.M.B., Ed., *Chambers Dictionary of Science and Technology*. Chambers Harrap Publishers Ltd: Edinbourgh, UK, 1999.
69. Eriksson, M.; Nystroem, C.; Alderborn, G. The use of air permeametry for the assessment of external surface area and sphericity of pelletized granules. Int. J. Pharmac. **1993**, *99* (2–3), 197–207.
70. Montillet, A.; Comiti, J.; Legrand, J. Determination of structural parameters of metallic foams from permeametry measurements. J. Mater. Sci. **1992**, *27* (16), 4460–4464.
71. Kadlec, O.; Varhanikova, A.; Zikanova, A.; Zukal, A. Investigation of the porous structure of solids. Porozimetrie a Jeji Pouziti **1969**, (no volume/issue), 1–69.
72. Weishauptova, Z. Principles of the porous structure description of coals, semicokes, and activates. Acta Montana, Series B: Fuel, Carbon, Mineral Process. **1994**, *4* (4), 55–63.
73. Domingo-Garcia, M.; Fernandez-Morales, I.; Lopez-Garzon, F. J.; Moreno-Castilla, C. Glassy carbons for application in selective adsorption. Cur. Topics Colloid Interface Sci. **1997**, *1* (no issue), 137–155.

74. (No author) Microporous ink-jet recording material. (Fuji Photo Film B.V., Neth.). Eur. Patent Appl. EP 1285772 A1 20030226, 2003.
75. Fathima, N.N.; Dhathathreyan, A.; Ramasami, T. Mercury intrusion porosimetry, nitrogen adsorption, and scanning electron microscopy analysis of pores in skin. Biomacromolecules **2002**, *3* (5), 899–904.
76. Leofanti, G.; Padovan, M.; Tozzola, G.; Venturelli, B. Surface area and pore texture of catalysts. Catalysis Today **1998**, *41* (1–3), 207–219.
77. Zalameda, J.N. Measured through-the-thickness thermal diffusivity of carbon fiber reinforced composite materials. J. Composit Technol. Res. **1999**, *21* (2), 98–102.
78. Ferrari, F; Rossi, S; Bonferoni, M.C.; Caramella, C. Rheological and mechanical properties of pharmaceutical gels. Part I: Nonmedicated systems. Boll. Chim. Farmac. **2001**, *140* (5), 329–336.

3
Thermodynamics of Microporous Structure Formation

I. WHY THERMODYNAMICS IS CHOSEN FOR PORE MODELING

In the previous analysis presented in Chapter 1 we have shown that most of specific properties of porous structures are determined by micropores: first of all, the adsorptive and selective separation properties that are exceptional and very important in the technical aspect features of microporosity. Hence, knowledge of micropores and microporous structures is extremely important for us. On the other hand, as we have shown in Chapter 2, experimental studies of microporosity are very complicated: direct methods (scattering by electromagnetic rays or elementary particles) are related to very significant errors of measurement, whereas indirect methods need careful theoretical treatment. It seems very perspective to combine different experimental methods for the obtainment of more information about microporosity, but in such situation the role of theoretical treatment of experimental results obtained by different techniques becomes even more important. This contradiction—between the enormous practical significance of studies of micropores and the serious problems in their experimental research—makes us to conclude about a great importance of theoretical studies of microporosity.

Theoretical models of microporosity regard processes of synthesis, structural characteristics of microporous media and measurable properties of microporous materials. The role of theories of microporosity in the hierarchy of theories of porous materials was generally considered in Chapter 1 (see the classification by Table 1.10). In this chapter we will consider models of processes of synthesis of microporous materials: applicability of these models to various processes, their advantages and shortcomings, mathematical form, and practical usefulness.

Methodologically, theoretical models of micropore formation should be considered as private applications of theories of matter. Theories of matter can be divided into phenomenological and molecular (phenomenological concepts deal with macroscopic characteristics of matter, without accounting for the structure on the microscopic level, whereas molecular theories are based on the consideration of the microscopic level). In their turn, phenomenological concepts can be divided into thermodynamic and statistical theories, while molecular concepts can be divided into classical and quantum approaches.

Let us consider briefly these four concepts.

A. Principles of Thermodynamics

Thermodynamic approach deals with parameters related only to the current state of the system (integrals of motion) and not related to the way this state is obtained. Thermodynamics uses the following parameters of system:

- Measurable parameters: mechanical job A_{mech} (J), heat Q_h (J), pressure P (measured in Pa or ata), volume V (m^3), temperature T (°C or K), electrical potential ϕ (V), electrical charge q (C), amounts of chemical reagents ν_i (mol), surface tension σ (N/m), surface area A (m^2), etc.
- Nonmeasurable parameters that can be derived and evaluated from the measurable ones: entropy S (J/K), enthalpy H (J), internal energy U (J), free energy by Gibbs G (J), free energy by Helmholtz F (J), chemical potential μ (J/mol), etc.

Let us consider an example when nonmeasurable parameter is derived from measurable parameters.

In a system providing no job all incoming heat is consumed only to the increase of entropy:

$$dQ_h = T\,dS \tag{3.1}$$

Obviously, integration of Eq. (3.1) always provides the current value of the entropy (assuming that the entropy of perfect one-component crystal at $T = 0$ K is zero). In the similar manner, one can always obtain values for all nonmeasurable parameters used in thermodynamics.

However, thermodynamic parameters can be divided not only into measurable and nonmeasurable but also into intensive and extensive (additive).

Intensive parameters are defined as thermodynamic parameters, values of each in any equilibrium system are equal for all parts of the system:

$$X_k = X_l \qquad \text{(any } k, l\text{)} \tag{3.2}$$

where X is an intensive parameter and k and l are indices of kth and lth parts of the system, respectively.

For example, pressure in any equilibrium system is equal for all parts of the system; the same is true also for temperature. Moreover: the equal values of each intensive parameter through all the system (for all intensive parameters) is the necessary and sufficient condition for equilibrium of the system.

Extensive (additive) parameters are defined as thermodynamic parameters, values of which are equal to the total sums of the values of these parameters for all parts of the system

$$Y_\Sigma = \sum_{k=1}^{n_{\text{part}}} Y_k \tag{3.3}$$

where Y is an extensive parameter of the considered system, Y_k is its value for kth part of the system, Y_Σ is its total value measured for the system, and n_{part} is the number of parts of the system.

For example, volume of a system is the sum of volumes of all parts of the system; the same is true for entropy, energy, etc.

The list of thermodynamic parameters most largely used in theoretical studies of pores is given in Table 3.1.

TABLE 3.1 Classification of Thermodynamic Parameters

Parameter	Intensive	Extensive	Measurable	Nonmeasurable
Pressure	Yes	No	Yes	No
Temperature	Yes	No	Yes	No
Volume	No	Yes	Yes	No
Amount of substance	No	Yes	Yes	No
Internal energy	No	Yes	No	Yes
Entropy	No	Yes	No	Yes
Enthalpy	No	Yes	No	Yes
Electrical charge	No	Yes	Yes	No
Chemical potential	Yes	No	No	Yes
Electrical potential	Yes	No	Yes	No
Surface tension	Yes	No	Yes	No
Surface area	No	Yes	No	Yes

Let us note that not all parameters largely used in chemical thermodynamics have the sense of thermodynamic parameters: e.g., concentration, heat capacity, and electrical conductivity do not have a sense of thermodynamic parameters.

Among the above-listed parameter, one needs to be careful with chemical potential. To be correct, not chemical potential but *generalized* chemical potential (accounting such specific factors as the influence of various fields and surface phenomena) is intensive parameter.

Equations of the relationship between the mentioned parameters (in the absence of mass transfer and specific fields) are

$$dU = T\,dS - P\,dV \tag{3.4}$$

$$dH = T\,dS + V\,dP \tag{3.5}$$

$$dG = V\,dP - S\,dT \tag{3.6}$$

$$dF = -P\,dV - S\,dT \tag{3.7}$$

Let us note that each term in Eqs. (3.4)–(3.7) consists of two physically compatible parameters: one intensive and one extensive. The same is true when all specific phenomena related to mass transfer are accounted.

The set of Eqs. (3.1), (3.4)–(3.7) is consistent: five unknown parameters in five independent equations.

Thermodynamic approach in description of various processes is based on two laws of thermodynamics:

First law of thermodynamics: balance of energy
Second law of thermodynamics: nondecrease of entropy in isolated systems

Balance of energy means that the total energy of an isolated system never changes. Nondecrease of entropy means that in isolated systems entropy may increase (in nonequilibrium, irreversible processes) or stay constant (in equilibrium, reversible processes) but never may decrease.

Let us notice that processes of pore formation and evolution are always irreversible; hence, they are always accompanied with increase of entropy.

Both laws are valid only for isolated systems, and any exterior interactions make these laws nonapplicable.

B. Quantum Approach

The fundamental base for the quantum approach in studying condensed systems is quantum physics, assuming corpuscular-wave dualism of elementary particles and their simple combinations. Mathematically, quantum approach is based on Schrödinger equation for stationary quantum system:

$$\hat{H}\psi_k = E_k\psi_k \tag{3.8}$$

where ψ_k is wave function, $\hat{H}$ is quantum Hamiltonian operator, and E_k is kth value of energy eigenvalue in the stationary system [1].

Theoretically, quantum approach is the most correct solution for microscopic systems, but in real practice that is applicable only to systems containing few number of particles or noninteracting particles, because of difficulties in computing. Both cases are absolutely *not* applicable to micropores. On the contrary to thermodynamics, quantum approach does not allow taking into account irreversibility of processes.

C. Statistical Approach

In its physical sense, the statistical approach, regarding microscopic processes, is very similar to the quantum approach. However, the principle advantage of statistics consists in the opportunity of counting large-number ensembles of particles. Mathematically, statistical dynamics operates with several distribution functions, applicability of which to studied processes is a separate problem [2].

Statistical approach cannot take into account irreversibility of pore formation.

D. Classical Approach

This is based on classical (Newtonian) mechanics and three well-known laws (equations) of Newton [2]. The classical approach, if applied to micropores, combines shortcomings of all other methods: ignorance of the quantum nature of particles, ignorance of irreversibility of pore formation, problems in counting large-number ensembles.

The comparison of fundamental methods of study of matter is illustrated by Table 3.2.

E. Compatibility of Approaches

As we notified above, the considered approaches are fundamental, principal, and their derivatives may get significantly differ from them. Is it possible, in the methodological point of view, to modify some of the above methods that a modified version combines advantages of these methods with minimizing

TABLE 3.2 Comparison of Fundamental Methods of Modeling of Irreversible Processes in Condensed Phase

Criterion for comparison	Quantum approach	Molecular dynamics	Statistical dynamics	Thermodynamics
Fundamental base for the approach	Principles of quantum physics	Classical mechanics	Brown motion	First and second laws of thermodynamics
Mathematical base for the approach	Schrödinger equation	Newton and Hamilton equations	Equations of statistics	Prigogine theorem, nonlinear theory
Applicability to low number of particles	Applicable	Applicable	Not applicable	Not applicable
Applicability to ensembles of particles	Too difficult	Too difficult	Applicable	Applicable
Applicability to irreversible processes	Not applicable	Not applicable	Not applicable	Applicable
Applicability to random processes	Applicable	Not applicable	Applicable	Applicable
Applicability to determinist processes	Not applicable	Applicable	Not applicable	Applicable
Accounting quantum nature of particles	Yes	No	Possible	Problematic
Factors of errors	Quantum errors	Ignorance of randomness	Ignorance of determinism	Ignorance of features of single particle

TABLE 3.3 Compatibility of Methods of Modeling of Irreversible Processes in Condensed Phase

Method	Quantum	Molecular dynamics	Statistical dynamics	Thermodynamics
Quantum	X	No	Yes	Problematic
Molecular dynamics	No	X	No	No
Statistical dynamics	Yes	No	X	Yes
Thermodynamics	Problematic	No	Yes	X

their shortcomings? This question is equivalent to the problem of methodological compatibility of the fundamental approach.

First of all, the classical approach presents no interest: as it has been noted, this approach has all shortcomings of other methods without their advantages. Statistical approach is compatible with both quantum and thermodynamic approaches, but the compatibility of thermodynamics with the traditional quantum mechanics is very difficult, because of low number of particles counted by the Schrödinger equation. In any case, the decisive advantage of thermodynamics consists in taking into account of irreversibility.

The problem of compatibility of various fundamental approaches is illustrated by Table 3.3.

Thus, in our further theoretical studies of pore formation we will use a combination of statistics and thermodynamics that is called *statistical thermodynamics*. This form of thermodynamics was derived namely for linking microscopic studies to macroscopic (thermodynamic) parameters [3–5].

F. Equilibrium and Reversibility

The notions of reversibility and equilibrium are very close but not identical in their sense. Whereas equilibrium means the absence in the changes of thermodynamic parameters with time, because of the minimum of potentials (U, H, G, and F above), reversibility means that there are some processes in the system accompanied with changes of the potentials, but the divergence between the values of an intensive parameter inside and outside the tested system is very little, negligibly short (for each and every intensive parameter), and a little change of the value of this parameter outside the system changes the direction of changes inside the system.

All equilibrium processes are reversible, but not all reversible processes are equilibrium.

The reversibility of processes is very important for the practical performance of processes near equilibrium—otherwise, such processes would be carried out for infinitely long time.

G. Measurement of Thermodynamic Parameters in Nonequilibrium and Method of Thermodynamic Cycles

According to the definition of thermodynamic parameters, they have sense not only in equilibrium but also in nonequilibrium. Measurement of measurable thermodynamic parameters in equilibrium is not difficult:

For measuring intensive parameters in equilibrium, one just introduces a measuring device into the tested system, waits for equilibrium, and then observes the results of the measurement.

For measuring extensive parameters in equilibrium, one just compares them to those of a measuring device.

Of course, there is an error of measurement, mostly for intensive parameters, because the measured values of intensive parameters are not exactly the same that the tested system had have *before* the measurements began (a new kind of the hysteresis problem analyzed in Chapter 2), but this error can be minimized, due to the use of energy measuring devices. An example of such measurement with the least error is the measurement of the temperature and the pressure of the environment: of course, traditional thermometers and manometers are so little in comparison with the environment that it is not serious to talk about their influence onto the measured parameters.

Although in the case of measuring of the same parameters in a system in nonequilibrium, the problem is much more difficult because it is impossible to wait for the equilibrium between the tested system and the measuring device just because of the absence of equilibrium in the system itself. We get a technical problem: the thermodynamic parameters have physical sense, but their measurement has no sense!

The above problem is partly solved due to the assumption of local equilibrium in a nonequilibrium system. The physical sense of this assumption consists in the opportunity, in several situations, to accomplish measurements of thermodynamic parameters in several spot so fast that the changes in this spot, caused by nonequilibrium processes in the system, can be neglected. A similar problem was analyzed in Chapter 2, regarding tests of a system with aging and/or periodic changes. As in that case, the condition of the correctness of such measurements consists in the upper rate of measurement process, in comparison with the rates of all changes inside the tested system.

The above analysis is valid also for such kind of nonequilibrium systems as metastable systems. Metastability is defined as a specific state of a system, in which its parameters may significantly change because of small exterior actions [6].

If the applicability of fast measurements of measurable thermodynamic parameters in some systems is possible, the sense of nonmeasurable parameters is even more problematic. Let us illustrate that on a classical example: measurements of parameters of superheated liquid boiling in nonequilibrium and irreversibly transformed to vapor. Its pressure and temperature are measured in accordance to the above recommendations, very carefully, to avoid the spontaneous evaporation (and we may suggest that the temperature and the pressure do not significantly change during the evaporation), but the evaluation of the entropy of the superheated liquid seems too difficult. One thing is obvious: this value is different from that for both liquid in equilibrium and vapor. In such situations, the value of nonmeasurable parameters can be estimated due to the method of thermodynamic cycles [6]. This method is based on the definition of thermodynamic parameters: their value depends only on the current state of the system but not on the way, by which this state has been gained. Therefore, one may "decompose" the way of evaporating the superheated liquid into the following reversible stages:

Superheated liquid is initially cooled down to the boiling point.
The liquid at the boiling point is boiled in equilibrium.
The obtained vapor is heated to the temperature of the initial superheated liquid.

Each of these stages is obviously reversible and can be carried out very close to equilibrium, hence, evaluations of entropy changes on each of the stages do not present a problem. After all, the resulting change of entropy (the sum of such changes on each stage) is equal to the entropy change in the initially considered evaporation of superheated liquid.

In Table 3.4 we analyze the methodological similarity and difference between notions of equilibrium, metastability, and reversibility.

H. Continuous and Discrete Systems

If the considered system is very small and can be adequately characterized by averaged values of intensive parameters, such system is defined as *discrete system*; otherwise, the system is defined as *continuous system*. Let us give some examples for discrete and continuous systems.

TABLE 3.4 Comparison of Equilibrium, Metastability, and Reversibility

Factor of comparison	Equilibrium	Metastability	Reversibility
Reversible or not	Reversible	Irreversible	Reversible
Stability with time	Stable	Not stable	Not stable
Applicability to isolated system	Applicable	Applicable	Not applicable
Applicability to open system	Not applicable	Applicable	Applicable
Minimum of thermodynamic potentials	Yes	No	No
Maximum of thermodynamic potentials	No	Yes	No

Discrete systems:
- Elementary particles
- Low-number ensembles of elementary particles
- Interface between two different phases

Continuous systems:
- Large-number ensembles of elementary particles
- Continuous condensed phase
- Continuous fluid flow

In most of situations, porous systems should be considered as discrete (first of all because of the interface between the continuous solid phase and the interior part or pores), but in several situations, continuous models are applicable.

I. Systems with Regular and Random Fluxes

All existing systems in nonequilibrium, in which thermodynamic parameters change due to some processes of heat and mass transfer, can be divided into two groups:

Systems with regular fluxes
Systems with random fluxes

Regular flux can be defined as a heat- or mass-transfer flux, direction, and value of which does not change with time, neither with spatial location, or changes with time and/or with spatial location according to a nonstatistical law.

Random flux can be defined as a heat- or mass-transfer flux, direction, and value of which changes with time and/or spatial coordinate randomly, and its changes are described by laws of statistics.

System in which all fluxes are regular is defined as *system with regular fluxes*.

System in which one flux at least is random is defined as *system with random fluxes*. In real practice, all systems contain only random or only regular fluxes.

The sufficient condition for the regularity of a flux is its predictability: its value and direction should be predictable always.

Using the above definitions, one needs to take into account that randomness can be false, and the considered system contain regular fluxes, regularity of which is very complex. A very important symptom is the accordance of the system with laws of statistics: if this condition is not satisfied, it shows that there is a law, according to which the fluxes in the system change, while this law is very complex.

Let us give some examples for typical systems with random and regular fluxes.

1. Systems with regular fluxes:
   a. Mass-transfer through a membrane, under the specified temperature, pressure and electrostatic potential
   b. Electrolysis without stirring
   c. Heat transfer in a noncontact heat exchanger
2. Systems with random fluxes:
   a. All turbulent systems
   b. Fluids containing dispersed phase
   c. Synthesis in fluidized bed
   d. Polymerizing systems
   e. Heat/mass-transfer in devices with stirring

Most of systems with pore formation are systems with random fluxes: there is no spot in such systems, where any flux (its value and direction) could be predicted more or less exactly.

J. Steady State and Non-Steady State

Steady state (*stationary state*) can be defined as such a state of a thermodynamic system, in which all parameters of the system do not change with time. If a system in steady state is isolated, it is certainly in equilibrium. A steady-state system can be in nonequilibrium only if this system is open.

If a system is *not in* steady state, that is, *non-steady-state* (*nonstationary*) *system*.

Both steady state and non-steady state are applicable to regular or random fluxes, as the researcher performing the experiments likes that.

Let us give some examples of steady state and non-steady state.

Steady state:
- All kinds of equilibrium
- All kinds of fluxes with specified values of intensive parameters
- Technological processes after the start and before the finish

Non-steady state:
- All kinds of metastability
- All periodical processes
- Start and finish in technological processes

Let us notice that a steady-state macroscopic process may comprise non-steady-state elementary processes: e.g., a turbulent flow can be performed in steady state (if its measurable parameters do not change with time), but, in any case, some local fluxes inside such flow appear, get changes, and disappear, which is a combination of nonstationary elementary processes.

Micropore formation processes can be steady state or non-steady state.

K. Methodology of Systems with Random Fluxes

Following the above methodological analysis, let us estimate which theoretical tools are applicable to systems with random fluxes.

It was shown above that thermodynamics of nonequilibrium is the best tool for the description of such irreversible process as micropore formation. However, there is a variety of thermodynamic method applicable to nonequilibrium. Among them, the most important are

- Onsager approach
- Langevin approach
- Liouville approach
- Method of maximum entropy (Prigogine theorem, in the case of linear systems)

The *Onsager approach* is based on the consideration of forces and fluxes in a nonequilibrium system. Each flux in such system is considered as a product of interactions of various driving (thermodynamic) forces. For the linear case, Onsager equations are written in the following form:

$$J_n = \sum_{k}^{N_f} g_{kn} X_k \tag{3.9}$$

where J_n is the flow of nth kind, X_k is the driving force of kth kind, N_f is the number of accounted kinds of forces (or flows), and g_{kn} is Onsager coefficient presenting the influence of driving force of kth kind onto flow of nth kind. Onsager coefficients form a square matrix that can be transformed to a symmetric form ($g_{nk} = g_{kn}$).

In a nonlinear situation, Onsager equations are much more complex and include nonlinear terms for the accounted driving forces.

A very important sequence of Onsager equations, Eq. (3.9), is the mutual influence of forces and flows having the different nature: e.g., temperature gradient (thermal driving force) can cause diffusion (thermo-diffusion), which should be normally caused by the gradient of concentrations.

Onsager approach is the base for nonequilibrium thermodynamics. Everyone knows such particular cases of this approach as Ohm's law in electricity, Fick laws for diffusion, etc.

The principal drawback of the Onsager approach is the difficulty in accounting random fluxes: for their description, one should write and solve an infinite number of Onsager equations with infinite number of unknown parameters.

L. The Langevin Approach

This approach considers systems with a very large (maybe infinite) number of particles in a fluid, motion of which is described by a combination of systematic and stochastic forces, comprising friction. The resulting equation for the ensemble of the particles is

$$\left\langle \vec{v}(t) \cdot \vec{v}(t) \right\rangle_{\vec{v}_0} = v_0^2 e^{-2\xi f} + \frac{C_{\vec{v}_0}}{2\xi}(1 - e^{-2\xi f}) \tag{3.10}$$

where $v(t)$ is the vector of particle's velocity, t is time, ξ is the friction ($\xi = 6\pi\eta a/m$), m is the particle's mass, η is the viscosity of the fluid, a is the characteristic size ("radius") of particle. For large t this should be equal to $3kT/m$, then one obtains the fluctuation–dissipation theorem:

$$\langle F(t)F(t') \rangle = \frac{6k_B T \xi \delta(t - t')}{m} \tag{3.11}$$

The Langevin approach allows accounting the infinite number of particles and the randomness of the system, but its applicability to interface continuous phase interior of pore will be related to serious difficulties.

M. Liouville Approach

This is derived for continuous flows and has the following form for the relationship between rotation rate of *i*th particle ω_i and its inertial moment I_i:

$$\frac{d(I_i\omega_i)}{dt} + \varepsilon_{ijk}\omega_j\omega_k I_k = \frac{dl_i}{dt} + \varepsilon_{ijk} l_j \omega_k = \tau_i \tag{3.12}$$

where $l_i = \int \rho(r) e_{ijk} x_j v'_k \, d^3 r$.

Liouville approach is good for the description of such systems with weak interactions as gases, but for interface like micropores this approach provides too complex equations to solve.

N. Maximum Entropy Approach

This approach is applicable only to steady-state and quasi-steady-state situations. In the linear case, this approach is supported by Prigogine theorem based on Onsager approach (and, in this sense, that is a modification of Onsager approach for infinite number of driving forces). The proof of the entropy maximum for steady state in nonlinear systems is given below.

The maximum entropy approach deals mostly with thermodynamic potentials and, therefore, is applicable to randomized systems with uncountable number of driving forces, influence of which results only in the evolution of thermodynamic potentials. The restriction in the application of this approach to non-steady state may seem very painful, but, as results from above analysis, in any case non-steady state is too difficult for all methods of modeling. Therefore, in the below theoretical analysis we will base all derivations on the maximum entropy approach.

In Table 3.5 we illustrate the comparison of the above-considered methods of thermodynamics of nonequilibrium, in the light of their applicability to micropore formation description.

O. Entropy in Steady-State Systems

As follows from above, all theoretical considerations of systems with micropore formation have physical sense only in steady-state systems (because the local values of intensive thermodynamic parameters in such systems do not change with time) or systems close to steady state (because in such systems the changes of intensive parameters with time is much slower than the measurements). Let us note that all equilibrium systems are steady state, but not all steady-state systems are equilibrium. For example, all open systems, through which some fluxes penetrate in a constant regime, are obviously steady state but certainly nonequilibrium. Aiming to derive the

TABLE 3.5 Comparison of Thermodynamic Methods in the Light of Their Applicability to Micropore Formation Description

Criterion for comparison	Onsager approach	Langevin approach	Liouville approach	Maximum entropy approach
Physical base for the method	Relationship between flows and driving forces in the system	Random perturbation in the system	Continuance of fluid flow	Second law of thermodynamics
Mathematical base of the method	Onsager equations	Langevin equations	Liouville equations	Prigogine theorem
Applicability to condensed phase	Applicable	Applicable	Not applicable	Applicable
Applicability to interface	Applicable	Not applicable	Not applicable	Applicable
Applicability to regular fluxes	Applicable	Not applicable	Applicable	Applicable
Applicability to random fluxes	Not applicable	Applicable	Applicable	Applicable

necessary theoretical tools for the description of above-considered processes of micropore formation, let us analyze the problem of entropy in a steady-state system. Let us prove that the value of entropy in such systems is always maximum, while the maximum is counted on the spectrum of parameters of freedom, meaning the thermodynamic parameters, value of which is not specified by the conditions of experiment. For example, in the process of synthesis of active carbon by oxidation of organic pollutants on active carbon matrix (item 1), the pressure and the temperature are specified, while the rates of fluxes through the reactor and the chemical potentials of the reagents may eventually change and can be considered as parameters of freedom.

1. Entropy in Linear Steady-State System

In such system, the rate of increase of entropy is minimum according to the Prigogine theorem [5]. However, that means that the entropy in such system is maximum on the spectrum of parameters of freedom; otherwise, an eventual rising of entropy would cause new fluxes increasing the rate of increase of entropy. That also means that the Prigogine theorem is equivalent to the principle of maximum of entropy for the system.

2. Entropy in Nonlinear Steady-State Systems

Let us consider a nonlinear steady-state system; this system and its interactions with its surrounding are characterized by a number n_0 of intensive thermodynamic parameters, among which n_1 parameters do not change in space except for fluctuations (let us call this the *pseudoconstant*; their gradients are zero) and n_2 parameters do (let us call them *variable*; their gradients are not zero). Obviously, the nonequilibrium features of the considered system are determined only by the variable parameters. The total entropy in the system is determined first of all by pseudoconstant parameters and, much less, by such variable parameters, gradients of which are enough short to allow fluctuations. Variable parameters with large gradients are specified and cannot have many available microstates, hence, their disorder is negligible. The term of entropy determined by pseudoconstant parameters does not differ from that for the equilibrium; hence, it has a maximal value measured on the spectrum of pseudoconstant parameters. The variable parameters having short gradients correspond to the linear situation considered above, hence, the entropy term related to such parameters is maximal too. Fluctuations of variable parameters are constrained by large thermodynamic forces in the system and do not influence the system's disorder. Those are significant only in the case of small thermodynamic forces, but this is the linear situation considered above.

Thus, we have proved the principle of maximum of entropy on a limited spectrum of parameters (which are considered as pseudoconstant or variable with short gradients) for the general case of nonequilibrium steady-state systems.

The above proof would not be valid for non-steady-state systems, because their intensive parameters cannot be divided into pseudoconstant and variable ones. However, some non-steady-state systems can be considered approximately as quasi-steady state. Such an approximation is acceptable if two following conditions are satisfied:

During the measurement of the intensive parameters of the system, the change of each parameter is negligible in comparison with its absolute value.

The duration of the measurement is much larger than the time of relaxation of every process inside the system (compare both conditions to these for tests of systems changing with time; see Chapter 2).

Let us note that the entropy of any steady-state system in non-equilibrium is always less than the entropy of the same system in equilibrium, because all intensive parameters of system in equilibrium are not specified. Moreover, the value of entropy of a steady-state system, compared to the same in equilibrium, can be considered as a characteristic parameter of the degree of nonequilibrium of the system.

P. Negentropy Analysis

Since entropy characterizes the *inability* of a thermodynamic system to provide useful work, one may characterize the *ability* of a thermodynamic system to provide useful work by a parameter having the sense contrary to entropy. This parameter is called *negentropy*. In principle, negentropy is very similar to potential Gibbs (for specified pressure and temperature) or Helmholtz (for specified volume and temperature), but these increase with temperature, while negentropy increases only with "nonrandomness" of the system. One may also define negentropy as the quantity of entropy that can be created by the considered system if this gets to be isolated.

Having opposite sense to entropy, negentropy has all other properties similar to entropy:

Negentropy is an extensive (additive) thermodynamic parameter.

In an insulated system, negentropy never increases.

Negentropy is the measure of nonrandomness of system.

Why do we need negentropy? The above-proved maximum of entropy in steady state is just a tool for description of very complex systems with infinite number of regular or random fluxes. Entropy is not an appropriate parameter for the characterization of the free energy of the system, while negentropy is. This fact is illustrated by the method of negentropy balance presented below. This method is especially effective for systems with random fluxes.

(a) Negentropy balance method. Let us consider an open continuous system with random fluxes, in steady state. This system is characterized by the maximum of entropy on the limited spectrum of intensive parameters, as it has been proved above. Let us divide the system into a very large number of subsystems. Each of these subsystems is itself an open system characterized by thermodynamic parameters, among which we are interested in negentropy. Each subsystem is characterized by negentropy corresponding to unit of substance contained in this subsystem, and it is very low probability that two different subsystems have the same negentropy per 1 mol of substance. Hence, the characterization of subsystems by negentropy is unique, and one may suggest that a value of negentropy U corresponds only to one subsystem. Since the initial system is stationary, we may suggest its statistical equilibrium (this is absolutely not the thermodynamic equilibrium); this means that eventual changes in the negentropy distribution of subsystems cannot influence the thermodynamic parameters of the whole system. The following equations describe the considered system [7]:

$$\Delta S = -R_g Q_u \int_0^{U_{\max}} f(U) \ln f(U)\, dU \tag{3.13}$$

$$\nu_0 = Q_u \int_0^{U_{\max}} f(U)\, dU \tag{3.14}$$

$$\Delta\Phi = \nu_0 U_{\text{av}} = Q_u \int_0^{U_{\max}} U f(U)\, dU \tag{3.15}$$

where U is the negentropy of a subsystem, $U_{\max}$ is the maximal attainable negentropy per 1 mol of substance in the considered system, ν_0 is the amount of the substance in the system, U_{av} is the average negentropy, and Q_u is a normalization coefficient. The application of the condition of the statistical equilibrium to Eqs. (3.13)–(3.15) provides [7]

$$\delta S = 0 \qquad \delta\nu_0 = 0 \qquad \delta\,\Delta\Phi = 0 \tag{3.16}$$

$$\delta \int_0^{U_{\max}} f(U) \ln f(U)\, dU = 0 \tag{3.17}$$

$$\delta \int_0^{U_{\max}} f(U)\, dU = 0 \tag{3.18}$$

$$\delta \int_0^{U_{\max}} Uf(U)\, dU = 0 \tag{3.19}$$

where δB is a very short variation of the parameter B.

Equations (3.17)–(3.19) can be transformed to

$$\int \delta f(U) \ln [f(U)]\, dU = 0 \tag{3.20}$$

$$\int \delta f(U)\, dU = 0 \tag{3.21}$$

$$\int [\delta f(U)] U\, dU = 0 \tag{3.22}$$

Since Eqs. (3.20)–(3.22) should be valid for all values of the variations of $f(U)$, one may multiply Eq. (3.21) by β_U and Eq. (3.22) by α_U, and then add to Eq. (3.20), then we obtain

$$f(U) = \exp\left(\frac{-U}{\alpha_U}\right) \tag{3.23}$$

From Eqs. (3.13)–(3.15) and (3.23) one obtains

$$Q_u = \frac{\nu_0/\alpha_U}{1 - f(\alpha_U, U_{\max})} \tag{3.24}$$

$$U_0 = \frac{\alpha_0}{1 - b_U/(\exp b_U - 1)} \tag{3.25}$$

$$\Delta S_0 = \frac{R_g}{1 - b_U/(\exp b_U - 1)} \tag{3.26}$$

where $b_U = U_{\max}/\alpha_U$.

Equations (3.23)–(3.26) give the subsystem distribution in negentropy. Now, if we like to obtain the value of any additive parameter Π for the system, one needs to estimate the relationship between the values of an extensive parameter Π for various values of U. Since both U and Π are extensive parameters, one can write

$$\Pi(U) = \Pi(U=0) + \frac{U * [\Pi(U=U_{\max}) - \Pi(U=0)]}{U_{\max}} \tag{3.27}$$

and one does not need to evaluate Π every time for all values of U.

Equation (3.27) can be called *principle of superposition* for subsystems in system.

Now, if the system gets negentropy with the rate dV/dt, the negentropy balance is determined by the negentropy obtained from the exterior system and the negentropy consumption for the increase of the entropy inside the system:

$$dU_0 = dV - \sigma_s\, dt \tag{3.28}$$

In the case of $dV=0$ and $\sigma_s = \gamma_0 U_0$ (isolated system with linear dissipation of negentropy), one obtains from Eq. (3.28)

$$U_0(t) = U_0(t=0)\exp(-\sigma_s t) \tag{3.29}$$

Equation (3.29) is valid not only for negentropy but also free energy (because these two parameters are analogues).

Equation (3.28) allows the following order of evaluations:

U_0 is evaluated from $V(t)$ and $\sigma_s(t)$;
Π_0 for the system is evaluated from the equation:

$$\Pi_0 = \int_0^{U_{\max}} f(U)\Pi(U)\, dU \tag{3.30}$$

After all, the only thing remains unknown for us: $U_{\max}$. How can we evaluate it? In principle, it is the simplest thing, that is, the negentropy of a subsystem in which all changes are very slow; hence, this subsystem retards in getting closer to the equilibrium. Let us consider two examples of the evaluation of $U_{\max}$ for subsystems appearing in processes of micropore formation.

1. Pyrolytic preparation of active carbon from a porous organic material in which the energy distribution of pores is $f_{\text{in}}(\varepsilon)$; Every current state of the system is characterized by a new hypothetical

equilibrium with energy distribution of pores $f_{\mathrm{cur}}(\varepsilon)$. The negentropy of the subsystem having maximal negentropy is

$$U_{\max} = R_g \int_{\varepsilon_{\min}}^{\varepsilon_{\max}} [f_{\mathrm{in}}(\varepsilon) \ln f_{\mathrm{in}}(\varepsilon) - f_{\mathrm{cur}}(\varepsilon) \ln f_{\mathrm{cur}}(\varepsilon)]\, d\varepsilon \tag{3.31}$$

2. Polymerization of liquid phase containing initially monomers only: The initial weight distribution of molecules in the system is given by the condition

$$C_N(t=0) = \begin{cases} 1 & \text{if } N = 1 \\ 0 & \text{if } N \neq 1 \end{cases} \tag{3.32}$$

The current equilibrium weight distribution would be given by

$$C_{N,\mathrm{eq}} = C_{N,\mathrm{eq}}(t) \tag{3.33}$$

The maximal negentropy subsystem corresponds to $C_N(t=0)$, because in this subsystem the negentropy is kept without changes; therefore, in such a case,

$$U_{\max} = R_g T \sum_{N=1}^{\infty} \left[C_N(t=0) \ln C_N(t=0) - C_{N,\mathrm{eq}}(t) \ln C_{N,\mathrm{eq}}(t)\right] \tag{3.34}$$

and one obtains from Eqs. (3.32) and (3.34)

$$U_{\max} = -R_g T \sum_{N=1}^{\infty} C_{N,\mathrm{eq}}(t) \ln C_{N,\mathrm{eq}}(t) \tag{3.35}$$

Thus, one may always evaluate $U_{\max}$ from elementary suggestions about the properties of the considered system.

II. THERMODYNAMICS OF PORE FORMATION

A. Stages in Processes of Micropore Formation

In Chapter 1 we considered most important processes of formation of microporous materials. We divided all them into two principal groups:

Micropore formation in solid phase [8,9]
Micropore formation simultaneously with solid phase appearance [8]

Let us now consider the mechanisms of micropore formation in both situations, and estimate the theoretical tools necessary for their modeling.

B. Micropore Formation in Solid Phase Prepared Beforehand

Though the processes of synthesis of microporous materials are very multiple, the list of principal theoretical tools needed for their modeling is well short. Let us show that by the consideration of some widespread processes. We are going to consider the following processes of microporous material synthesis from solid raw material:

Synthesis of active carbon by oxidation of organic contaminants on existing active carbon
Synthesis of active carbon by oxidation of organic polymer
Synthesis of active silica and active alumina by dehydration of silica and alumina gels, respectively
Synthesis of zeolite by dehydration of alumosilicates
Synthesis of metal foam by oxidation—decomposition of polymeric matrix covered with metal coating
Synthesis of porous silicon by decomposition of silicon-organic compounds.

1. Active Carbon Prepared by Oxidation of Organics on Active Carbon Matrix

This process includes the following stages:

1. Initiation of radical formation in the organic medium by means of heating to the temperature of radical formation (also possible to use initiating rays for the same purpose):

$$C_mH_{2n} + O_2 \rightarrow C_{m_1}H_{n_1}OH^{\bullet} + C_{m_2}H_{n_2}OH^{\bullet} \quad (3.36)$$

   where $m = m_1 + m_2$, $2n = n_1 + n_2 + 2$, $n \le m$. The products of reaction (3.36) are free radicals characterized by very high free energy and able to easily react with both organics and oxygen. In principle, also active carbon can be attacked by free radicals, but, if the process is carried out in conditions of low content of oxygen (or air), particles inside the solid matrix (active carbon itself) are more or less effectively protected.

2. Chain radical reactions of oxidation of the organic components by air oxygen with intermediate attacks by free radicals, e.g.,

$$C_mH_{2n} + C_{m_1}H_{n_1}OH^{\bullet} \rightarrow C_{m_3}H_{n_3}OH^{\bullet} \tag{3.37}$$

$$C_{m_3}H_{n_3}OH^{\bullet} + O_2 \rightarrow C_{m_4}H_{n_4}OH^{\bullet} + C_{m_5}H_{n_5}O_2H^{\bullet} \tag{3.38}$$

where $m + m_1 = m_3$, $2n + n_1 = n_3$, $m_3 = m_4 + m_5$, $n_3 = n_4 + n_5 + 1$. On this stage, carbon can be eventually delivered, e.g.,

$$n_6O_2 + C_{m_6}H_{2n_6}OH^{\bullet} \rightarrow m_6C + n_6H_2O + OH^{\bullet} \tag{3.39}$$

3. Dilution of the organic phase and extinction of the radical processes, e.g.,

$$C_mH_{2n}OH^{\bullet} + C_{m_1}H_{n_1}OH^{\bullet} \rightarrow C_{m_3}H_{n_7} + H_2O \tag{3.40}$$

where $m + m_1 = m_3$, $2n + n_1 = n_7$.

Among all products of reactions (3.36)–(3.40):

Carbon formed by reaction (3.39) can participate in building of the porous structure.
Volatile components (e.g., water vapor) partly leave the reaction zone and partly are adsorbed inside the new formed microporous structure when its temperature decreases.

The following factors determine the growth of the microporous structure:

1. Features of the previously formed microporous structure (first of all the carbon matrix, on which the oxidation is performed)
2. The chemical composition of the oxidized organics
3. The regime of the oxidation

The previously formed microporous structure stimulates the growth of the solid phase with micropores, if the organic components are adsorbed so strongly they cannot leave the pores when the temperature gets rising.

The chemical composition of the oxidized organics influences the rate of the growth of micropores: more is the content of carbon in the organics, more reactions like reaction (3.39) take place, faster is the process of carbon aggregation to already existing micropores [9–11].

The regime of the oxidation is extremely important and can even change the mechanisms of the processes: only low content of oxygen is favorable for reactions like reaction (3.39), while the high content of oxygen causes the oxidation of carbon—not the only product of reaction (3.39) since also carbon is in the primary matrix.

In the case that three above factors are so chosen they favor the carbon matrix growth, carbon particles formed by reactions like reaction (3.39), join the existing primary carbon matrix and preferably occupy free positions in accordance to requirements of thermodynamics. This process can be considered as a specific form of adsorption:

$$C_x + C(\text{matrix}) \longleftrightarrow C(\text{matrix})^{\bullet}C_x \tag{3.41}$$

Though the reactions (3.36)–(3.40) are obviously irreversible, interactions between carbon particles and the microporous matrix under high temperatures [reaction (3.41)] are close to equilibrium. In such conditions, the matrix growth is very slow because of intense motion of particles, whereas the decrease of temperature accelerates the matrix growth. Assuming the simple first-order reversible reaction for Eq. (3.41) and [C(matrix)] = 1, without taking into account the permanent formation of carbon particles, we obtain the kinetic equations:

$$\frac{d[A]}{dt} = k_2[B] - k_1[A] \tag{3.42}$$

$$\frac{d[B]}{dt} = k_1[A] - k_2[B] \tag{3.43}$$

where k_1 and k_2 are constants of the rates for the direct and the reverse reactions (3.41), t is time, A is the carbon aggregating to the matrix, and B is the unstable new matrix, [X] is the concentration of component X (mole parts). The solution of Eqs. (3.42) and (3.43) is given by

$$[A] = [A]_0 * \frac{k_2 + k_1 * \exp\,[-(k_1 + k_2)t]}{(k_1 + k_2)} \tag{3.44}$$

The total rate of micropore formation is given by

$$W_{\text{mf}} = k_1[A] - k_2[B] \tag{3.45}$$

The relationship between the temperature and the total rate of micropore growth is given by Fig. 3.1.

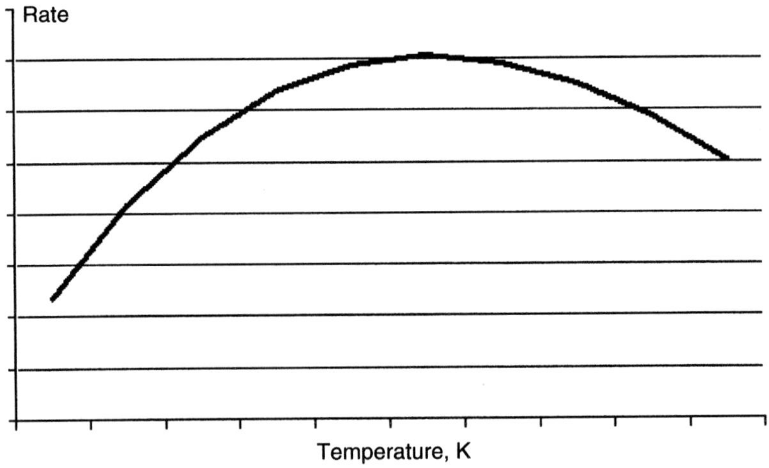

FIG. 3.1 Total rate of micropore growth vs. temperature for active carbon synthesis on active carbon matrix.

Equations (3.42)–(3.44) are valid on the stage of dilution of free radicals, when the carbon particle formation is negligible. However, we can notify that taking into account of the permanent formation of carbon particles does not significantly change the situation.

Equations (3.42)–(3.44) bring us information about the amount of substance in fresh carbon matrix but do not allow us to estimate the structural characteristics of the prepared microporous carbon. To solve this problem, we need to use methods of linear thermodynamics. This aspect will be analyzed below.

2. Active Carbon Preparation by Oxidation of Polymeric Matrix

This process includes stages similar to those in the above case (item 1), but the mechanism of micropore formation includes a stage of the destruction of the polymeric matrix. The freshly formed micropore coexists not only with the aggressive gas phase but also with macromolecules (this problem does not exist in item 1, where the condensed phase is the same that the freshly formed phase). Equations of the growth of the carbon matrix are given by

$$\sum_{k=1}^{N} C_{nk} \longleftrightarrow C_{\Sigma nk} \tag{3.46}$$

Reaction (3.46) means that a carbon aggregate is formed reversibly from N different aggregates. The direct reaction (3.46) provides a nucleus of the carbon matrix, the further growth of which is described by Eq. (3.41). As follows from Eq. (3.46), the nucleus can, in principle, be destroyed. This can happen because of

The reverse reaction (3.46)
Just oxidation

Comparing both options to item 1, we find that the existence of a ready carbon matrix greatly reduces the chances for both these options.

The free energy of the matrix nucleus is determined also by interactions with the polymeric medium, for which the nucleus becomes a factor of structural defect increasing the resulting free energy of the medium.

To build a model for such case, we need not only linear thermodynamic model for nucleus growth, but also model for a polymeric structure with defect—nucleus of a carbon matrix.

3. Metal Foam from Coated Polymer

That is Example 2.13 (see Chapter 2). The process of metal foam preparation includes two steps:

Coating of a polymeric matrix with filling its pores by a metal powder
Decomposition–oxidation of the polymeric fraction on the metal matrix, in which former macromolecules are replaced with micropores

The resulting microporous structure of the obtained metal foam is exactly the same as the initial continuous phase of the polymeric matrix, which is described by polymer theory. Hence, for the description of the structure of the product (metal foam) we need the following theoretical tools:

Theory of adsorption: for modeling of the process of coating the polymeric matrix with the metal powder
Theory of polymeric structures: for the estimation of zones of the concentration of the metal inside voids in the polymeric matrix [8,12–25]

Let us notify that the polymeric theory sufficient for our needs should be able to describe not only linear macromolecules (that is not difficult) but also branched and cross-linked polymeric structures.

4. Porous Silicon from Silicon Organics

Careful decomposition of silicon-organic compounds results in the formation of high-porosity silica containing organic groups:

$$2\text{Si}^{\bullet}\text{R} \rightarrow {}^{\bullet}\text{Si} - \text{Si}^{\bullet}\text{R} + \text{R} \qquad (3.47)$$

$$\text{Si}^{\bullet}\text{R} + {}^{\bullet}\text{Si}_n - \text{Si}^{\bullet}\text{R} \rightarrow {}^{\bullet}\text{Si}_{(n+1)} - \text{Si}^{\bullet}\text{R} + \text{R} \tag{3.48}$$

where R is an organic radical removed by vaporization, degradation, or very careful oxidation. The reactions (3.47) and (3.48) are similar to Eq. (3.46), but the silicon nucleus can be much more stable if the organic radical is removed without oxidation. The principal thermodynamic tools needed for the description of process Eqs. (3.47) and (3.48) are the same that in item 2.

5. Active Silica Preparation by Dehydration of Silica Gel

This process is described by the following equation [8]:

$$n\text{H}_2\text{SiO}_3 \rightarrow (\text{SiO}_2)_n + n\text{H}_2\text{O} \tag{3.49}$$

Though the temperature of process Eq. (3.49) can be very high (300–600°C), the structure of the product is determined by the aggregation under low temperatures, when water molecules just begin to leave the solid phase. This process is very similar in all aspects to the traditional polymerization. As it is concluded in Ref. 8, there is no principal difference between the formation of organic macromolecules in the traditional polymerization processes and the aggregation of silica in the process of the preparation of active silica. Hence, the above-mentioned polymeric approach will be applicable not only to the considered organic polymers but also to active silica synthesis. Let us note that the preparation of active alumina is very similar to that for active silica, and the necessary theoretical approach, of course, will be the same [8,12].

6. Preparation of Zeolite by Removal of Aluminum Oxide from Alumosilicate

(See Example 2.10.) This process is regular: that is performed with a regular raw material (crystalline alumosilicate) with regular and selective changes of its internal structure. Such process can be described by theory of crystals [10], and that is not necessary to elaborate any special thermodynamic theory for such processes.

Thus, after the analysis of typical processes of synthesis of microporous solids from solid raw materials, we can conclude that though the rich variety of such processes, the list of theoretical tools needed for their description is well limited:

Linear thermodynamic model for fast irreversible reactions of aggregation of particles with formation of solid phase

Model for polymers with branching (and maybe cross-linking) applicable not only to traditional polymers but also various inorganic aggregates

Model of solid nucleus contacting with two phases: raw material (organic polymer or amorphous organics) and high-temperature gas
Theory of adsorption

The above analysis is illustrated by Table 3.6.

C. Micropore Formation Simultaneously with Synthesis of Solid Phase

Now, let us consider some typical processes of micropore formation, in which the solid phase does not initially exist but is formed with simultaneous appearance of micropores.

(a) Polymerization. The general reactions of polymerization are written as

$$2\,\text{Mon} \rightarrow \text{Pol}(2) \tag{3.50}$$

$$\text{Mon} + \text{Pol}(2) \rightarrow \text{Pol}(3) \tag{3.51}$$

$$\ldots$$

$$\text{Mon} + \text{Pol}(N-1) \rightarrow \text{Pol}(N) \tag{3.52}$$

where Mon is monomer and Pol(N) is a polymer (macromolecule) consisting of N monomeric units. Reactions (3.50)–(3.52) are reversible, and the process reversible to polymerization is defined as *depolymerization* (also called as *destruction* or *degradation* of polymer, while these terms may also mean not reversing of polymerization but irreversible breaking of polymeric structures by oxidation).

The theory of polymerization and its application to micropore formation processes—not only in organic polymers (polyethylene, polystyrene, proteins, etc.) but also in inorganic aggregates (silica, alumina)—is considered in more details below.

1. Preparation of Active Carbon by Oxidation of Organic Matrix

This process is traditional for the preparation of active carbon [9]. The chemical reactions do not significantly differ from Eqs. (3.36)–(3.41). The physical difference from the cases 1 and 2 (preparation of active carbon by oxidation of organics adsorbed on the ready carbon matrix or by oxidation of organic polymer) consists in the absence of a matrix, in which carbon particles could form micropores. The carbon matrix begins with a nucleus in a liquid organic phase, contacting with gas–vapor phase, in which radical processes lead to oxidation of volatile components from the organic phase.

TABLE 3.6 Processes and Theoretical Tools Needed for Their Description in Micropore Formation in Solid Phase

Process[a]	Stages	Needed theoretical tools
Active carbon by oxidation of adsorbed organics	Radical formation, chain radical reactions, chain breaking	Linear thermodynamic model
Active carbon by oxidation of polymer	Radical formation, depolymerization, chain radical formation, chain breaking	Polymeric model, linear thermodynamic model, model of micropore nucleus
Metal foam from coated polymer	Coating of polymer matrix, polymer decomposition	Theory of adsorption, polymeric model
Porous silicon from silicon organics	Silicon-organics decomposition, silicon particle aggregation	Model of micropore nucleus, polymeric model
Active silica–alumina by dehydration of silica–alumina gel	Water removal	Polymeric theory
Zeolite by removal of aluminum oxide	Removal of aluminum oxide	Existing theory of crystal growth

[a]Processes determining micropore formation are outlined.

Methodologically, this process needs a thermodynamic model for the nucleus formation and growth.

2. Preparation of Microporous Materials by Simple (Random) Sedimentation

Sedimentation allows the preparation of numerous microporous materials, comprising silica and alumina gels. The mechanism of sedimentation comprises random processes of aggregation of particles to macromolecules. The relevant theoretical model is very close to polymerization theory, while some heterogeneous effects (e.g., events caused by gravitation) should be also taken into account by linear thermodynamics tools.

3. Preparation of Crystals by Regular Sedimentation (Crystallization)

Regular sedimentation allows preparation of crystalline products that become raw materials in zeolite synthesis. As in item 6, the existing theory of crystal growth is sufficient for modeling regular sedimentation, and we do not need to elaborate new theoretical methods for such cases.

The above analysis is illustrated by Table 3.7.

As follows from Table 3.7, the theoretical tools needed for the description of micropore formation simultaneously with the synthesis of the solid phase are, in principle, the same that in the case of micropore formation in already existing solid phase (items 1–6).

TABLE 3.7 Processes and Theoretical Tools for Their Modeling in Micropore Formation Simultaneously with Solid Material Synthesis

Process[a]	Stages	Needed theoretical tools
Polymerization	Initiation, chain reactions, chain breaking	Polymeric model
Active carbon by oxidation of amorphous or liquid organics	Radical formation, chain radical reactions, chain breaking	Linear thermodynamic model, model of micropore nucleus
Gels by sedimentation	Random aggregation of particles	Polymeric model, linear thermodynamic model
Crystallization	Regular aggregation of particles	Existing theory of crystallization

[a]Processes causing micropore formation are outlined.

Thus, for the adequate modeling of processes of micropore formation we need the following theoretical models:

Linear thermodynamic model for high-temperature aggregation
Linear thermodynamic model accounting for gravitation effects in aggregation by random sedimentation
Model for polymerization with branching and cross-linking
Model for nucleus formation in amorphous and/or liquid organics, contacting with high-temperature gas
Model for nucleus formation in organic polymeric structure, contacting with high-temperature gas

In addition to the above models, we will also need accounting restructuring process in homogeneous and heterogeneous microporous structures.

D. Evaluation of Pore's Energy and Entropy

All thermodynamic analyses begin with evaluations of energy and entropy (nonmeasurable thermodynamic parameters) of the studied objects. Therefore, first of all we need to evaluate energy and entropy of micropore on purpose of understanding what is the driving force for micropore formation. In our considerations, we will use the discrete model of solid phase, accounting for the corpuscular structure of the condensed matter.

Let us consider a structure including N_s cells occupied by a substance (continuous phase) and N_p voids (the micropore). The shape of the micropore does not interest us. The number of microstates available for the considered structure is [10]

$$W_r = \frac{(N_p + N_s)!}{N_p! N_s!} \tag{3.53}$$

$$\ln W_r \approx N_s \ln \frac{N_0}{N_s} + N_p \ln \frac{N_0}{N_p} \tag{3.54}$$

where $N_0 = N_s + N_p$.

According to Bolzmann comprehension of entropy of a thermodynamic system, it is related to the number of microstates available for the system:

$$S_s = R_g \ln W_r \tag{3.55}$$

where R_g is gas constant. Based on Eqs. (3.53)–(3.55), we obtain a formula analogous to entropy of mixing:

$$S_p = -R_g[x_p \ln x_p + (1 - x_p) \ln (1 - x_p)] \tag{3.56}$$

where $x_p = N_p/(N_s + N_p)$ is the volume fraction of voids. Equation (3.56) means that the entropy of pores is always positive value; hence, the entropy rising is the driving force for pore formation, the increase of the internal energy.

For a porous structure containing numerous pores with different energy, the balances of substance and energy are given by, respectively,

$$v_\Sigma = Q_0 \int_{\varepsilon_{min}}^{\varepsilon_{max}} f(\varepsilon)\, d\varepsilon \tag{3.57}$$

$$E_\Sigma = Q_0 \int_{\varepsilon_{min}}^{\varepsilon_{max}} \varepsilon f(\varepsilon)\, d\varepsilon \tag{3.58}$$

where ε_{min} is the minimal energy of substance inside pores, ε_{max} is the maximal energy of substance inside pores, and Q_0 is a normalization coefficient. Let us note that, as follows from Eqs. (3.57) and (3.58), pores are accounted in a continuous spectrum, while above we assumed the discrete model for a micrporous structure; that is just a mathematical approximation aiming to use integration from the minimal energy to the maximal energy of pore's substance, and we always mean in physical models the discreteness of porous structures.

Taking into account the energetic divergence between micropores, one may write the following equation for the entropy of a microporous system (using Bolzmann's definition of entropy) [10]:

$$S_{p\Sigma} = -R_g Q_0 \int_0^{\varepsilon_{max}} \ln f(\varepsilon) f(\varepsilon)\, d\varepsilon \tag{3.59}$$

Equation (3.59) mean that the energetic degeneracy of pores is not considered; that seems physically reasonable, because two different micropores, having the same energy of the substance inside them, almost always differ or in the amount of the included substance, or in shape, or both.

Let us note that Eqs. (3.57)–(3.59) bring us the relationship between the energy distribution of micropores and their thermodynamic parameters, but we do not still know the energy distribution of micropores in concrete situations.

Now, let us derive the energy distribution of micropores for various situations that can be realized in processes considered above (Tables 3.4 and 3.5).

E. Linear Stationary System

Such system is realized when the values of intensive thermodynamic parameters of the considered system do not change with time or their change if negligible. Practically, such model is applicable for

Growth of active carbon matrix when the concentration of free radicals is close to constant

Vertical coordinate of a system formed by random sedimentation in gravitation field

According to the proof presented above and regarding systems in steady state, the entropy of such system is maximal on the available spectrum of parameters of freedom. Let us assume that the system with maximal entropy and specified energy and amount of substance is stable against eventual variations of the form of the energy distribution function. Applying the variation of the energy distribution function to Eqs. (3.57)–(3.59), we obtain

$$\delta v_{\Sigma} = Q_0 \int \delta f(\varepsilon)\, d\varepsilon = 0 \tag{3.60}$$

$$\delta E_{\Sigma} = Q_0 \int \varepsilon \delta f(\varepsilon)\, d\varepsilon = 0 \tag{3.61}$$

$$\delta S_{p\Sigma} = -R_g Q_0 \int \delta f(\varepsilon) \ln f(\varepsilon)\, d\varepsilon = 0 \tag{3.62}$$

where the symbol δ means the variation of the corresponding parameters. Taking into account that Eqs. (3.60)–(3.62) should be valid for all short variations, we obtain from Eqs. (3.60)–(3.62)

$$f(\varepsilon) = \exp \frac{-\varepsilon}{\alpha} \tag{3.63}$$

where the parameter α shows the degree of nonequilibrium of the system: for the system in equilibrium, $\alpha = R_g T$ (because pores allow one-dimensional motion only for fluid inside the porous medium). If $\alpha \neq R_g T$, Eq. (3.63) becomes the Gibbs energy distribution for porous structures. The parameter α can be, theoretically, found from preparation conditions of the considered porous material, but in real practice that is a semiempirical parameter found from experimental measurements, e.g., from adsorption isotherms as it was done in Ref. 10. We can also note that theoretical curves built on the base of Eq. (3.63) provided a very good correlation with the experimental data treated in Ref. 10.

In the particular case of micropore formation by sedimentation, $\alpha \approx R_g T$ and $\varepsilon = \varepsilon^o + \mu_p g h$, where μ_p is the molar of mass of the particles in the sedimentation process.

F. Non-Steady-State Formation of Micropore

Such situation is very widespread in industry and corresponds (among items 1–9) to the active carbon matrix formation on the stage of dilution of free radicals and cooling of the system. As it was shown above, the stage of cooling of microporous system allows the resulting rate of micropore growth even larger than the main reaction itself.

Since we are concerned with non-steady state, we have several difficulties in the application of the above-proved principle of maximal entropy in steady-state systems.

To obtain the solution of the non-steady-state problem, we will initially consider a thermally homogeneous system, in which the temperature depends on the time but not on the space coordinate. Such situation is realized in the interior part of the reaction zone, far from the reactor walls.

In such case, the process of pore formation can be considered as a chemical reaction written in the following form:

$$\text{Solid} \longleftrightarrow \text{Solid} \bullet \text{pore}(\varepsilon, T) \tag{3.64}$$

The constant of the rate of the direct reaction (3.64) is $k_1(\varepsilon, T)$, while the constant of the rate of the reverse reaction $-k_2(\varepsilon, T)$, and the sense of both constants is the same as in item 1, while respecting *not* to the ensemble of micropores but to a single micropore, substance in which has energy ε.

The process of pore formation changes both the total volume of the pores and their energy distribution. The corresponding kinetic equation is [11]

$$\frac{\partial X(t,\varepsilon,T)}{\partial t} = k_1(\varepsilon,T)[\text{solid}] - k_2(\varepsilon,T)X(t,\varepsilon,T)[\text{solid}] \tag{3.65}$$

where X and [solid] are the volume fractions of the micropores with the energy close ε and of the solid, respectively. Taking into account that the total volume of micropores is much less than that of the solid, we assume [solid] = const ≈ 1, then Eq. (3.65) is transformed to

$$\frac{\partial X(t,\varepsilon,T)}{\partial t} = k_1(\varepsilon,T) - k_2(\varepsilon,T)X(t,\varepsilon,T) \tag{3.66}$$

If the system was in equilibrium, we would obtain

$$X(t=\infty,\varepsilon,T) = X^s(\varepsilon,T) = \frac{k_1(\varepsilon,T)}{k_2(\varepsilon,T)} = K_r(\varepsilon,T) \tag{3.67a}$$

$$K_r(\varepsilon,T) = \zeta(\varepsilon)\exp\frac{-\Delta H^o(\varepsilon)}{R_g T} \tag{3.67b}$$

$$\zeta(\varepsilon) = \exp\frac{\Delta S^o(\varepsilon)}{R_g} \tag{3.67c}$$

where $\Delta H^o(\varepsilon)$ and $\Delta S^o(\varepsilon)$ are the changes of enthalpy and of entropy in reaction (3.65), respectively [11].

For a thermally homogeneous system that passes from temperature T_1 to temperature T_2 one obtains

$$X^s(\varepsilon,T_1) = \zeta(\varepsilon)\exp\frac{-\Delta H^o(\varepsilon)}{R_g T_1} \tag{3.68}$$

$$X^s(\varepsilon,T_2) = \zeta(\varepsilon)\exp\frac{-\Delta H^o(\varepsilon)}{R_g T_2} \tag{3.69}$$

If the transfer from T_1 to T_2 is very fast and temperature T_2 does not change, the solution of Eqs. (3.66)–(3.69) is [11]

$$X(t,\varepsilon) = X^s(\varepsilon,T_2)\{1 - \exp[-k_1(\varepsilon,T_2)t]\} \tag{3.70}$$

If the rate of the change of the temperature is much slower than the rate of pore formation, the system can be considered as quasi-steady state, and

$$X(t,\varepsilon,T) \approx X^s[\varepsilon,T(t)] \tag{3.71}$$

For a process, in which the rate of the pore formation is much slower than the change in the temperature, one obtains from Eq. (3.66)

$$\frac{\partial X(t,\varepsilon,T)}{\partial t} \approx k_1(\varepsilon,T)[X^s(\varepsilon,T(t)) - X(t,\varepsilon,T)] \tag{3.72a}$$

Using the Arrenius equation for the constant of the rate of chemical reaction, we can write Eq. (3.72a) as [11]:

$$\frac{\partial X(t,\varepsilon,T)}{\partial t} \approx A_0 \exp\frac{-E_a\varepsilon}{R_g T}[x^s(\varepsilon,T(t)) - X(t,\varepsilon,T)] \tag{3.72b}$$

where A_0 is a constant, and $E_a(\varepsilon)$ is the activation energy of the formation of pore, inside which the substance has energy ε(J/mol). The solution of Eq. (3.72a) is given by

$$X(t,\varepsilon,T) = X^0(\varepsilon,T) + [X^s(\varepsilon,T) - X^0(\varepsilon,T)]\{1 - \exp[-k_2(\varepsilon,T)t]\} \tag{3.73}$$

1. Zone of Transfer Regime

According to the equations above, in many situations one can find two main areas of energy distribution of micropores: for a very slow heating, the distribution is close to that of equilibrium, and in the case of a very fast heating, the distribution does not change. In both of these cases, the energy distribution is given by Eq. (3.63), but the value of α is different. Between these areas, we find a transfer zone where the distribution function is not Gibbs.

Let us consider a process of heating for which the temperature changes linearly: $T(\tau) = T_1 + a\tau$. If τ is given in minutes, a and A_0 are given in 1/minutes. In the zone of the transfer regime the distribution function can be approximately given by the empirical equation:

$$X(\varepsilon) \approx A'(1 + \beta\varepsilon)\exp\frac{\varepsilon}{\alpha} \tag{3.74}$$

where A' is the normalization coefficient, while

$$\begin{aligned} &\alpha = \alpha_1 = R_g T_1 && \text{when } a \to 0 \\ &\alpha_1 < \alpha < \alpha_2 && \text{for intermediate values of } a \\ &\alpha_2 = R_g T_2 && \text{when } a \to \infty \end{aligned} \tag{3.75}$$

In the linear approximation, the activation energy of pore formation is proportional to the pore energy:

$$E_a = B\varepsilon \tag{3.76}$$

where B is a semiempirical coefficient [11].

2. Thermally Inhomogeneous System

All above considerations are valid in thermally homogeneous systems, in which the temperature gradient is negligible. Such situation exists in the interior parts of most of chemical reactors. However, near the walls the situation changes, and one needs to take into account the thermal inhomogeneity of the system.

Let us consider a thermally inhomogeneous system, in which the oxidizing agent is injected into the raw material in a nonstationary regime. Let us neglect the reaction of carbon matrix destruction. Then, the thermoconduction equation contains the convection term and the heat effect of the chemical reaction:

$$\frac{\partial T}{\partial t} = \alpha_t \, \Delta T + \frac{W_G C^0_{pg}(T - T_G)}{C^0_{ps}} - \beta_t A_0 \int_{\varepsilon_{\min}}^{\varepsilon_{\max}} \Delta H^0(\varepsilon) \exp\left[\frac{-E_a(\varepsilon)}{(R_g T(\tau))}\right] d\varepsilon \tag{3.77}$$

where T_G is the temperature of the injected agent, α_t and β_t are the thermal coefficients for the continuous and porous phases, respectively, and C^0_{pg} and C^0_{ps} are the heat capacities of the gas and solid phases. In the case of slab symmetry, one obtains from Eq. (3.77):

$$\begin{aligned} \frac{\partial T}{\partial t} &= \alpha_t \frac{\partial^2 T}{\partial r^2} + \frac{W_G C^0_{pg}(T - T_G)}{C^0_{ps}} \\ &\quad - A'_0 \int_{\varepsilon_{\min}}^{\varepsilon_{\max}} \Delta H^0(\varepsilon) \exp\left[\frac{-E_a(\varepsilon)}{(R_g T(\tau))}\right] d\varepsilon \end{aligned} \tag{3.78}$$

where $A'_0 = A_0 \beta_t$. Taking into account that $\Delta H^0(\varepsilon) = \varepsilon$, one obtains from Eq. (3.78):

$$\frac{\partial T}{\partial t} = \frac{\alpha_t \partial^2 T}{\partial r^2} + \frac{W_G C^0_{pg}(T - T_G)}{C^0_{ps}} - A'_0 \int_{\varepsilon_{\min}}^{\varepsilon_{\max}} \varepsilon \exp\left[\frac{-E_a(\varepsilon)}{(R_g T(\tau))}\right] d\varepsilon \tag{3.79}$$

The boundary conditions are determined by the fact that the values of the temperature at the borders of the system are specified (or measured) by the experimenter, whereas the initial temperature is measured before the experiment. Those are as follows:

$$T(0,t) = F_1(t)$$
$$T(R,t) = F_2(t),$$
$$T(r,0) = f_0(r) = \text{constant}$$

where R is the reactor's size, functions $F_1(t)$ and $F_2(t)$ are results of the measurements at the borders, and $f_0(r)$ is the initial temperature.

To solve the above-defined problem, let us use the perturbation method. The nonperturbed solution is given by $A_0' = 0$ and can be written as

$$\frac{\partial(T - T_G)}{\partial t} = \alpha_t \frac{\partial^2(T - T_G)}{\partial r^2} + \frac{W_G C_{pg}^0 (T - T_G)}{C_{ps}^0} \tag{3.80}$$

The form of equation (3.80) is well known and has a standard solution.

If $T^{(n)}$ is the reference solution at nth iteration of perturbation, the perturbed solution is given by

$$T^{(n+1)} = T^{(n)} + \delta T \tag{3.81}$$

Then, the value of δT is found from the following differential equation:

$$\frac{\partial \delta T}{\partial t} = \alpha_t \frac{\partial^2 \delta T}{\partial r^2} + \frac{W_G C_{pg}^0 (T - T_G)}{C_{ps}^0} + \psi_p(r,t) \tag{3.82}$$

Here

$$\psi_p = -A_0' \int_{\varepsilon_{\min}}^{\varepsilon_{\max}} \varepsilon \exp\left[-E_a\left(R_g T^{(n)}(r,t)\right)\right] d\varepsilon$$

The boundary conditions formulated above are written for the perturbed problem as

$$\delta T(0,t) = 0 \qquad \delta T(R,t) = 0 \qquad \delta T(r,0) = 0$$

The solution of the perturbed problem is given by Fourier representation:

$$\delta T(r,t) = \sum_{n=0}^{\infty} \left[a_n(t) \sin \frac{n\pi r}{R} \right]$$
$$\phi_p(r,t) = \frac{1}{\sqrt{2\pi}} \sum_{n=0}^{\infty} \left[b_n(t) \sin \frac{n\pi r}{R} \right] \qquad (3.83)$$
$$b_n = \frac{1}{\sqrt{2\pi}} \int_0^R \varphi_p(r,t) \sin \frac{n\pi r}{R}\, dn$$

where the unknown functions are found from the following condition:

$$\frac{da_n}{dt} = \left[-\left(\frac{n\pi}{R}\right)^2 \alpha_t + \frac{W_G C^0_{pg}}{C^0_{ps}} \right] a_n + b_n(t) = \lambda_n a_n + b_n(t) \qquad (3.84)$$

where

$$a_n(t) = c_n(t) \exp(\lambda_n t)$$
$$c'_n(t) = b_n(t) \exp(\lambda_n t)$$
$$a_n(t) = \int_0^t b_n(t) \exp \lambda_n (t - \tau)\, d\tau$$

Thus, we have obtained a solution for the problem of evaluation of temperature distribution in thermally inhomogeneous system. When the temperature distribution is found from Eqs. (3.83) and (3.84), the pore formation process gets a description in accordance to the thermodynamic approach developed above. This solution is applicable to many of processes of high-temperature synthesis of microporous materials but first of all to manufacturing of active carbon—not only in laboratory but also industrial conditions.

Now, let us analyze the completeness of the solutions obtained above. After all, we can evaluate the pore energy distribution functions of microporous media prepared in various processes under high temperatures, but we do not still know the limits of energy for Eqs. (3.57) and (3.58). It is normal to assume $\varepsilon_{min} = 0$, but the value of ε_{max} remains unknown. One way of its estimation consists in energy estimations for processes of restructuring of porous structures (that will be discussed below).

G. Substantial and Local Considerations of Active Carbon Synthesis by Pyrolysis

Let us consider a typical technological process of pyrolytic preparation of active carbon with burning off the organic volatile components (see Fig. 3.2).

The raw material moves inside a tube heated with exterior heat. The length of the tube is L, the coordinate of a considered fragment of the raw material inside the tube is l $(0 < l < L)$, the exterior temperature $T(l)$ changes from $T(l=0)$ to $T(l=L)$. In real technology, the stationary regime of the treatment is optimal; in such case, Eq. (3.63) is valid, and the value of α depends on l. If we observe the system from a spot with fixed l, we have the *local consideration* of the system. In this local consideration, the system is observed as steady state, Eq. (3.63) is applicable, and the only problem consists in the change of values Q_0 and α with l.

On the other hand, the involved system moves inside the tube, gets varied temperatures and follows some reactions analyzed above. The velocity of the involved system is v_f, and the time of the treatment is $\tau_t = L/v_f$. One may relate the observation with the moving raw material; this consideration is defined as *substantial*. In this substantial consideration, the system is not of all steady state (because the observer finds changes of

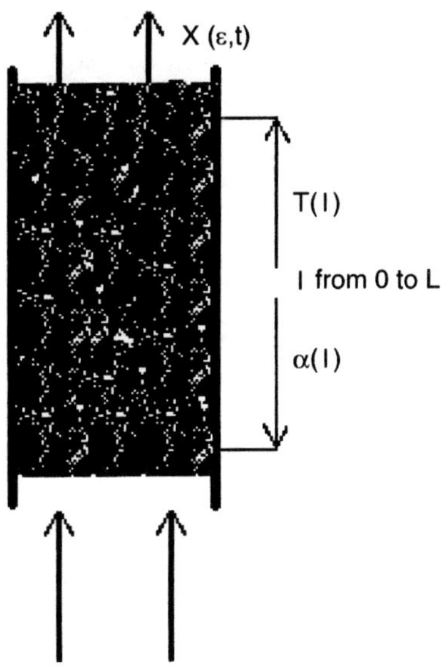

FIG. 3.2 Schematic presentation of the motion of organic raw material in synthesis of active carbon.

exterior and interior parameters with time), and the solution is obtained from Eq. (3.66).

Thus, we have a paradox: depending on the choice of the observation spot, we obtain formally different regimes of treatment (steady state or non-steady state) and the absolutely different mathematical models.

However, this paradox is easily solved: of course, both forms of considerations are valid, both mathematical models are valid, and one needs just to write the condition of their conjugation.

Accounting that the substantial notion $X(t, \varepsilon, T)$ means the volume fraction of pores having energy ε per 1 mol of internal substance under temperature T at the moment t, we may write

$$X(t,\varepsilon,T) = \xi f(\varepsilon) = \xi \exp \frac{-\varepsilon}{\alpha(t)} \tag{3.85}$$

where ξ is porosity and $f(\varepsilon)$ and α are parameters applicable to the local consideration. Writing $X(t, \varepsilon, T)$, we mean that the pore volume fraction may change with

- Pore energy ε, because X is related to the pore energy distribution function
- Temperature T, because this influences the rate of formation or destruction of pores
- Time t, because X may change with time even under a specified temperature T, if the system is not in equilibrium

Equation (3.66) is applicable without any changes, while Eq. (3.63) should be transformed to the following form:

$$\frac{\partial X}{\partial \varepsilon} = XZ(T,t) \tag{3.86}$$

where the function $Z(T, t)$ takes into account the dependence of α on all factors except ε. Equations (3.66) and (3.86) are transformed to

$$\frac{\partial^2 X/\partial \varepsilon}{\partial t} = \frac{\partial k_1}{\partial \varepsilon}(\varepsilon,t) - \frac{\partial k_2}{\partial \varepsilon}(\varepsilon,t)X(t,\varepsilon,T) - k_2(\varepsilon,t)\frac{\partial X}{\partial \varepsilon}(t,\varepsilon,T) \tag{3.87}$$

$$\frac{\partial^2 X/\partial \varepsilon}{\partial t} = \frac{\partial X}{\partial t}\, Z + X(t)\, \frac{\partial Z}{\partial t} \tag{3.88}$$

Equations (3.87) and (3.88) are transformed to

$$Z\frac{\partial X}{\partial t} + X\left(\frac{\partial Z}{\partial t} + k_2 Z + \frac{\partial k_2}{\partial \varepsilon} - \frac{\partial k_1}{\partial \varepsilon}\right) = 0 \tag{3.89}$$

For a very slow process, when $\tau_t \to \infty$, one obtains from Eq. (3.89)

$$\frac{\partial X}{\partial t} = 0 \qquad \frac{\partial Z}{\partial t} = 0 \qquad X^e = \frac{\partial k_2/\partial \varepsilon}{k_2 + (\partial k_2/\partial \varepsilon)} \tag{3.90}$$

that is the equilibrium solution!

Let us notify that Eq. (3.89) is valid only in condition that the local consideration provides the steady-state image of the process of treatment.

H. Pattern Formation and False Heterogeneity

Analyzing pore formation processes (Chapter 1), we notified that the homogeneous conditions applied to a homogeneous raw material always provide the homogeneity of the product. Now, let us consider a situation when this conclusion seems wrong.

Let us consider a process of active carbon synthesis on a device presented by Fig. 3.2 when the velocity of the material motion increases. First, when the velocity v_f is moderate, the locally observed system is steady state, while the process of burning off becomes short, the amount of burned-off organics decreased, the product after synthesis is cooled rapidly, and the pore formation on the final stage of the synthesis is very fast, while the product is still homogeneous. The further increase of v_f results in the following phenomenon: the process of burning off is very short, and only its central zone (source of inflammation) gets the sufficient oxidation. Around the central zone, the volatile organics are not well removed, and its rest concentration increases with the distance from the source of inflammation. However, that is less than the initial concentration; therefore, its inflammation is more difficult than for the initial content. We observe four zones:

The central zone (source of inflammation), where the degree of burning off of volatile organics is close to the normal

The zone of moderate burning off of volatile organics, where the rest concentration is not high but much over the desired level

The zone of low burning off of volatiles, where the rest concentration of organics is high while less than the initial one

The zone where no burning took place, where the concentration of volatile organics is the initial

The central zone leaves the reactor. Then, the zone of moderate burning off comes to the zone of upper temperatures but cannot get inflammation—not only because of low content of volatile organics but also because the motion through the reactor is too fast. This fraction, though not sufficient for the needs of user of porous material, also leaves the reactor and is mixed to the well-treated fraction (central zone—already kept as product). Thus, we obtain a heterogeneity: though all raw material passed the same exterior conditions, it contains at least two fractions characterized by different burning off, different porosity, different chemical composition, pore energy distribution, etc. Let us notify that this heterogeneity is false: each separate fraction is well homogeneous.

What happens to the other two fractions? The fraction of low burning off has a chance to get inflammation in the zone of upper temperatures, depending on the combination of these factors—internal temperature determined by the exterior temperature and the rate of heat conduction. However, there is also a chance that this fraction does not get burned.

The last fraction—nonburned part of the raw material—should get inflammation, due to the high content of volatile organics, and form the good fraction of the product.

After all, if the velocity v_f continues to increase, the zones of high and moderate burning off are reduced, and then the inflammation becomes impossible because the motion of the raw material through the reactor is too fast.

Let us describe the above system with a simple model: We assume that

The considered part of the raw material is in homogeneous conditions.
The inflammation takes place only in the gaseous phase.
The gaseous phase is heated due to two factors (heat from the walls and from the burning process) and cooled because of loosing heat with the removed material.
The raw material is heated due to thermal conductivity.
The criterion of inflammation consists in the temperature of gaseous phase T_{ex} coming over the temperature of walls T_b.
The relevant equations are (very roughly) written as

$$dh_+ = [\lambda_{gas} Q_{gas}(T_b - T_{ex}) + C_{vol,\,gas}|\Delta H^0_{inf}| Q_{gas} - v_f(T_{ex} - T_0) Q_{gas}]\, dt \tag{3.91}$$

$$dT_{ex} = \lambda_{gas} Q_{gas}(T_b - T_{ex})\, dt \tag{3.92}$$

$$dT_{in} = \lambda_{gas\text{-}solid} Q_{gas}(T_{ex} - T_{in})\, dt \tag{3.93}$$

$$dQ_{gas} = Q_{solid} \exp\left(\frac{\Delta S^0_{vol}}{R_g} - \frac{\Delta H^0_{vol}}{R_g T_{in}}\right) dt \tag{3.94}$$

$$C_{vol,gas} Q_{gas} + C_{vol,solid} Q_{solid} = C_{vol,0} Q_{solid,0} \tag{3.95}$$

where ΔS^0_{vol} is entropy of volatile delivering, ΔH^0_{vol} is the enthalpy of this process, ΔH^0_{inf} is the heat effect of burning [that is negative, therefore, in Eq. (3.91) figuration modulus], Q_{gas} is the amount of substance in the gas phase, Q_{solid} is the same for solid phase, $Q_{solid,0}$ is the initial amount of the raw material, $C_{vol,gas}$ is the concentration of volatiles in the gas phase, $C_{vol,solid}$ is the same for the solid phase, $C_{vol,0}$ is the same for the initial raw material, λ_{gas} is heat conductivity for the gas phase, and $\lambda_{gas\text{-}solid}$ is the same for gas–solid interface.

Equations (3.91)–(3.95) present no interest in the mathematical aspect; it is much more important to analyze the results of their numerical solution on the computer. We have taken the following parameters: $\Delta S^0_{vol} = 100\,\text{J/mol/K}$, $\Delta H^0_{vol} = 60\,\text{kJ/mol}$, $\Delta H^0_{inf} = -500\,\text{kJ/mol}$, $Q_{solid,0} = 1$, $\lambda_{gas} = \lambda_{gas\text{-}solid} = 0.1(\text{mol·s})^{-1}$.

The results of computer simulations on the base of Eqs. (3.91)–(3.95) are presented on Fig. 3.3.

As follows from Fig. 3.3, over some value of v_f (5 for the conditions of the given calculations) the inflammation is absolutely impossible. Below this

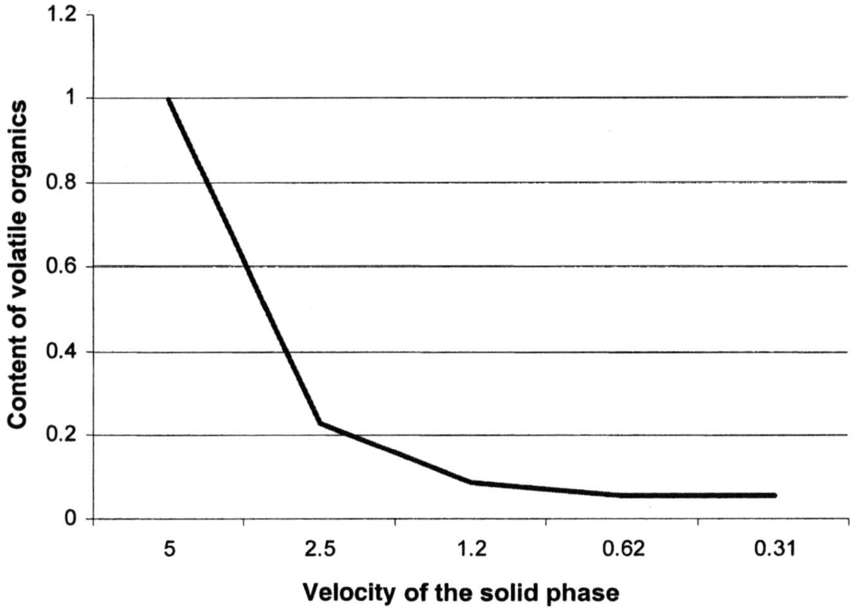

FIG. 3.3 Minimal content of volatile organics needed for inflammation vs. the velocity of the solid-phase flow.

value, the minimal concentration allowing the inflammation decreases with the decrease of v_f. When this minimal concentration becomes enough low, the process of burning off becomes continuous, and the pattern (false heterogeneity) disappears.

Let us notice again that Eqs. (3.91)–(3.95) do not pretend to the correctness but just aim to illustrate the physical phenomenon: the relationship between the rate of treatment of the raw material and the appearance of patterns though the homogeneity of treatment conditions, in the sense as formulated in Chapter 1.

III. PORE THERMODYNAMICS IN LINEAR AND BRANCHED STRUCTURES

Linear (*chain*) *structures* are defined as structures whose configuration is characterized by the opportunity of joining of two fragments (structural units) to the same fragment. An example of chain structure is presented by Fig. 3.4.

The ability of a fragment to join one, two, or more other fragments is defined as *functionality*. If functionality is over two, each fragment may join three or more other fragments, this phenomenon is called *branching*. *Branched structures* are defined as structures whose configuration is characterized by the joining of three or more fragments (structural units) to the same fragment. Some examples of branched structures are given on Fig. 3.4a–d.

As we can see from Fig. 3.4a–d, branched structures may have

Identical or different fragments
Cross-linking (also called *ringing* or *cycling*) or no cross-linking
Various branching
Various angles between the fragments (not shown on the figures)

If a structure has number of functionality 3, it does not mean that each of its fragments is connected to three other fragments. *Number of functionality* just means the maximum of possible connections, whereas the minimum is always 0. The same is true for cross-linking: the existence of *cross-linking* means that some but not all fragments are connected into closed configurations. In principle, cross-linking is possible also in nonbranched structures (meaning branching = 1), forming closed rings. If branching is 1 and no cross-linking is found, such structure is defined as *linear structure*.

Branched structures of various configurations are called *graphs* in mathematics and are described by the mathematical theory of *graphs* (or *graph theory* [12]).

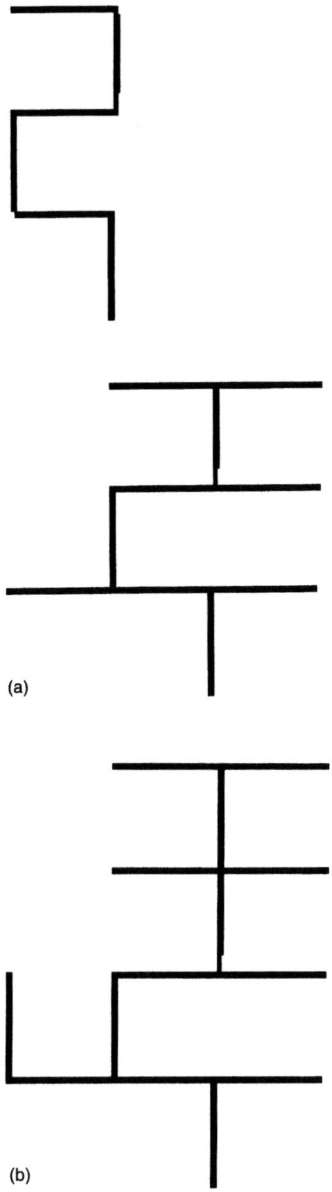

FIG. 3.4 Linear (chain) structure. (a) Branched structure without cross-linking, functionality = 2. (b) Branched structure without cross-linking, functionality = 3. (c) Branched structure with cross-linking, functionality = 3. (d) Branched structure with cross-linking, functionality = 3, constructed from various fragments.

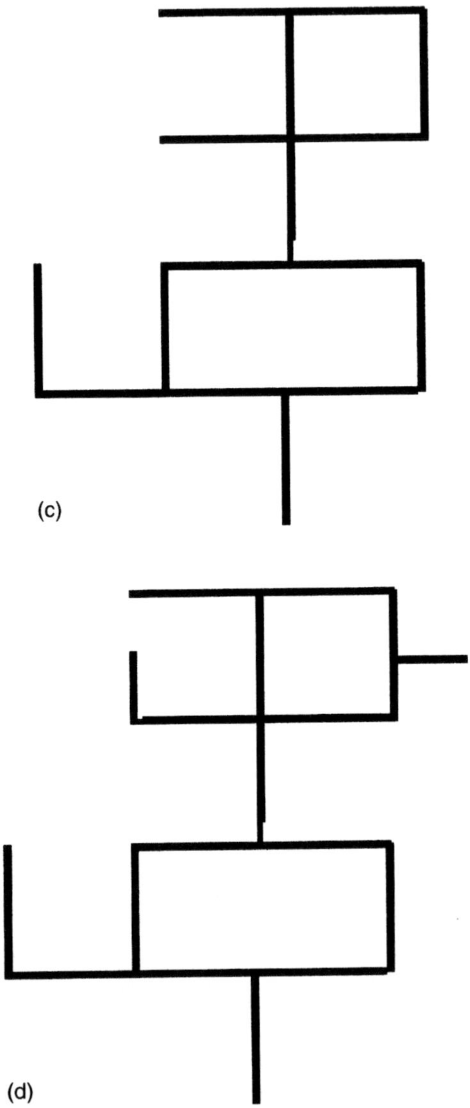

FIG. 3.4 Continued.

Branched structures find multiple applications, but we are interested only in their use for simulations of macromolecular structures: polymer, gels, and various molecular aggregates. All macromolecules can be described as branched or linear structures, with or without cross-linking.

A. Polymers

Polymer can be defined as a heavy-mass electroneutral organic macromolecular structure constructed of a number of molecular fragments, connected by chemical bonds. Let us consider some examples of polymers:

Polyethylene: worldwide known polymer constructed of ethylene molecular fragments and having linear structure; ethylene fragments are connected by chemical bonds $-CH_2-CH_2-CH_2-\cdots$ that are formed due to breaking double bonds in ethylene molecules in the polymerization process.

Polybutadiene: the simplest rubber constructed of butadiene molecular fragments and having branched cross-linked structure with branching = 3; butadiene fragments are connected by chemical bonds formed due to breaking double-double bonds in butadiene molecules in the polymerization (vulcanization) process.

Polybutadiene rubber: a widespread rubber similar to polybutadiene but containing sulfur, due to which the process of vulcanization is performed much faster.

Proteins: natural polymers composing all kinds of living tissue, constructed of amino-acid molecular fragments connected by ionic bonds.

Now, let us consider some structures that are not polymeric while they have similar features:

Concrete: this has a macromolecular structure but is not an organic compound.

Molecule of hexane: this is organic and consists of some fragments, but its weight is absolutely not sufficient to be considered a polymer.

Petroleum: this consists of organic substances, but their molecules are not connected by chemical bonds.

B. Polymerization

Polymerization is defined as the process of synthesis of polymer from separate molecules (monomers). In principle, the chemical composition of the resulting polymer may differ from the sum of all monomers—e.g., if the polymerization reaction is accompanied by delivering water (polycondensation reaction). According to its mechanism, polymerization can be free-radical or ionic reaction. For example, polyethylene is formed due to free-radical reaction, while proteins are due to ionic reaction between bipolar amino acids. Polymerization is usually initiated by a little amount of a substance that does not need to be included into the resulting polymer: e.g., polymerization of ethylene can be initiated by metal sodium that easily delivers electrons to double bonds of ethylene, which excites ethylene monomers and stimulates their interactions. In some cases, polymerization is initiated by high-energy irradiation that stimulates the formation of free radicals.

Polymerization can be regular or random. Regularity of polymerization means a selection in the way of polymer formation: not all thermodynamically possible structures can be formed; polymerization degree is limited, etc. Regular polymerization is very important in technologies requiring pure chemicals, e.g., in pharmacy and laser technology.

The randomness of polymerization means that all thermodynamically allowed structures can be formed and all levels of polymerization degree are available. Random polymerization is typical for the chemical technology. Unless the regularity is imposed by technical means, polymerization gets a random mechanism.

If the product of the polymerization is linear polymer, such polymerization is called *linear polymerization*. In the case of branching, that is *branched polymerization*.

(a) Model of Linear Polymerization. The theory of linear polymerization was developed by Nobel Laureate P. Flory [13].

Polymerization can be defined as a process of formation of polymer (macromolecule) consisting of N units (fragments), where $N > 1$. Polymers can be

One-component polymers consisting of fragments of the same chemical composition, formation of which is described by the same equation of chemical reaction between an unstable intermediary product (for example, radical) and a neutral molecule:

$$-(CH_2)_n{-}CH_2^{\bullet} + CH_2{=}CH_2 \rightarrow -(CH_2)_{n+2}{-}CH_2^{\bullet} \tag{3.96}$$

Multicomponent polymer consists of some different kinds of fragments having the different chemical composition, each of them being presented many times in the resulting structure, e.g.,

$$-(CH_2)_n{-}(CHCH_3)_m^{\bullet} + CH_2{=}CH_2 \rightarrow -(CH_2)_n{-}(CHCH_3)_m{-}(CH_2)_2^{\bullet} \tag{3.97}$$

Multicomponent polymers are also called *copolymers.*

The normal polymeric system contains monomers and macromolecules of different weight. The content of heavy polymers in such a system can be characterized by degree of polymerization:

$$d_p = \frac{\sum_{N=1}^{\infty} NC_N}{\sum_{N=1}^{\infty} C_N} \tag{3.98}$$

where C_N is the concentration of N-mer.

Let us notify that there are two options for the characterization of the concentration of polymers: the traditional manner meaning the molar part of N-mers among all molecules (C_N) and the specific for polymeric chemistry style, when the concentration means the weight part of the macromolecules (NC_N). The second form of designing polymeric concentration is more applicable to polymerization processes, when

$$\sum_{N=1}^{\infty} NC_N = 1 \tag{3.99}$$

Let us consider a linear polymerization as bimolecular reaction, meaning that polymer consisting of N monomeric units (N-mer) can be formed by (1) reaction of a monomer with $(N-1)$-mer or (2) reaction of a M-mer to $(N-M)$-mer, $1 < M < (N-1)$. The kinetics of such processes is very complicated, because obviously properties of macromolecules having different weight should differ. However, the procedure of modeling of linear polymerization can be significantly simplified, based on the elementary logic: in two similar macromolecules having the same chemical composition, the reaction ability of end groups should be approximately the same because the near neighbor fragments "isolate" the end groups from the rest part of the polymer. The suggestion that the main influence on each fragment (comprising end groups) is due to its nearest neighbors (first environment), while the influence of other fragments decreases fast with the distance from the considered fragment, is the sense of *Flory principle.*

Flory principle simplifies much all calculations related to macromolecular systems. Flory principle is applicable not only to one-component chains but to all kinds of polymers, aggregates, gels, etc., while one must always take into account that the 1st environment in each case can be different.

C. Radical and Ionic Polymerization

Depending on the form of the unstable intermediate product in the polymerization process, this can be radical or ionic. For example, reaction

TABLE 3.8 Initiation of Polymerization by Various Mechanisms

Factor of initiation	Mechanism of polymerization	Cause of nonstability of intermediate product
Metallic sodium	Ionic	Getting electron from sodium
X-rays	Radical	Getting energy by irradiation
Scattering by electrons	Ionic	Getting energy and electrons by irradiation
Scattering by neutrons	Both radical and ionic	Energy, electrons, and protons from neutrons
Ultrasound irradiation	Radical	Energy by acoustics

(3.96) is radical, while ionic mechanism of initiation, e.g., due to an interaction with metallic sodium, can provide electrically negative charge onto the unstable intermediary macromolecular product, then the polymerization gets an ionic mechanism. Processes of gels formation have ionic mechanism.

Table 3.8 gives examples of radical and ionic initiation of traditional polymerization of organic monomers.

Comparing Table 3.8 to Tables 2.3–2.5, we can notify that high-energy scattering of polymeric porous materials is not acceptable as the technique of their study—even initiating the further polymerization. On the other hand, for many of gels formed by ionic polymerization, x-ray and ultrasound scattering techniques cannot cause any changes in the polymerization degree.

D. Irreversible Chain Polymerization Far from Equilibrium

Based on Flory principle, let us build a theoretical model for irreversible chain polymerization having the following mechanism [14]:

$$\mathrm{M} + \mathrm{M} \rightarrow \mathrm{M}_2 \tag{3.100}$$

$$\mathrm{Pol}(2) + \mathrm{M} \rightarrow \mathrm{Pol}(3) \tag{3.101a}$$

$$\mathrm{Pol}(m) + \mathrm{Pol}(n) \rightarrow \mathrm{Pol}(m+n) \tag{3.101b}$$

where M is monomer, Pol(m) is linear polymer containing m identical fragments, and $\mathrm{M}_2 = \mathrm{Pol}(2)$.

Kinetics of reactions (3.100)–(3.101) is described by the following equations:

$$dC_1/dt = -k_{11}C_1^2 - k_{12}C_1C_2 - \cdots - k_{1(m-1)}C_1C_{m-1} - \cdots$$
$$= -\sum_{m=1}^{\infty} k_{1m}C_1C_m \tag{3.102}$$

$$\frac{dC_2}{dt} = k_{11}C_1^2 - \sum_{m=1}^{\infty} k_{2(m-2)}C_2C_{(m-2)} \tag{3.103}$$

$$\frac{dC_n}{dt} = \sum_{m=1}^{\leq n/2} k_{m(n-m)}C_mC_{(n-m)} - \sum_{l=1}^{\infty} k_{nl}C_nC_l \tag{3.104}$$

where k_{ij} is the constant of the rate of the reaction between Pol(i) and Pol(j) with formation of ($i+j$)-mer and C_i is the concentration of i-mer. Based on Flory principle, let us assume that $k_{i_1j_1} = k_{i_2j_2} = k_0$ (for all i_1, i_2, j_1, j_2), then Eqs. (3.102) and (3.104) are transformed to the following:

$$\frac{dC_1}{dt} = -k_0C_1\sum_{l=1}^{\infty} C_l = -k_0C_1C_\Sigma \tag{3.105}$$

$$\frac{dC_2}{dt} = k_0C_1^2 - k_0C_2\sum_{l=1}^{\infty} C_l = k_0C_1^2 - k_0C_2C_\Sigma \tag{3.106}$$

$$\frac{dC_n}{dt} = k_0\sum_{m=1}^{\leq n/2} C_mC_{(n-m)} - k_0C_nC_\Sigma \tag{3.107}$$

where C_Σ is the total concentration [that is not 1, because of condition (3.99)].

Equation (3.105) for monomers has the obvious solution:

$$C_1 = C_0\exp(-k_0t) = \exp(-k_0t) \tag{3.108}$$

where $C_0 = 1$ is the initial concentration of monomers. Neglecting the change of C_Σ with time, for dimers and trimers we obtain from Eqs. (3.106)–(3.108), respectively:

$$C_2 = 1 - \exp(-C_\Sigma k_0t)]\exp(-k_0t) \tag{3.109}$$

$$C_3 = [1 - \exp(-C_\Sigma k_0 t)]^2 \frac{\exp(-k_0 t)}{2} \tag{3.110}$$

E. Reversible Polymerization

Reversible polymerization takes place very close to equilibrium, then Eqs. (3.100)–(3.101) are written as

$$\mathrm{M} + \mathrm{M} \longleftrightarrow \mathrm{M}_2 = \mathrm{Pol}(2) \tag{3.111}$$

$$\mathrm{Pol}(2) + \mathrm{M} \longleftrightarrow \mathrm{Pol}(3) \tag{3.112}$$

$$\mathrm{Pol}(m) + \mathrm{Pol}(n) \longleftrightarrow \mathrm{Pol}(m+n) \tag{3.113}$$

Kinetics of reactions (3.111)–(3.113) is described by the following equations:

$$\frac{dC_1}{dt} = -k_+ C_1 + k_- \sum_{m \neq 1} C_m \tag{3.114a}$$

$$\frac{dC_1}{dt} = -k_+ C_1 + k_- C_2 + k_- C_1^e \sum_{m \geq 3} \frac{C_m}{\sum_{i \neq 1,\, i \neq (m-2)}^{m-1} C_i^e} \tag{3.114b}$$

$$\frac{dC_2}{dt} = k_+ C_1^2 - k_+ C_2 - k_- C_2 + k_- C_3 + k_- C_2^e \sum_{m \geq 3} \frac{C_m}{\sum_{i \neq 2,\, i \neq (m-2)}^{m-1} C_i^e} \tag{3.115}$$

$$\frac{dC_n}{dt} = k_+ \sum_{i=1}^{\leq (n/2)} C_i C_{n-i} - k_+ C_n - k_- C_n + k_- C_n^e \sum_{m \geq (n+1)} \frac{C_m}{\sum_{i \neq n,\, i \neq (m-n)}^{m-1} C_i^e} \tag{3.116}$$

where k_+ is the constant of rate of synthesis, k_- is that of depolymerization, C_m^e is the equilibrium concentration of m-mer.

Equations (3.114)–(3.116) mean that each n-mer in the system with reversible chain polymerization:

- Is formed due to synthesis from components having lower weight, for which $m < n$.
- Is formed due to decomposition of macromolecules having more weight, for which $m > n$.

Is consumed to synthesis of macromolecules having more weight ($m > n$).
Is decomposed to components having lower weight ($m < n$).
The probability of its formation is proportional to its concentration in equilibrium.

Solution of Eqs. (3.114)–(3.116) needs, first of all, the estimation of equilibrium concentrations of macromolecules.

F. Equilibrium Distribution of Chain Polymers

On the contrary to nonequilibrium processes, equilibrium polymerization does not need to take into account all kinds of possible reactions like Eqs. (3.111)–(3.113). Since thermodynamic parameters depend only on the current state of the system and not on the way by which the system has came to this state, one may base the consideration only on reactions with participation of monomers (principle of independent reactions):

$$\mathrm{M} + \mathrm{M} \longleftrightarrow \mathrm{Pol}(2) \tag{3.117}$$

$$\mathrm{M} + \mathrm{Pol}(m-1) \longleftrightarrow \mathrm{Pol}(m) \tag{3.118}$$

Each of the above reactions is characterized by constant of equilibrium:

$$K_m = \frac{C_m}{C_1 C_{m-1}} \tag{3.119}$$

Using again Flory principle, we obtain from Eq. (3.119)

$$K_m = K_0 = \frac{k_+}{k_-} \quad \text{(for all } m\text{)} \tag{3.120}$$

Hence,

$$C_m = (K_0)^{m-1}(C_1)^m \tag{3.121}$$

G. Differential Equations for Concentrations of Chain *N*-mers Close to Equilibrium

The general equations are given by Eqs. (3.114)–(3.116), these are too complicated for solution. However, these can be simplified due to the use of perturbation method.

Let us present the current concentration of N-mer close to equilibrium as

$$C_N = C_N^e + \delta C_N, \tag{3.122}$$

where C_N^e is the equilibrium concentration of N-mer and δC_N is its perturbation because of nonequilibrium processes. Obviously

$$dC_N = d\delta C_N \tag{3.123}$$

because C_N^e is constant, and

$$|\delta C_N| \ll C_N^e \tag{3.124}$$

because the concentration is close to equilibrium. We obtain from Eqs. (3.116)–(3.124)

$$\frac{d\delta C_1}{dt} = -k_+\delta C_1 + k_-\delta C_2 + k_- C_1^e \sum_{m\geq 3} \frac{\delta C_m}{\sum_{i\neq 1,\, i\neq(m-1)}^{m-1} C_i^e} \tag{3.125}$$

$$\frac{d\delta C_2}{dt} = 2k_+ C_1^e \delta C_1 - k_+\delta C_2 + k_-\delta C_3 + k_- C_2^e \sum_{m\geq 4} \frac{\delta C_m}{\sum_{i\neq 2,\, i\neq(m-2)}^{m-1} C_i^e} \tag{3.126}$$

$$\frac{d\delta C_n}{dt} = k_+\left[\sum_{i=1}^{\leq(n/2)} \left(C_i^e \delta C_{n-i} + \delta C_i C_{n-i}^e\right)\right] - k_+\delta C_n - k_-\delta C_n + k_- C_n^e \times \sum_{m\geq(n+1)} \frac{\delta C_m}{\sum_{i\neq n,\, i\neq(m-n)}^{m-1} C_i^e} \tag{3.127}$$

Equations (3.122)–(3.127) are linear and can be solved numerically on computer. The question of their solution stands out of this book. (Why? Because the thermodynamic solution presented below allows the description of nonequilibrium due to using a much more simple way.)

H. Thermodynamic Functions of Linear Macromolecules

The above-presented kinetic method for modeling linear polymerization just illustrates how is it difficult to describe even so simple system directly.

What will happen in the case of branched polymerization, much more complicated? One may suggest that branching will not be described of all. However, the indirect modeling of polymerization by thermodynamic method, as it is shown below, allows the solution of very difficult problems related to polymerization. Let us consider the facilities of thermodynamics, first of all, for linear (chain) polymerization.

As usually in thermodynamics of pores, thermodynamic functions of polymers correspond to the equilibrium situation, independently on the real way of polymer synthesis. The equilibrium situation is described by Eqs. (3.119)–(3.121). The thermodynamic condition for equilibrium is given by the equal values of total chemical potentials for both sides of Eq. (3.119):

$$\mu_1 + \mu_{m-1} = \mu_m \tag{3.128}$$

Equation (3.128) can be rewritten with using activities of the components:

$$\mu_1^0 + R_g T \ln a_1 + \mu_{m-1}^0 + R_g T \ln a_{m-1} = \mu_m^0 + R_g T \ln a_m \tag{3.129}$$

where a_i is the activity of ith component. Assuming $a_i \approx C_i$, we obtain from Eq. (3.129)

$$\mu_1^0 + \mu_{m-1}^0 - \mu_m^0 = R_g T \ln \frac{C_m}{C_1 C_{m-1}} \tag{3.130}$$

Comparing Eqs. (3.119) and (3.129), we obtain

$$\mu_1^0 + \mu_{m-1}^0 - \mu_m^0 = R_g T \ln K_0, \tag{3.131}$$

or

$$\mu_m^0 = \mu_1^0 + \mu_{m-1}^0 - R_g T \ln K_0 = m\mu_1^0 - (m-1) R_g T \ln K_0 \tag{3.132}$$

I. Evaluation of Nonequilibrium Concentrations of Chain Polymers, Using Principle of Superposition

Let us consider a process of linear (chain) polymerization, in which monomers are presented initially in concentration $C_1(t=0) = C_0 = 1$, then partly transformed to polymers. The system since starting with polymerization becomes a mixture of monomers and various polymers in non-equilibrium. The initial free energy of the system is equal; obviously,

$$\Delta F(t=0) = \mu_1^0 C_1(t=0) = \mu_1^0 \tag{3.133}$$

Of course, $\mu_1^0 > 0$, otherwise the polymerization would be impossible.

If the system gains equilibrium, its free energy is

$$\Delta F(t=\infty) = \Delta F^e = \sum_{N=1}^{\infty} C_N^e \mu_N^e \tag{3.134}$$

where index e means the equilibrium. From Eqs. (3.132) and (3.134) we obtain

$$\Delta F^e = \sum_{N=1}^{\infty} (C_1^e)^N K_0{}^{N-1}(N\mu_1^0 - (m-1)R_g T \ln k_0) \tag{3.135}$$

Each available state of the system is characterized by a value of ΔF:

$$\Delta F^e \leq \Delta F(t) \leq \Delta F(t=0) \tag{3.136}$$

The same is true for $C_N\mu_N$, which is an additional parameter too:

$$C_N^e \mu_N^e \leq C_N \mu_N(t) \leq \delta_{N1} \tag{3.137}$$

where $\delta_{N1} = 0$ if $N \neq 1$ but $\delta_{N1} = \mu_1^0$ if $N = 1$. Both ΔF and $(C_N\mu_N)$ are additive parameters; hence, they can be treated with Eq. (3.27), meaning the principle of superposition; therefore,

$$(C_N\mu_N)(t) = (C_N\mu_N)(t=\infty) + [\Delta F(t) - \Delta F^e]\frac{\delta_{N1} - (C_N\mu_N)(t=\infty)}{\Delta F(t=0) - \Delta F^e} \tag{3.138}$$

Equation (3.138) provides the value of $(C_N\mu_N)(t) = C_N(t)\mu_N(t)$ for any value of ΔF. If the system in polymerization does not get negentropy from exterior and is described by linear law, Eq. (3.29) is applicable, and one obtains the sequence of operations providing the nonequilibrium concentrations:

1. The current value of ΔF is found from Eq. (3.29).
2. The current value of $(C_N\mu_N)$ is found from Eq. (3.138).
3. The current value of C_N is found from the condition $(C_N\mu_N) = (C_N\mu_N^0) + R_g T C_N \ln C_N$.

Thus, we have obtained the solution for the general case of nonequilibrium in system with chain polymerization, without solving differential equations, Eqs. (3.125)–(3.127).

J. Ring Formation in Processes of Chain Polymerization

End groups in linear macromolecular radicals like Eq. (3.96) may react not only with groups in other macromolecules (that causes the growth of the macromolecule) but also with other end groups in the same radicals, with breaking the radical reaction chain. The product of such reaction is not a macromolecular chain but a ring (sometimes also called *cycle*):

$$\begin{array}{lcc} {}^{\bullet}CH_2\text{–}CH_2\text{–}CH_2\text{–}CH_2{}^{\bullet} \longleftrightarrow & CH_2\text{–}CH_2 \\ & | \quad\quad | \\ & CH_2\text{–}CH_2 \end{array} \tag{3.139}$$

Reactions like Eq. (3.139) are restrained because of loss in entropy but stimulated due to loss of internal energy (since two radicals are simultaneously lost); therefore, such reactions are allowed mostly under low temperatures, when the polymerization process is mainly achieved.

The kinetics of ring formation is described by the following equations:

$$\frac{dC_N}{dt} = -k_{\mathrm{rf}} C_N + k_{\mathrm{rd}} C_{Nr} \tag{3.140}$$

$$\frac{dC_{Nr}}{dt} = k_{\mathrm{rf}} C_N - k_{\mathrm{rd}} C_{Nr} \tag{3.141}$$

where C_{Nr} is the concentration of N-meric rings, k_{rf} is the constant of the rate of ring formation, and k_{rd} is the constant of the rate of ring destruction. The solution of Eqs. (3.140) and (3.141) is given by

$$C_N(t) = C_N^0(t) - C_{Nr}(t) \tag{3.142}$$

$$C_{Nr}(t) = f_{Nr}(t)\exp[-(k_{\mathrm{rf}} + k_{\mathrm{rd}})t] \tag{3.143}$$

$$f_{Nr}(t) = k_{\mathrm{rf}} \int_0^t C_N^0(\tau)\exp[(k_{\mathrm{rf}} + k_{\mathrm{rd}})\tau]\,d\tau, \tag{3.144}$$

where $C_N^0(\tau)$ is the total concentration of chain and ring N-mers at moment τ, while $C_N^0(\tau)$ changes only due to polymerization–depolymerization processes, independently on ring formation–destruction.

As we have noted above, ring formation does not significantly influence the polymerization process until this is achieved. However, another process very similar to ring formation—cross-linking in branched macromolecules—is thermodynamically allowed and influences the process of branched polymerization and the resulting properties of branched macromolecules, as is shown below.

K. Chain Polymerization in Multicomponent System

Multicomponent polymerization (copolymerization) is very similar to the above-considered chain polymerization, but the principal difference consists in the possibility of including different monomeric units into the same macromolecule:

$$-M_1-M_2-M_3- \cdots -M_n- + M_{n+1} \longleftrightarrow -M_1-M_2-M_3- \cdots -M_{n+1}- \tag{3.145}$$

where monomeric units M_1, M_2, $M_3, \ldots, M_n$ correspond to chemically different substances. The kinetic and thermodynamic equations for multicomponent chain polymerization are much more complex, but the main problem is the difficulties in the application of the Flory principle. Indeed, for example, monomeric units M_2 in the groups $-M_1-M_2-M_1-M_1-M_2-M_1-$ and $-M_1-M_2-M_2-M_1-M_1-M_1-$ have obviously a different neighborhood, and one cannot describe such macromolecules without taking into account all possible kinds of environment; that is too difficult.

Another way for solving this problem is the use of a statistical approach. This is used below for a more general problem: branched polymerization in a multicomponent system.

L. Pore Formation in Chain Polymeric Systems

As it was noted in Chapter 1, linear polymerization is accompanied by formation of narrow pores, mostly supermicropores, also mesopores and ultramicropores, but found less. Thermodynamics of these pores does not significantly differ from that analyzed above, but pores in chain macromolecules have some specific features:

The voids in polymeric materials are due to the eventuality of configurations of polymeric chains, allowing the appearance of voids without delivering volatile components.

The energy of pores is due to the reactability of the ends of the chains (respectively, polymeric rings do not contribute the pore energy).

Hence, the estimation of the pore volume in polymers needs modeling of their mechanical properties, analysis of which stands out of this chapter.

The relationship between the total energy of pores and that of a linear macromolecular system is given by

$$Q_0 \int_{\varepsilon_{\min}}^{\varepsilon_{\max}} \varepsilon f(\varepsilon)\, d\varepsilon = 2 \sum_{N=1}^{\infty} C_N e_0 \nu_{\Sigma p} \tag{3.146}$$

where $\nu_{\Sigma p}$ is the total amount of substance in the macromolecular system, C_N here is the concentration of nonringed structures only, and e_0 is the energy of the active end of a nonring macromolecule: obviously, either monomers and all nonring polymers possess (each) two active centers, as is accounted in Eq. (3.146).

Let us consider, just as an elementary approximation, a pore as a cylinder having radius r_p. Its surface area is proportional to r_p, while its volume is proportional r_p^2, respectively. *Very roughly* one may assume that energy of pore is proportional to $1/r_p$. However, the pore energy should be proportional to the local concentration C_l of active centers located at the pore wall. Two active centers cannot be neighbors—otherwise they react and are lost. Hence, the maximal local concentration of active centers at the pore wall is 50% of the total concentration of the monomeric units there, that is, proportional the surface area of the pore. The local concentration of active centers (let us call them *vacancies* in the further analysis because they can be replaced with monomeric units) can differ from the mean concentration [this is $2\Sigma C_N$, compare to Eq. (3.146)] only due to eventual fluctuations of concentration. The probability of such fluctuation that allows the local concentration C_l of active centers located at the pore wall is given by the normal gaussian distribution:

$$F_G(\mu) = \frac{1}{\sqrt{2\pi}\sigma_G} \exp\left[\frac{-(\mu - \mu_m)^2}{2(\sigma_G)^2}\right] \tag{3.147}$$

where μ is the local chemical potential determined by the local concentration of active centers ($\mu = R_g T \ln C_l$), μ_m is the mean chemical potential determined by the mean concentration of active centers (that is, equal to the current free energy of the polymeric system, as it has been discussed above), σ_G is the statistical dispersion of the distribution Eq. (3.147). The function in Eq. (3.147) is also called *error function* and *designed Er*. The total energy of the system, while found from Eq. (3.146), is also equal to

$$E_\Sigma \sim \int_0^{C_{l,\max}} Er(\mu(C_l)) C_l \, dC_l \int_{r_{\min}}^{r_{\max}} \frac{dr}{r} \tag{3.148}$$

Let us take into account that the probability of finding a pore having radius r and energy ε is proportional to the probability that its local concentration of energetic centers (ending groups) is C_l. Then, $r \sim (1/C_l)$, and the probability of such event is $\mathrm{Er}(C_l)\varphi_r(1/C_l)$. The total probability that a pore has energy ε is the integral of all probabilities related to a different one results in the relationship between the pore energy distribution and the polymer appearance:

$$f(\varepsilon) = \int_0^{C_{l,\max}} Er(C_l)\varphi_r\left(r \sim \left(\frac{C_l}{\varepsilon}\right)\right) dC_l \tag{3.149}$$

where the pore radius r is expressed through the relationship between pore energy and pore size, and $\varphi_r(r)$ is the size distribution of pores.

Equation (3.149) provides, in principle, the relationship between pore-size distribution and pore-energy distribution for a porous structure formed by polymerization. Let us note, however, that the applicability of Eq. (3.149) is very problematic, for the following reasons:

The derivation of Eq. (3.149) needs assumptions about the pore shape (in the example above, it was assumed that pore is cylinder); as we noticed in Chapter 1, such assumptions are almost always nonsense.

The assumption of the relationship between the pore energy ε and its size r like $\varepsilon \sim 1/r$ is incorrect not only for noncylinder but also for very narrow pores, e.g., micropores.

These two limitations make Eq. (3.149) too exotic for serious theoretical studies. However, it is possible that several modifications may allow using of Eq. (3.149) for empirical treatment of experimental data. In any case, Eq. (3.149) gives some comprehension about the relationship between pore-size and -energy distributions.

M. Branched Polymerization

Branched polymerization is defined as a polymerization process, products of which have branched structure, where each monomeric unit can be connected to three or more other monomeric units. Contrary to the theory of linear polymerization, branched polymerization does not have any unified description. That is because the enormous complexity of modeling branched structures—the possible number of such structures, their interactions, etc. The number of possible structures of N-mers only for a one-component macromolecular system having branching $m = f_p - 1$ (f is the functionality of monomeric unit) can be approximately estimated as m^N (obviously, for chain structures that is always 1, and no problem related to the configuration aspect appears). Hence, each model of branching needs, first of all, to reduce the number of accounted structures. The manner in which this problem is solved makes the physical sense of each model.

Let us consider the most important models of branched polymerization.

1. Models of Branched Polymerization: Classification

Existing models of branched polymerization can be classified by

The role of computing: Is it the decisive factor or just a tool for performing the calculations?

Assumptions about any specification of treated macromolecular structures: Does the model deal with all possible structures or only with some specified shapes?

According to the role of computing, models are divided into analytical and numerical ones. Numerical models do not need analytical solution but replace it with numerical simulations on computer. Analytical models intend to find more or less relevant analytical solutions that can be later simplified and simulated on computer; however, in most situations, analytical models provide also particular solutions simple enough in their form to be obtained without computer. Numerical models can never avoid using the computer. The merit of numerical models is absolutely determined by the use of a computer, while analytical models can have merit even without any use of a computer.

According to the main assumption about several specifications of treated structures, existing models can be divided into fixing structure (architecture) and nonfixing structure. The fixation of considered structures, based on several physical considerations, reduces significantly the number of treated structures but may sometimes lose the correctness of the description of branched macromolecular systems.

A very important requirement to all models of branching is their applicability also to linear polymerization, because this can be considered as branching with functionality = 2.

2. Structure-Fixing Models

Absolute majority of existing models of branching is based on several fixations of treated structures: stars, quasichains, etc. [15,16]. That is very important for technologies using regular polymerization but absolutely not applicable to random polymerization.

3. Numerical Models

This approach to the solution of problems of branching avoids numerous complicated equations, due to the limitation of the accounted degree of polymerization. Numerical models allow, in principle, evaluation of every characteristic parameter of a considered system.

Numerical simulations are very useful in studies of conformational properties of branched polymers having a low degree of polymerization. Monte Carlo, molecular dynamics, and Brownian dynamics methods are employed to simulate the equilibrium and dynamic behavior and also to reproduce hydrodynamic properties. The simulations are performed on several polymer models. Different Monte Carlo algorithms are devised for lattice and off-lattice models. Numerical models are applicable to both regular and random kinds of polymerization. Numerical models are effective in the description of uniform homopolymer stars as single chains, or in nondiluted solutions and melts, employing a variety of techniques, models, and properties. Other important structures, such as stars with different types of monomer units, combs, brushes, dendrimers, and absorbed branched polymers have also been the subject of specific simulation studies [15–17].

Nevertheless, numerical models have the following shortcomings:

- Numerical models are very sensitive to initial data, error of calculations, and selection of the random number generator.
- The physical sense of obtained numerical results is not always clear.
- It is difficult to distinguish physical tendencies, as they are calculated, from eventual errors.

4. Statistical Models

The statistical approach is based on the Flory–Stockmayer concept of branching trees. The mathematical techniques of this approach use probability generating functions. The building (monomer) units and molecules are represented by graphs. The collection of branched molecules in the system is represented by a collection of molecular trees composed of monomers. This collection is transformed into a collection of rooted trees by choosing every monomer unit (node) for the root with the same probability and placing it on generation zero [18]. The covalently bound units appear in the first, second, etc., generations with respect to the unit in the root. This transformation provides two principal consequences: (1) the distribution of units in the root represents the distribution of units in the system, and (2) an N-mer is rooted N times so that it appears in the collection of rooted trees—the rooted forest—N times. The transformation into the rooted trees is performed in order to be able to generate the trees using simple probabilistic considerations [18]. The monomer units differ in the number of reacted functional groups. For a single-component system, this distribution is sufficient for building trees. It is assumed that the reactivity of a group in a unit does not depend on the state of groups in neighboring units [18].

Thus, in the mathematical sense, the main characteristic of a polymeric system (in the statistical approach) is a probability generating function (PGF), which is defined for the number of bonds issuing from a f_p functional monomer in the root $F_c(z)$ as

$$F_0(z) = \sum_{i=1}^{f} p_i z^i \tag{3.150}$$

In Eq. (3.150), p_i is the probability of finding a monomer unit in the root having i issuing bonds. This probability is equal to the fraction of units with i reacted functional groups; z is a variable of the generating function through which the operations with PGF are performed [18].

The statistical approach is applicable to two opposite kinds of systems: polymers in equilibrium, for which the sequence of reactions leading to their formation is not important, and polymers formed due to eventual interaction of radicals. Obviously, for real practice, the equilibrium situation is much more important, especially for aggregates.

One of the principal shortcomings of the statistical approach consists in the assumption that structures having various shapes are formed with the same probability. Other drawbacks of this approach will be considered below.

Another statistical method for the study of branched structures, based on the assumption of a given configuration of branched polymers on the lattice uniquely characterized by specifying the total number of monomers, number of chemical bonds, the total number of k functional sites ($k = 1,2,3,\ldots$), the total number of loops in all the polymers, and the number of polymers was developed in Refs. 17 and 19–23. In many cases, the equilibrium distribution of the polymers is assumed [19–22]. This approach allows solution of some problems regarding the behavior of polymers on interface. However, not all assumptions in this approach seem acceptable (for example, the applicability of the lattice model is sometimes doubtable).

N. Kinetic Approach

This approach was developed by S. Kuchanov [24]. The polymeric system is described by an infinite set of differential equations for the concentration of each N-mer. There are two principal schemes realizing this approach.

1. Random Irreversible Step Polyaddition of an f-Functional Monomer

Each macromolecule is characterized by its degree of polymerization N and the number of unreacting groups $V_p(N)$ (the notion of unreacting group was introduced above, in the analysis of chain polymerization, as an end group or vacancy; of course, "end group" is not well applicable to branched structures, because they may include vacancies in any part of the structure). For a tree structure, V is found from the following equation [18,24]:

$$V_p(N) = N(f_p - 2) + 2 = N(m - 1) + 2 \tag{3.151}$$

The formation of N-mer having V_p unreacting groups ($A_{N,V}$) is described by the following equation:

$$A_{k,j+1} + A_{N-k,V-j+1} \rightarrow A_{N,V} \tag{3.152}$$

The system is described by corresponding kinetic equations for the change of $C_{N,V}$ with time [18,24].

2. Initiated Living Polymerization

This process is important for such substances as epoxy resins and described by the following reaction scheme:

$$\mathrm{M} + \mathrm{Pol}(N) \rightarrow \mathrm{Pol}(N+1) \tag{3.153}$$

$$\mathrm{M} + \mathrm{Pol}(N+1) \rightarrow \mathrm{Pol}(N+2), \tag{3.154}$$

where Pol(N) means N-mer. Also for this case, kinetic equations in the differential form are written and solved [18,24].

The kinetic approach is applicable only to systems in nonequilibrium, for which the reverse reaction (decomposition of macromolecules) is neglected. As a result, this approach cannot be used for derivation and/or evaluation of thermodynamic functions. Also, in this approach, the problem of equivalence of all shapes of macromolecules (as it was mentioned for the statistical approach) stands out.

Both statistical and analytical approaches can be employed not only in analytical but also numerical forms.

Complexity and a large number of equations in the kinetic approach make it less used by researchers, in comparison with the statistical approach.

For both statistical and kinetic approaches, we mention the methodological problem: they have no conjugation between them or with the model describing chain structures. A system transformed from nonequilibrium to equilibrium cannot be described by the statistical approach (which is invalid for most forms of nonequilibrium processes) or the kinetic approach (which is not applicable to equilibrium). Hence, before the application of any approach, the researcher must know whether the system deals with equilibrium.

O. Statistical Polymer Method

Statistical polymer is defined as the average structure including all possible (taking into account their probability) structures of polymers containing the same number of monomers. This definition allows formulation of all processes in polymeric mixtures in terms of statistical polymers.

The concept of statistical polymer method comprises three levels of modeling:

Single macromolecules with branching (no cross-links)

Simple macromolecular systems (equilibrium mixtures of macromolecules having branched or cross-linked structure), estimation of their additive parameters

Complex macromolecular systems: nonequilibrium, nonadditive parameters

The level of modeling single macromolecules means the consideration of statistical polymer itself as an averaged macromolecule of a specified weight, analysis of its properties related to all possible macromolecular structures.

The level of modeling simple macromolecular systems means the consideration of macromolecular mixtures as combinations of statistical polymers having different weight: their weight distribution in equilibrium, thermodynamic parameters, and related macroscopic properties. On this level, cross-linking is accounted.

The level of modeling complex macromolecular systems means the consideration of nonequilibrium processes in polymeric mixtures with branching and evaluation of nonadditive parameters. All solutions obtained on this level are indirect and based on direct solutions obtained on the previous levels of study.

The statistical polymer method was experimentally tested indirectly (against existing curves of adsorption isotherms measured on microporous materials having branched cross-linked structure [25]) and directly (with reproduction of Trommsdorf effect [26]).

The statistical polymer method is easily applicable to linear structures: in such case, a statistical polymer is just the chain macromolecule of the same weight.

The principal shortcoming of the statistical polymer method consists in its applicability to random polymerization only; otherwise the physical sense of the statistical polymer is lost. However, for purposes of modeling pore formation in polymeric structures this disadvantage has no importance because the absolute majority of porous polymeric materials have the random nature.

The results of the comparative analysis of existing methods for study of branched polymerization are presented in Table 3.9.

As follows from Table 3.9, the statistical polymer method has the absolute advantage in the description of porous materials, for which regular polymerization is not typical.

Thus, in the analysis of branched polymerization and related pore formation described below we will use the statistical polymer method. Let us consider it in more detail.

P. Equilibrium Polymerization in Terms of Statistical Polymer Method

In the statistical polymer method, all processes in the system are considered as reactions between statistical polymers and monomers. The reactions of polymerization–destruction (decomposition, de-polymerization) are written as:

$$\mathrm{SP}(N) + \mathrm{M} \longleftrightarrow \mathrm{SP}(N+1) \tag{3.155}$$

$$\mathrm{SP}(N_1) + \mathrm{SP}(N_2) \longleftrightarrow SP(N_1 + N_2) \tag{3.156}$$

where M is monomer, SP(N) is an N-meric statistical polymer [27]. As in the consideration of linear polymerization in equilibrium, we do not need to account for all possible reactions but may just consider independent reversible reactions between statistical polymers and monomers, Eq. (3.155).

The capability of N-mer to accept an additional monomer is characterized by occurrence of vacancies (unreacting bonds in the kinetic model, end groups in the linear polymerization), their number being found (for a non-cross-linked situation) as $V_p(N) = (f_p - 2)\times N + 2 = (m-1)N + 2$, where f_p is the functionality, and $m = (f_p - 1)$ is the number of branching.

TABLE 3.9 Comparison of Various Methods of Modeling Branched Polymerization

Factor of comparison	Fixation (specification) of structure	Numerical models	Statistical models	Kinetic approach	Statistical polymer method
Main idea	Limitation of treated shapes	Limitation of treated weight of polymers	Treatment of branching trees, using probability-generating functions	Consequence of kinetic equations for branched growth of polymers	Averaged description of ensembles of macromolecules
Existence of analytical solution	Yes	No	Yes	Yes	Yes
Applicability to chain polymerization	Yes	Yes	No	No	Yes
Applicability to regular polymerization	Yes	Yes	No	Yes	No
Applicability to random polymerization	No	Yes	Yes	Yes	Yes
Applicability to equilibrium	Yes	Yes	Yes	No	Yes
Applicability to nonequilibrium	Yes	Yes	No	Yes	Yes
Applicability to cross-linking	Yes	Yes	No	No	Yes

Independent reactions Eq. (3.155) are written as reactions of occupation of vacancies:

$$\mathrm{SP}(N) \times \mathrm{Vac} + \mathrm{M} \longleftrightarrow \mathrm{SP}(N+1) \quad (3.157)$$

The reverse reaction [transformation of statistical $(N+1)$-mer to statistical N-mer] is characterized by removal from $(N+1)$-mer of a monomeric unit having one single bond only with the rest of polymer (otherwise N-mer is not obtained but two new polymers). Such monomeric units are defined as *extreme units* (let us note: also they would be defined in the linear polymerization concept as end groups). Their number (in the noncross-linked case) is found from the following considerations:

In a process like Eq. (3.155), the capture of a monomeric unit by statistical N-mer results in increasing the number of extreme units by one—because the newcomer unit becomes obviously an extreme unit—but there is a chance that this newcomer occupies a vacancy belonging to another extreme unit, which may reduce the number of extreme units proportionally to the number of vacancies they have. Hence, we obtain a recursive equation:

$$U_\Sigma(N+1) = U_\Sigma(N) + 1 - \frac{mU_\Sigma(N)}{V_\Sigma(n)} \quad (3.158)$$

where $U_\Sigma(N)$ is the number of extreme units in statistical N-mer (and $V_\Sigma(N)$, of course, is the number of vacancies). Obviously,

$U_\Sigma(1) = 1$, because free monomer is automatically extreme unit.
$U_\Sigma(2) = 2$, because in dimer both monomeric units are extreme units.
$U_\Sigma(3) = U_\Sigma(2) = 2$, because in trimer two opposite monomeric units are extreme units, while the unit intermediary between them cannot be extreme.
For chains having $N > 1$, $U_\Sigma(N) = 2$, because chains have two end groups working as extreme units.

For statistical polymers with $m > 1$ and $N > 3$, the number of extreme units is not so simple but found from Eq. (3.158).

As follows from Fig. 3.5, over $m = 5$ the distance between curves corresponding to different m is negligible. Hence, any errors in the estimation of m, when the functionality is large, do not significantly influence the result in the estimation of the number of extreme units. Below we will see that this tendency is found also for other parameters of polymeric systems with branching: the effect of branching is especially important, the

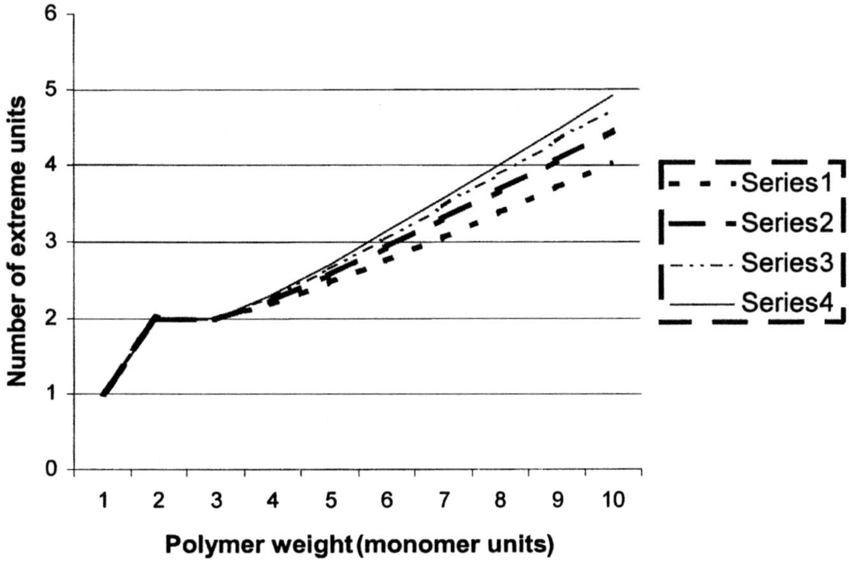

FIG. 3.5 Number of extreme units vs. polymer weight for different branching: $m=2$ (functionality 3) series 1; $m=3$ (functionality 4) series 2; $m=5$ (functionality 6) series 3; $m=9$ (functionality 10) series 4.

value of branching from 2 to 5 is significant but not decisive, whereas the further increase of m plays a negligible role, and one may always assume for large values of functionality that m is approximately 5.

The rates of direct and reverse reactions in Eq. (3.157) are found, respectively, from

$$W_+(N+1) = K_+ V_\Sigma(N) C_1 C_N; \quad W_-(N+1) = K_- U_\Sigma(N+1) C_{N+1} \tag{3.159}$$

where C_1 and C_N are concentrations of monomers and N-mers, respectively, and K_+ and K_- are constants of direct and reverse reactions (3.157), respectively. The condition of equilibrium is given by [27]

$$W_+(N+1) = W_-(N+1) \Longrightarrow \tag{3.160}$$

$$\frac{C_{(N+1)}}{C_N C_1} = \frac{K_+ V_\Sigma(N)}{K_- U_\Sigma(N+1)} = \frac{K_0 V_\Sigma(N)}{U_\Sigma(N+1)} \tag{3.161}$$

where K_+ and K_- are the constants of reaction rates for the direct and reverse reactions (3.157), respectively, and K_0 is the constant of equilibrium. The previous derivations let us assume that K_0 does not depend on N. This is the formulation of Flory principle in the statistical polymer method.

Equation (3.161) can be transformed if the temperature changes:

$$\frac{C_{(N+1)}}{C_N C_1} = \exp\left[\frac{\Delta S_e^0}{R_g} - \frac{\Delta H_e^0}{R_g T}\right] \frac{V_\Sigma(N)}{U_\Sigma(N+1)} \tag{3.162}$$

where ΔS_e^0 and ΔH_e^0 are the changes of entropy and enthalpy in reaction (3.157). Of course, $\Delta H_e^0 < 0$ (because reaction (3.157) is exothermal) and $\Delta S_e^0 < 0$ (because the system loses degrees of freedom due to the reduction of nonconnected particles).

Equations (3.161) and (3.162) provide the weight distribution of branched polymers in equilibrium.

Figures 3.6–3.8 provide the weight distribution curves for different values of m and K_0, comprising the chain case ($m = 1$).

As follows from Fig. 3.6, the factor of branching significantly changes the weight distribution of polymers. On the other hand, the further increase of m does not significantly influence the weight distribution.

Figure 3.7 illustrates the well-known fact that the value of polymerization constant influences very much the weight distribution of polymers. The influence of K_0 onto the weight distribution of polymers is always very significant (for low, moderate, or high values of K_0). Now, it is interesting to compare the influence of both factors, polymerization constant and branching. Let us do it for the total weight fraction of hard polymers (for which $N \geq 11$) (see Fig. 3.8).

As follows from Fig. 3.8, the influence of branching on the weight distribution of polymers is definitely comparable to that of the polymerization constant. It is very interesting to note that high values of branching compensate the low values of polymerization constant.

Q. Cross-Link Formation

The process of cross-link formation in branched macromolecules is very similar to ring formation in chain polymers (that was considered above). As for chains we assumed ring formation due to interactions of end groups rich in energy also for branched polymers cross-linking is due to interactions of monomer units possessing vacancies. However, there are some factors of difference:

Ring formation in chain macromolecules is possible only between the end groups, with the transformation of the whole macromolecule to

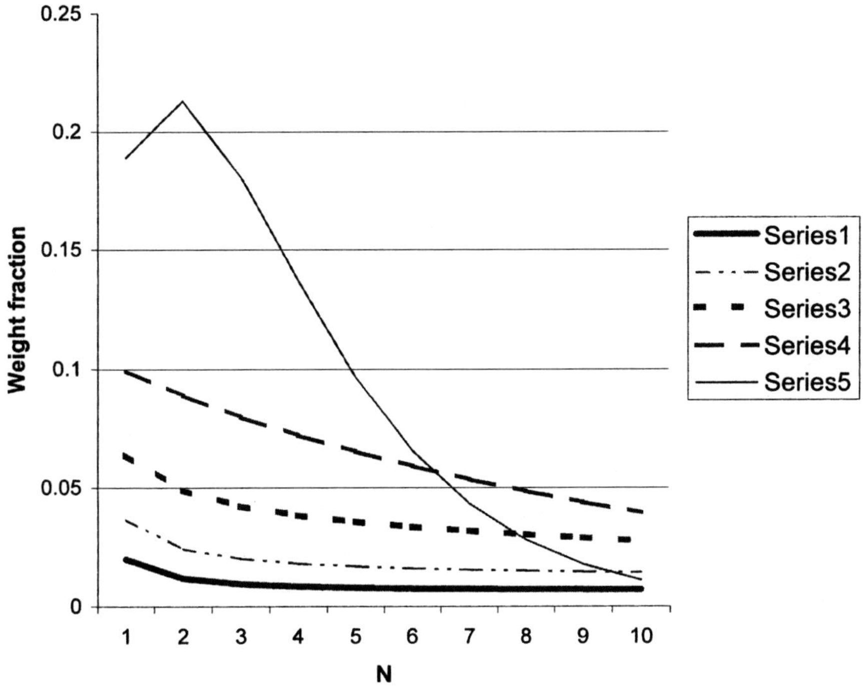

FIG. 3.6 Weight distribution curves for statistical N-mers (≤ 10) for $K_0 = 3$. Series 1: $m = 9$; Series 2: $m = 5$; Series 3: $m = 3$; Series 4: $m = 2$; Series 5: $m = 1$ (chain polymerization).

a ring, while cross-linking is possible inside branched polymers, and its result is a macromolecule containing one or more internal rings.

Ring formation in a chain neutralizes the active groups of the polymer, and the resulting ring has the minimal free energy, whereas cross-linkage just reduces the number of vacancies, and the resulting structure remains able to participate in polymerization–cross-linking reactions.

The probability of ring formation in chains is very low, because of the significant loss in entropy of the ringed structures, while cross-linking in branched polymers can become very probable in several conditions, and the entropy loss in such cases is less because of the variety of options for cross-linking.

The statistical polymer method describes cross-links as internal bonds in polymers, hence, their formation is determined by the same vacancies

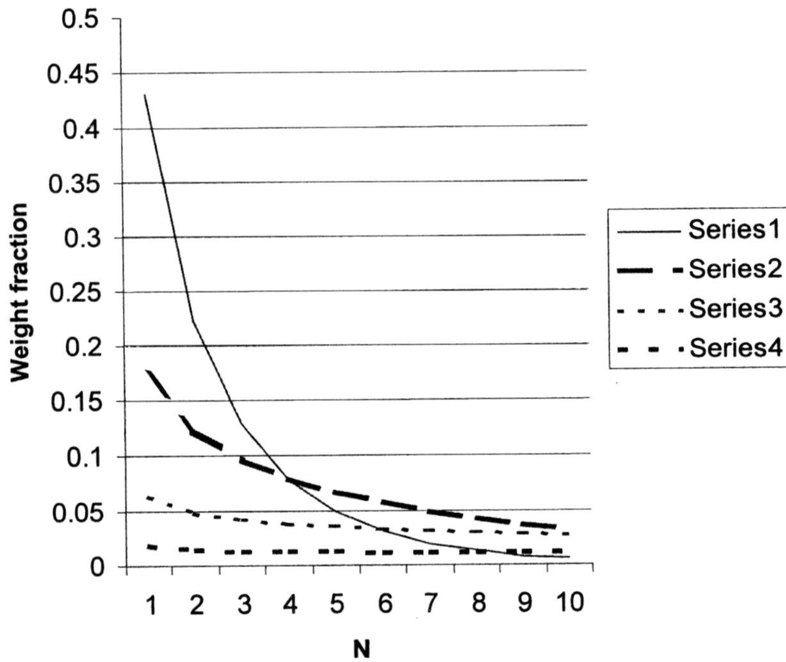

FIG. 3.7 Weight distribution curves for statistical N-mers (≤ 10) for $m=3$. Series 1: $K_0=0.3$; Series 2: $K_0=1$; Series 3: $K_0=3$; Series 4: $K_0=10$.

which cause polymerization. The number of possibilities for cross-link formation is found from [25,26]:

$$C_r = \frac{1}{2} V_\Sigma(N)[N - 1 - B_{0\Sigma}(N)] \tag{3.163}$$

where $B_{0\Sigma}$ is the number of monomeric units having no vacancies. Equation (3.163) is based on the assumptions that (1) a monomer unit cannot link itself, (2) the rate of cross-linking is proportional to the number of vacancies in N-mer, and (3) a monomer unit cannot join another unit whose vacancies are already occupied.

The value of $B_{0\Sigma}$ is found from the following considerations: Getting a cross-link, statistical N-mer may loose the number of monomer units having (each) j vacancies with the probability proportional the number of vacancies belonging them but also may get a number of the same units (if they are not

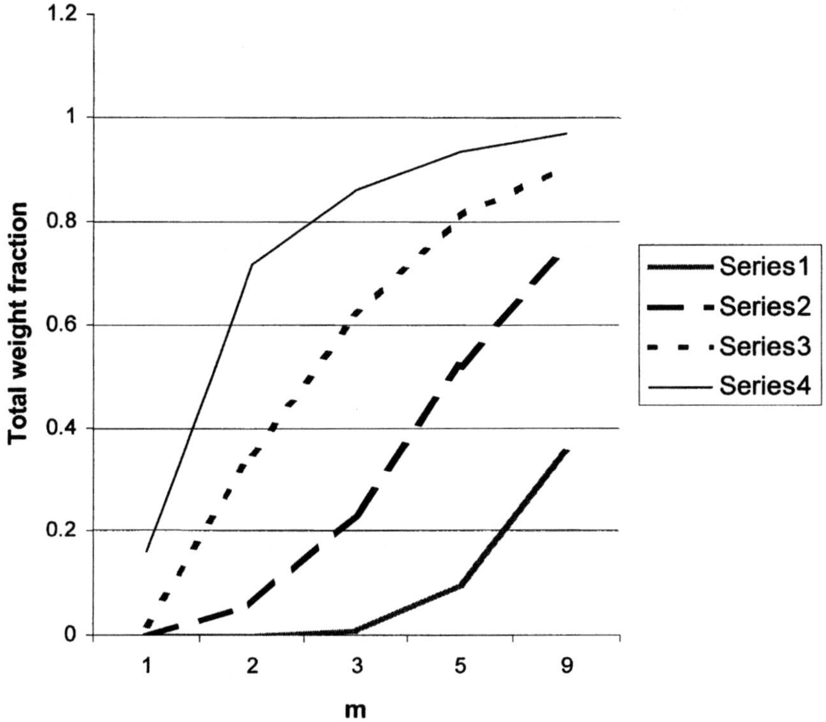

FIG. 3.8 Influence of polymerization constant and branching onto the total weight fraction of hard polymers. Series 1: Polymerization constant $K_0 = 0.3$; Series 2: Polymerization constant $K_0 = 1$; Series 3: Polymerization constant $K_0 = 3$; Series 4: Polymerization constant $K_0 = 10$.

extreme units) due to the occupation of vacancies belonging to units having (each) $(j+1)$ vacancies:

$$B_{j\Sigma}(C_r + 1) = B_{j\Sigma}(C_r) + \frac{B_{(j+1)\Sigma}(C_r)(j+1) - B_{j\Sigma}(C_r)j}{V_p(C_r)} \tag{3.164}$$

where $1 \le j \le (m-1)$. However, the number of units having no vacancies can only increase:

$$B_{0\Sigma}(C_r + 1) = B_{0\Sigma}(C_r) + \frac{B_{1\Sigma}(C_r)}{V_p(C_r)} \tag{3.165}$$

For $j=m$, obviously, $B_{m\Sigma}(C_r)=U_\Sigma(C_r)$, the number of extreme units. Their loss is given by the following equation:

$$B_{m\Sigma}(C_r+1) = U_\Sigma(C_r) = B_{m\Sigma}(C_r) - \frac{B_{m\Sigma}(C_r)m}{V_p(C_r)} \tag{3.166}$$

The rates of reaction of cross-link formation and destruction are given by the following equations, respectively:

$$W_{c+} = K_{c+}C_r \qquad W_{c-} = K_{c-}G_r \tag{3.167}$$

where G_r is the number of cross-links and K_{c+} and K_{c-} are constants of cross-link formation and destruction, respectively. In equilibrium, the left parts of Eqs. (3.167) are equal:

$$W_{c+} = W_{c-} \rightarrow K_{cr} = \frac{G_r}{C_r} = \frac{2G_r/V_\Sigma(N)}{[N - B_{0\Sigma}(N) - 1]} \tag{3.168}$$

The loss of vacancies because of cross-linkage is found from the following balance:

$$V_p(C_r) = V_p(C_r = 0) - 2C_r \tag{3.169}$$

Obviously, the system of Eqs. (3.168) and (3.169) has the single solution.

Let us note that Eqs. (3.165)–(3.169) are basic not only for equilibrium but also for nonequilibrium system description.

The evaluation of K_{c+}, K_{c-}, and K_{cr} is very problematic. Of course, K_{cr} cannot be over K_0 in Eq. (3.161), because energetically cross-linking has no preference against polymerization. However, K_{cr} must be *below* K_0, because of orientation problems (e.g., cross-linking a unit in the second environment is very difficult, because the formed ring is very sensitive and can be destroyed by a simple heat motion). Therefore, Eqs. (3.165)–(3.169) have mostly semiempirical sense and can be useful first of all for the treatment of experimental data obtained for cross-linked branched polymers, gels, and aggregates but not for fundamental predictions.

R. Thermodynamic Functions of Branched Structures

Now, let us derive the equations for thermodynamic functions of branched structures. Obviously, that can be done on the basis of the equilibrium solution obtained above.

Equation (3.161) can be rewritten as follows:

$$\Delta S^0(N) = \Delta S_e^0 + R_g \ln \frac{V_\Sigma(N)}{U_\Sigma(N+1)} \tag{3.170}$$

where $K_0 = K_+/K_-$. The heat effect of polymerization is $\triangle H_e^0 = \triangle E_a = 2\varepsilon_0$, where ε_0 is the free energy of vacancy. From Eq. (3.170) we obtain:

$$\Delta S^0(N+1) = R_g[\ln K_0 + \ln V_\Sigma(N) - \ln U_\Sigma(N+1)] \tag{3.171}$$

For the case of dimerization, Eq. (3.171) is transformed to

$$\Delta S^0(2) = R_g[\ln K_0 + \ln V_\Sigma(1) - \ln U_\Sigma(2)] \tag{3.172}$$

Comparing Eqs. (3.170) and (3.172), we obtain [26]

$$\Delta S^0(N+1) = \Delta S^0(2) + R_g \ln \frac{2V_\Sigma(N)}{(m+1)U_\Sigma(N+1)} \tag{3.173}$$

The chemical potential of statistical N-mer is found from

$$\mu(N) = \mu^o(N) + R_g T \ln C(N) \tag{3.174}$$

Now, we obtain from Eqs. (3.171)–(3.174) the equation of non-cross-linked statistical N-mer:

$$\mu^0(N) = \varepsilon_0 V_\Sigma(N) - M \tag{3.175}$$

$$M = T\left[NS^0(1) + N\Delta S^0(2) + \sum_{n=1}^{N} R_g \ln \frac{2V_\Sigma(n-1)}{(m+1)U_\Sigma(n)}\right] \tag{3.176}$$

where $S^0(1)$ is the entropy of monomer [28,29].

As in the case of chains, free energy of branched structures in non-equilibrium is estimated due to the superposition principle.

Equations (3.171)–(3.174) allow the description of branched non-cross-linked structures and chains. Some examples of such simulations are presented on Figs. 3.9–3.11.

As follows from Fig. 3.9, well-known Trommsdorf effect (acceleration of polymerization for short time and slowing of the reaction for large time) is reproduced for chains and branched polymers. In accordance with the previous results, more m corresponds to higher polymerization degree.

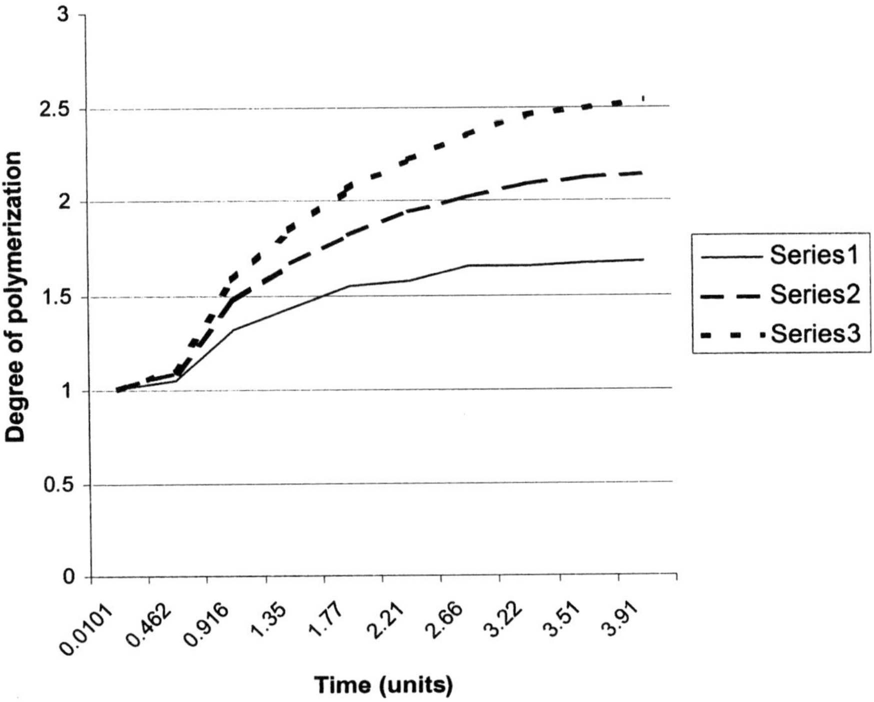

FIG. 3.9 Trommsdorf effect: Series 1: $m=1$ (chains); Series 2: $m=2$; Series 3: $m=3$.

Figure 3.10 presents the weight distribution of low-weight polymers for specified time (corresponds to $t=0.916$ on Fig. 3.9).

As follows from Fig. 3.10, the tendency of dominating weight fraction of hard polymers for more m is found since several moments after polymerization reaction. However, the tendency of high content of dimers for chains (see Fig. 3.6) is not observed—probably just because the studied moment of the reaction is very far of equilibrium.

As follows from Fig. 3.11, the general tendencies in evolution of weight distribution profiles with time correspond to the general laws found from experiments on polymerization.

S. Multicomponent Structures

The problem of multicomponent macromolecules is very important not only for copolymers (since we left without a solution the problem of

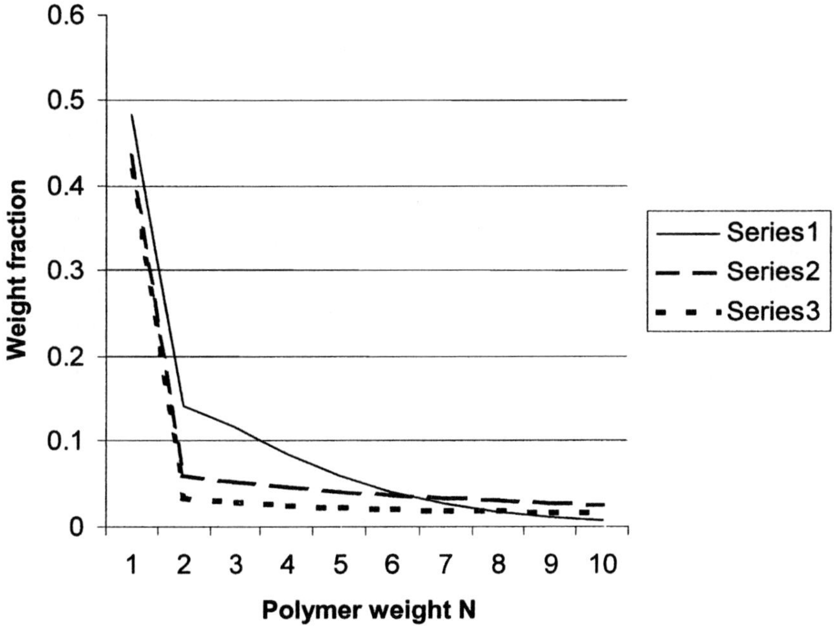

FIG. 3.10 Weight distribution of polymers ($N \leq 10$) for specified time: Series 1: $m = 1$ (chains); series 2: $m = 2$; series 3: $m = 3$.

multicomponent chains) but also for gels and other aggregates different from traditional polymers.

Let us consider a macromolecular system containing M components, in which multicomponent macromolecules coexist with the liquid phase containing monomers. Each component is characterized by its specific functionality f_j determining the corresponding branching $m_j = f_j - 1$. Among macromolecules belonging to this system, we consider N-mers, for which the following condition is satisfied:

$$N = \sum_{j=1}^{M} N_j \tag{3.177}$$

where N_j is the number of monomeric units of jth kind ($j \leq M$) belonging to the considered N-mer. Statistical polymer for multicomponent system is defined as an averaged structure accounting not only all possible structures of macromolecules but also their chemical composition; hence, the values of

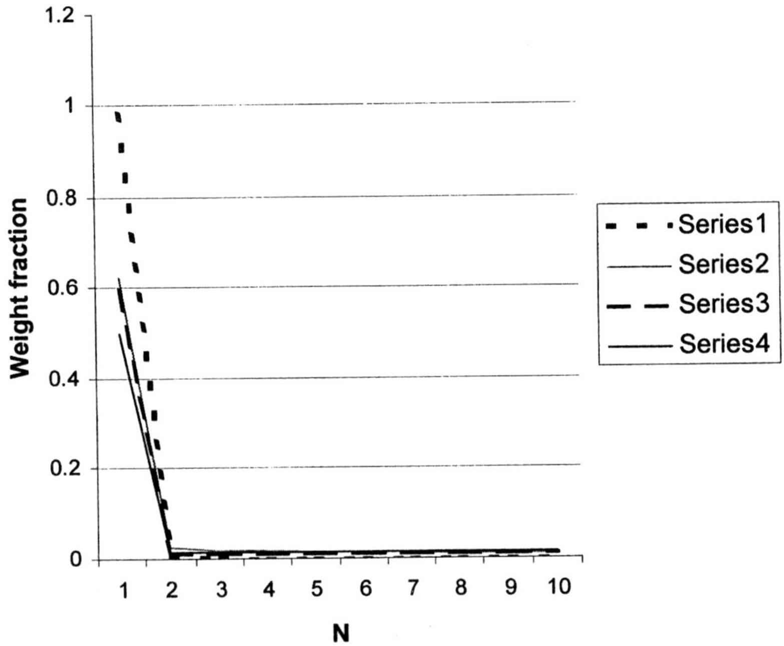

FIG. 3.11 Profiles of weight distribution with time for $N \leq 10$, $m = 3$: Series 1: moment $t = 0.01$ (by Fig. 3.9); Series 2: moment $t = 0.462$ (by Fig. 3.9); Series 3: moment $t = 0.916$ (by Fig. 3.9); Series 4: moment $t = 1.35$ (by Fig. 3.9).

N_j do not need to be integer. The specific interactions between the statistical polymers and monomers are characterized by the following parameter [30]:

$$s_j = \lim_{n \to \infty} \frac{N_j}{x_j N} \tag{3.178}$$

where x_j is the molar fraction of the jth component in the liquid phase; if the molar part of the jth component in macromolecules is $y_j = N_j/N$, $x_j = y_j/s_j$; hence, the parameter s_j introduced by Eq. (3.178) characterizes some preferences in joining monomers of some component by macromolecules. Based on Flory principle, we assume that y_j and s_j do not depend on N.

The number of vacancies for M-component statistical N-mer is found from the following equations:

$$V_\Sigma(M, N) = 2 + \sum_{j=1}^{M} N_j(m_j - 1) = \sum_{j=1}^{M} V_j(N) \tag{3.179}$$

where $V_j(N)$ is the number of vacancies belonging to the jth component in statistical N-mer; obviously, $V_j(N) = N_j(m_j - 1) + 2y_j$.

The reactions in the considered system in equilibrium are described, as in the one-component situation, as results of interactions between statistical polymers and monomers:

$$\mathrm{Pol}(N) + \sum_{j=1}^{M} y_j \mathrm{Mon}_j \longleftrightarrow \mathrm{Pol}(N+1) \tag{3.180}$$

The mechanism of the direct reaction (3.180) is given by the interaction between vacancies and monomers:

$$\mathrm{Pol}(N)\mathrm{Vac}(N) + \sum_{j=1}^{M} y_j \mathrm{Mon}_j \longleftrightarrow \mathrm{Pol}(N+1) \tag{3.181}$$

As in the one-component case, we define extreme units as monomers (not important which component) having one only bond to the main statistical polymer. Extreme unit formed by jth component obviously has m_j vacancies. As in the one-component case, the number of extreme units of jth component in statistical N-mer is found from the consideration of the change of the number of extreme units of jth kind due to joining monomer $(+y_j)$ and their loss because of possible joining monomer to existing extreme unit $[-y_j V_j / V_\Sigma(M,N)]$:

$$U_j(N+1) = U_j(N) + y_j - y_j \frac{V_j}{V_\Sigma(M,N)} = U_j(N) + y_j\left[1 - \frac{V_j(N)}{V_\Sigma(M,N)}\right] \tag{3.182}$$

$$U_\Sigma(M,(N+1)) = \sum_{j=1}^{M} U_j(N+1) = U_\Sigma(M,N) + 1 - \sum_{j=1}^{M} \frac{y_j V_j(N)}{V_\Sigma(M,N)} \tag{3.183}$$

$$U_j(1) = y_j \qquad U_j(2) = 2y_j \tag{3.184}$$

The resulting condition of equilibrium is exactly the same as Eq. (3.161), while the sense of parameters $U_\Sigma(M, N+1)$ and $V_\Sigma(M, N)$ is given by the above equations. Respectively, the thermodynamic potentials are given by Eqs. (3.170)–(3.176), with the relevant expressions for the parameters $U_\Sigma(M, N)$ and $V_\Sigma(M, N)$ as written above.

Obviously, the solution for multicomponent chains is obtained if all $m_j = 1$.

T. Pore Formation in Branched Polymeric Systems

The procedure of characterization of pores formed in polymeric systems with branching is very similar to the linear case. As was noticed in Chapter 1, branched polymerization is accompanied by formation of very narrow pores, mostly ultramicropores (less found also supermicropores and mesopores). Thermodynamics of these pores does not significantly differ from that analyzed above, but pores in branched macromolecules have some specific features:

The voids in polymeric materials are not only due to eventuality of configurations of polymeric chains but also because of vacancies remaining free.

The energy of pores is due to the reactability of vacancies in macromolecules.

Since energetic centers are determined by the number of vacancies, the maximal energy of pore walls corresponds to the situation when the wall is formed by the same macromolecule with minimal cross-linkage, in such case the number of vacancies at the wall of a cylinderlike pore, having radius r and length h, is about $(m-1)\,4\pi rh$. For branched macromolecular systems, the relationship between the total energy of pores and that of macromoles system is given by

$$Q_0 \int_{\varepsilon_{\min}}^{\varepsilon_{\max}} \varepsilon f(\varepsilon)\, d\varepsilon \approx (m-1) e_0 \nu_{\Sigma p} \sum_{N=1}^{\infty} C_N \tag{3.185}$$

where $\nu_{\Sigma p}$ is the total amount of substance in the macromolecular system, C_N here is the concentration of monomeric units that do not form cross-links, and e_0 is the energy of vacancy.

Based on the same assumptions that allowed us to derive Eq. (3.149), we obtain the following equation for the total energy of the system, found from Eq. (3.146):

$$E_\Sigma \sim \int_0^{C_{l,\,\max}} \mathrm{Er}(\mu(C_l)) C_l dC_l \int_{r_{\min}}^{r_{\max}} dr/r \tag{3.186}$$

Formally, Eq. (3.186) does not differ from Eq. (3.148), but in this case the averaged concentration of active centers (vacancies) is found from the following equation:

$$C_{v,\mathrm{av}} = \frac{\sum_{N=1}^{\infty} [NC_N V_\Sigma(N) - 2C_r(N)]}{v_{\Sigma p}} \tag{3.187}$$

Equation (3.187) means that cross-linkage reduces the number of vacancies, as it has been mentioned above.

The total probability of finding a pore in energetic state ε is estimated, in the general case, from Eq. (3.149). Solution of Eq. (3.149) means finding the orthogonal function for $\varphi_r(r)$. However, if the system is in steady state, one can analyze some properties of function $\varphi_r(r)$. For steady state, Eq. (3.149) is transformed to

$$\exp\left(\frac{-\varepsilon}{\alpha}\right) = \int_0^{C_{l,\max}} \mathrm{Er}(C_l)\varphi_r\left(r \sim \frac{C_l}{\varepsilon}\right) dC_l \tag{3.188}$$

Let us denote

$$I_0 = \int_0^{C_{l,\max}} \mathrm{Er}(C_l)\varphi_r\left(r \sim \frac{C_l}{\varepsilon}\right) dC_l \tag{3.189a}$$

$$I_n = \int_0^{C_{l,\max}} \mathrm{Er}(C_l)C_l^n\varphi_r^{(n)}\left(r \sim \frac{C_l}{\varepsilon}\right) dC_l \tag{3.189b}$$

where $\varphi_r^{(n)}$ is the nth derivative of the function $\varphi_r^{(n)}$ (C_l/ε). Now, we obtain from Eqs. (3.188) and (3.189)

$$\frac{\partial I_n}{\partial \varepsilon} = \frac{-I_{n+1}}{\varepsilon^2} \tag{3.190}$$

$$I_1 = \frac{\varepsilon^2 I_0}{\alpha} \tag{3.191a}$$

$$I_2 = \varepsilon^2\left(\frac{\varepsilon^2}{\alpha^2} - \frac{2\varepsilon}{\alpha}\right) I_0 \tag{3.191b}$$

$$I_3 = \varepsilon^3\left(\frac{\varepsilon^3}{\alpha^3} - \frac{6\varepsilon^2}{\alpha^2} + \frac{6\varepsilon}{\alpha}\right) I_0 \tag{3.191c}$$

$$I_n = \varepsilon^n \sum_{i=1}^{n} z_i \left(\frac{\varepsilon}{\alpha}\right)^i I_0 \tag{3.191d}$$

where coefficients z_i are found from Eqs. (3.189) and (3.190). Equations (3.189)–(3.191) are transformed to

$$\int_0^1 \mathrm{Er}(C_l)\left\{\sum_{j=1}^{\infty}\gamma_j\left[C_l^j\varphi^{(j)} - \varepsilon^l\varphi\sum_{i=1}^{j} z_i\left(\frac{\varepsilon}{\alpha}\right)^i\right]\right\} dC_l = 0 \qquad (3.192)$$

where γ_i are some coefficients. Since Eq. (3.192) should be valid for any values of γ_i, we obtain from Eq. (3.192):

$$\sum_{j=1}^{\infty}\gamma_j\left[C_l^j\varphi^{(j)} - \varepsilon^l\varphi\sum_{i=1}^{j} z_i\left(\frac{\varepsilon}{\alpha}\right)^i\right] = 0 \qquad (3.193)$$

or

$$\varphi = \frac{\sum_{j=1}^{\infty}\gamma_j C_l^j \varphi^{(j)}}{\sum_{j=1}^{\infty}\gamma_j\varepsilon^l\sum_{i=1}^{j} z_i(\varepsilon/\alpha)^i} \qquad (3.194)$$

Equations (3.188)–(3.194) provide the principal relationship between pore size distribution and pore energy distribution for a porous structure formed by linear or branched polymerization.

IV. SELF-ORGANIZATION OF MICROPOROUS STRUCTURES AND FRACTAL FORMATION

The above-considered models have dealt with separate micropore formation or ensembles of micropores formed due to chemical reactions in processes of synthesis of microporous materials. However, in addition to several laws of energy distribution of micropores, their ensembles are characterized also by restructuring phenomena having no direct relation to the chemical processes of micropore formation.

The process of pore formation comprises not only the appearance of pores themselves but also their restructuring, meaning the aspect of the location and the neighborhood of assemblages of pores. The phenomenon of restructuring can be considered also as macroscopic fragmentation of porous sub-structures. The same re-structuring processes may happen in macro-, meso-, and microporous structures.

Depending on conditions of performing the synthesis, these phenomena can be classified to heterogeneous and homogeneous restructuring. The general conditions for the formation of homo- and heterogeneous structures were analyzed in Chapter 1. Now, let us consider restructuring processes in both situations (meaning homogeneity and heterogeneity).

Let us note that all models describing restructuring of microporous structures are compatible will models of microporosity, because they are applied to absolutely different phenomena.

Heterogeneous restructuring was considered in detail in Ref. 26.

A. Homogeneous Restructuring: Formation of Random Fractals

The physical sense of random fractals was considered in Chapter 1. Random fractals are defined as structures, fragments in which are characterized by approximately the same values of extensive parameters (per 1 mol of substance). The physical cause for random fractal formation is the homogeneity of the preparation conditions, as they were defined in Chapter 1 (with exception for pattern formation in such specific situations as that considered above). That physical cause determines the thermodynamic cause for random fractal formation:

1. Stationary situation: This is characterized by conditions of statistical equilibrium, i.e., stability against short perturbations of distribution function:

$$\delta v_{\Sigma} = 0 \quad \delta U_{\Sigma} = 0, \quad \delta S_{\Sigma} = 0 \tag{3.195}$$

where δv_{Σ}, δU_{Σ}, and δS_{Σ} are variations of the amount of substance, of the total energy, and of the entropy of the system. As it was shown above, equations like Eq. (3.195) provide the Gibbs energy distribution of fragments of the system—compare to Eq. (3.115):

$$f(U) = \exp\left(\frac{-U}{\alpha_U}\right) \tag{3.196}$$

where the physical sense of parameters is the same as for Eq. (3.115).

Obviously, since the fragment energy distribution is the same through all the structure, all additive parameters of the fragments, being closely related to the energy distribution, are always proportional to the amount of substance inside them, which is equivalent to the above definition of random fractals.

2. Quasistationary situation: This is similar to the stationary case, under conditions analyzed above (see the proof of the principle of entropy maximum).

3. Nonstationary situation. In this case, the principle of entropy maximum is invalid. However, one can solve this problem if two principal conditions of homogeneity of preparation conditions are satisfied (see Chapter 1):
   a. The initial medium (raw material) is homogeneous.
   b. The treatment conditions are homogeneous (intensive parameters have the same values through all the system).

Let us consider the initial medium as a system of identical cells filled with the substance of the raw material. For such system, the process of pore formation can be considered as a random removal of substance from the cells of the initial medium. If this is described by a law of symmetry, the same law describes also the process of empty cell appearance. In the statistical point of view, that process means the increase of the probability that a cell becomes empty, independent of the location of this spot. In the described process, the random formation of pores results in the formation of two new substructures in the space of the initial medium: an empty substructure (comprising empty cells) and a filled substructure (the rest of the initially filled medium). Both fragments have the same symmetry as the initial medium. Moreover, every fragment has size much larger than cell size and also has the same symmetry, being not important whether the fragment is already empty. Hence, we obtain two random fractal structures: one empty structure and one filled structure. Each of them consists of similar fragments having the same symmetry (and, consequently, having proportional values of extensive parameters).

Note that the above proof is invalid in the case of homogeneous structure formation *though* the inhomogeneity of the initial medium (such situation was considered in Chapter 1), if the treatment process is not steady state or close steady state.

It is interesting that, in its physical sense, the above-described process of random fractal appearance is *not formation but transformation* of the initially existing random fractal substructures, symmetry of which repeats in two resulting fractal substructures, one being filled and the second being empty. Nevertheless, in the further analysis we will talk about fractal formation—always meaning fractal transformation.

B. Fundamental Aspects of Random Fractal Formation

Formation (transformation) of random fractals is an example of self-organization of matter. The phenomenon of self-organization of systems is well known; in 1977 Prof. Ilya Prigogine (Belgium–USA) got the Nobel Prize for studies of self-organization of systems. I. Prigogine largely used the notion of dissipative structures as open systems in nonequilibrium, observed properties of which repeat periodically in time (temporary dissipative structures), space (special dissipative structures), or both (temporary-spatial dissipative structures). In his estimation, dissipative structures do not obey Boltzmann's order principle, because a dissipative structure is not in equilibrium. This suggestion is well problematic; as we have shown above, many of nonequilibrium systems exhibit properties similar to equilibrium (mostly because the limited-spectrum maximum of entropy in an open system in nonequilibrium steady state has many common features with the recognized maximum of entropy in isolated system in equilibrium). Let us mention some examples of self-organization in nature:

Astrophysical macrosystems (hierarchy "down"): our Universe, Milky Way Galaxy, Solar system, Jupiter, Earth, etc.
Microsystems (hierarchy "up"): a gluon quanta, an electron quanta, elementary particles, atoms, molecules, biological cells, multicell organisms, society
Cosmological nuclear physics: the formation of assemblages of atomic nuclei in stars (that can be considered as a form of replication)
Magnetic phenomena: magnetization of ferromagnetic in magnetic field, due to selective orientation of fragments, while the removal of the exterior field does not cancel the magnetization effect
Chemistry: Beloussov–Zhiabotinsky reaction, in which the color of the reacting solution changes periodically (temporary and temporary-spatial dissipation structure)
Society: formation of nations inside the same population (spatial dissipative structure)
Microporosity: pattern formation in the high-velocity process of synthesis of active carbon (considered above temporary dissipative structure)

Many of dissipative structures appear in steady-state open systems, in which the entropy maximum on the limited spectrum of intensive parameters was proved above. That is a reason to suggest that all dissipative structures appear due to entropy mechanism—increase of entropy of the system. As we have shown above, random fractals in porous structures are the case. Moreover, many complex phenomena in nature are very similar to random fractals. Let us mention two such phenomena, importance of which is recognized, that can be interpreted as analogues of fractal formation (see Appendix 2):

Universe formation by big bang: The concept of big bang is the mainly accepted explanation for beginning of the Universe. However, the existing theories of big bang cannot explain why that happened? What was before big bang? The thermodynamic theory of big bang, proposed recently by the author, considers beginning of the Universe as self-organization of exterior "hyper-cosmos." Mathematical evaluations show that, according to the thermodynamic theory of big bang, this increased the entropy of hyper-cosmos. Hence, the Universe inside the hyper-cosmos can be compared to a porous fragment inside continuous phase.

Beginning of life on early earth. The early earth can be considered as a quasi-steady-state system, in which the entropy rising causes self-organization of protein structures. That becomes the thermodynamic cause for polymerization–destruction processes in protein systems. Also this situation is very similar to random fractal formation, while protein cell appearance can be compared to pore formation.

C. Thermodynamic Functions of Fractals

Though fractals do not distinguish different pores, those distinguish continuous and porous (empty) phases, due to the existence of interface. On the other hand, the volume and energy balance cannot depend on distinguishing or nondistinguishing particular properties of structural fragment. Therefore, the fractal formation (transformation) is characterized by changes in volume and free energy evaluated from the following equations:

$$E_{ff} = \sigma A_{ff} = Q_0 \int_{\varepsilon_{\min}}^{\varepsilon_{\max}} \varepsilon f(\varepsilon)\, d\varepsilon \tag{3.197a}$$

$$V_{ff} = (Q_0/\rho_0) \int_{\varepsilon_{\min}}^{\varepsilon_{\max}} f(\varepsilon)\, d\varepsilon = \nu_{f2} V_c \tag{3.197b}$$

where ρ_0 is the molar density of substance in pores (mol/m^3) and V_c is the volume of cell.

Let us use the discrete approach to the material. Then, the entropy of fractal formation is additional to the entropy of the pores themselves and is found from

$$S_{ff} = k_B[\nu_0 \ln \nu_0 - \nu_{f1} \ln \nu_{f1} - \nu_{f2} \ln \nu_{f2}] \tag{3.197c}$$

where ν_{f1} is the number of filled cells, ν_{f2} is that of empty cells, and $\nu_{f1}+\nu_{f2}=\nu_0$ is the number of cells in the initial structure. The pores' entropy is found as before:

$$S_{\text{pores}} = -R_g Q_0 \int_{\varepsilon_{\min}}^{\varepsilon_{\max}} f(\varepsilon) \ln f(\varepsilon)\, d\varepsilon \tag{3.198}$$

where the physical sense of all parameters is the same as in Eq. (3.59). In steady state, the conditions of the statistical equilibrium of the system are

$$\delta V_{ff} = 0, \qquad \delta E_{ff} = 0, \qquad \delta(S_{ff} + S_{\text{pores}}) = 0 \tag{3.199}$$

Equations (3.195)–(3.199) give us:

$$\begin{aligned} &\alpha_1 Q_0 \delta_f(\varepsilon) + \alpha_2 Q_0 \delta f(\varepsilon) + R_g Q_0 [\ln f(\varepsilon)] \delta f(\varepsilon) + R_g \delta f(\varepsilon) \\ &\quad + \delta\nu_{f1} \ln \frac{(\nu_0/\nu_{f1} - 1)}{\varepsilon_{\max} - \varepsilon_{\min}} = 0 \end{aligned} \tag{3.200a}$$

$$\delta f(\varepsilon) Q_0 [\alpha_1 + \alpha_1 \varepsilon + R_g [\ln f(\varepsilon)] = -\frac{\delta\nu_{f1} \ln (\nu_0/\nu_{f1} - 1)}{\varepsilon_{\max} - \varepsilon_{\min}} \tag{3.200b}$$

or

$$\frac{\delta f(\varepsilon)}{\delta \nu_{f1}} = -\frac{\ln (\nu_0/\nu_{f1-1})/Q_0/(\varepsilon_{\max} - \varepsilon_{\min})}{\alpha_1 + \alpha_2 \varepsilon + R_g \ln f(\varepsilon)} \tag{3.201}$$

Equation (3.201) means that the change of the numbers ν_{f1} and ν_{f2} influences the pore energy distribution $f(\varepsilon)$

In the case $\delta\nu_{f1}=0$, we reproduce Gibbs energy distribution Eq. (3.63).

What is the physical sense of Eq. (3.201) when $\delta\nu_{f1} \neq 0$? In such a case, Eq. (3.201) allows the estimation of the influence of the limit amount of the substance on the pore energy distribution. The practical significance of Eq. (3.201) is minor: just remember that finite structures with steady-state formation of pores obtain pore-energy distribution different from Gibbs (3.63).

Let us find the solution of Eq. (3.201) in the case when the influence of fractal growth onto the pore-energy distribution is small, using perturbation method.

1. The nonperturbed solution, obviously, is Gibbs distribution:

$$\alpha_1 + \alpha_2 \varepsilon + R_g \ln f^0(\varepsilon) = 0 \tag{3.202}$$

where $f^0(\varepsilon)$ is the nonperturbed distribution (Gibbs).

2. The perturbed solution is

$$f(\varepsilon) = f^{0}(\varepsilon) + Df(\varepsilon) \tag{3.203}$$

where $Df(\varepsilon)$ is the perturbation. Now, we obtain from Eqs. (3.202) and (3.203):

$$\ln[Df(\varepsilon)] = \frac{-\ln(\nu_0/\nu_{f1} - 1)/R_g Q_0/(\varepsilon_{\max} - \varepsilon_{\min})}{\delta f(\varepsilon)/\delta\nu_{f1}} \tag{3.204}$$

The derivative $\delta f(\varepsilon)/\delta\nu_{f1}$ shows the sensitivity of the pore-energy distribution to the growth of the pore fractal.

Note that Eqs. (3.195)–(3.204) are valid only under the condition that the rate of fractal growth is comparable to the rate of relaxation of energy distribution function. Otherwise (when the relaxation rate is much larger) the system is described just as quasi-steady state.

D. Discrete Consideration of Fractal Formation

Since the process of fractal formation may begin from zero, the factor of discreteness may influence the solution. On the other hand, for large number of cells belonging to each fractal, the continuous and the discrete models should give a conjugation. The discreteness involves the factor of randomness (when the applicability of thermodynamics is problematic), while the case of large numbers of cells in both fractal clusters (porous and filled) leads to the correct applicability of thermodynamics. When does thermodynamics start to influence the process of cluster growth? This problem should be solved numerically.

Let us consider two situations for discrete fractal cluster growth:

Cluster growth in limited volume
Cluster growth in unlimited volume

1. Cluster Growth in Finite Volume

Let us consider a microporous system as a net containing N_0-cells, N_p of which are empty spots corresponding to pores. Each cell has n_n neighbors, n_e of which are empty and $n_f = n_n - n_e$ are filled. Neglecting quantum effects in micropores, we assume that the surface energy of the system is proportional to the internal surface area A_p, the surface between two clusters [30]:

$$\Delta G_p \approx \sigma_p A_p \tag{3.205}$$

$$A_p = \sum_{n_e=1}^{n_n} n_e N_p(n_e) = \sum_{n_e=1}^{n_n} (n_n - n_e)[N_0 - N_p(n_e)] \qquad (3.206)$$

where $N_p(n_e)$ is the number of empty cells having (each) n_e empty neighbors. The situation $n_e = n_n$ corresponds to cells having the minimal mechanical bonds to their neighbors—only at the corners. If the system was in equilibrium, the maximum entropy would correspond to Gibbs energy distribution and would be described by Poisson distribution of filled cells having empty neighbors [30]:

$$\xi(n_e) = \frac{p_p^{ne}(1 - p_p^{nn-ne})(1 - p_p)}{1 - p_p^{nn+1}} \qquad (3.207)$$

where $p_p = N_{pe}/N_0$; N_{pe} is the equilibrium value of N_p. If the real distribution in the interface field is $\xi^*(n_e) = N_p(n_e)/(N_0 - N_p)$, the resulting free energy of surface is:

$$\Delta G_p \approx \sigma_p A_p + R_g T \sum_{n_e=1}^{n_n} [\xi^*(n_e) \ln \xi^*(n_e) - \xi(n_e) \ln \xi(n_e)] \qquad (3.208)$$

Variations of the free energy by Eq. (3.208) and surface energy by Eq. (3.206) provide

$$\delta A_p = 0 \qquad \delta \Delta G_p = 0 \qquad (3.209)$$

hence,

$$\sum_{n_e=1}^{n_n} \{\sigma_p n_e \delta\xi^*(n_e) - R_g T[\ln \xi^*(n_e)\delta\xi^*(n_e) + \xi^*(n_e)]\} = 0 \qquad (3.210a)$$

$$\sum_{n_e=1}^{n_n} \delta\xi^*(n_e) = 0 \qquad \sum_{n_e=1}^{n_n} n_e \delta\xi^*(n_e) = 0 \qquad (3.210b)$$

That results in the following cell distribution in the number of neighbors [30]:

$$\xi(n_e) = \alpha_2 \exp -\frac{(\sigma - \alpha_1)n_e}{R_g T} \qquad (3.211)$$

where

$$\alpha_1 = R_g T \ln \frac{p_p}{1 - p_p} \tag{3.212a}$$

$$\alpha_2 = \frac{1 - \exp[(\alpha_1 - \sigma)/R_g T)]}{[1 - \exp[(n_n + 1)(\alpha_1 - \sigma)/R_g T]} \tag{3.212b}$$

Equation (3.211) is very similar to the Gibbs distribution Eq. (3.63), but the changes between values of function $\xi(n_e)$ are discrete, and the number of its allowed values is limited (that is, $n_n + 1$).

The probability that a filled cell having n_e empty and n_f filled neighbors looses the substance and becomes an empty cell is proportional to $N_p(n_e)$ and depends on the number of filled neighbors $(n_n - n_e)$, more exactly on the work needed to cut the bond with these neighbors [30]:

$$\Pr(n_e) \sim N_p(n_e) \exp\left[-\frac{(n_n - n_e)\sigma}{(R_g T)}\right] \tag{3.213}$$

Let us consider a filled cell having n_e empty and n_f filled neighbors, among which n_{f0} have no empty neighbor, n_{f1} have one empty neighbor only, ..., n_{fK} have K empty neighbors:

$$\sum_{K=0}^{n_n} n_{fK} = n_f \tag{3.214}$$

When the considered filled cell transforms itself to an empty cell, also the environment of its neighbors changes, they loose one filled neighbor and get one empty neighbor. Hence [30]

1. The number of filled cells having n_e empty neighbors is reduced by one, because of the loss of the above cell.
2. The number of filled cells having (each) K filled neighbors decreases by n_{fK}, but the number of filled cells having $(K - 1)$ filled neighbors increases by n_{fK}.
3. The statistical criterion found from Eq. (3.213) should be estimated not for the initial but for the potential resulting distribution of filled cells in the number of their empty neighbors.

Hence

$$\Delta N_{eK} = \begin{cases} n_f \dfrac{\left(N_{f(K-1)} + \delta_2\right) - \left(N_{fK} + \delta_1\right)}{N_f - 1} & \text{if} \quad K \geq 1 \\ -n_f \dfrac{N_{fK} + \delta_1}{N_f - 1} \quad \text{if} \quad K = 0 & \end{cases} \tag{3.215}$$

$$\delta_1 = \begin{cases} 0 & \text{if} \quad K \neq m_e \\ -1 & \text{if} \quad K = m_e \end{cases} \tag{3.216a}$$

$$\delta_2 = \begin{cases} 0 & \text{if} \quad K \neq 1 + m_e \\ -1 & \text{if} \quad K = 0 \end{cases} \tag{3.216b}$$

Equations (3.211)–(3.216) allow the numerical modeling of discrete formation of porous cluster in finite-size system.

2. Cluster Growth in Unlimited Volume

The general physical assumptions about the growth of cluster in the infinite-size volume are the same as in the finite-size case:

The system is considered as a medium consisting of cells, filled or empty (while this time the number of these cells is infinite).

The transformation of a filled cell to an empty cell is determined by thermodynamic restrictions.

The kinetics of the transformation of a filled cell to an empty cell is determined by the prehistory of the system (the near environment of each cell: filled and empty neighbors), the rate of such transformation depending on the number of filled neighbors.

One of sequences of the transformation of a filled cell to an empty cell consists in the change of the distribution of filled cells having a number of empty neighbors and, hence, influences the further process of porous cluster formation.

The reactionability of each filled cell is determined by its availability for the pore cluster, therefore, the reaction rate of empty cell formation is proportional the probability of the event of filled cell becoming empty [31]:

$$W_r(n_e) = N_p(n_e) \exp\left[-\frac{(n_n - n_e)a_0\sigma}{(R_gT)}\right] \tag{3.217}$$

where a_0 is the surface area of each cell. Cells having filled neighbors only are permanently consumed, they have no source but only sinks (pore formation and the appearance of filled cells having one empty neighbor each). The change of their number is found from the following equation [compare to Eq. (3.215)]:

$$dN_{f0} = -dV\psi \frac{N_{f0}n_n}{\left(\sum_{K=0}^{n_n-1} (n_n - K)N_{f(n_n-K)}\right)} \tag{3.218a}$$

where ψ is the normalization coefficient and dV is the change of the volume of the porous cluster [31]

$$dV = \frac{-dN_f}{\rho_f} \tag{3.218b}$$

The change of the number of filled cells having maximum (n_n) of empty neighbors, that have one source (filled cells having $n_n - 1$ empty neighbors) and one sink (pore formation), is found from the following equation [31]:

$$dN_{fnn} = -W_r(n_n)\,dV + \frac{\psi dV}{\left[\sum_{K=0}^{n_n-1} KN_{f(n_n-K)}\right]} \tag{3.219}$$

For filled cells, number of empty neighbors for which is n_e from 1 to $(n_n - 1)$, there is a source (cells having $n_e - 1$ empty neighbors) and sink (void formation and formation of cells having $n_e + 1$ empty neighbors), hence, the change of their number is found from the following equation [31]:

$$dN_{fne} = -W_r(n_e)\,dV + \psi \frac{dV}{[N_{f(n_e-1)}(n_n - n_e + 1) - N_{fn_e}(n_n - n_e)]/\sum_{K=0}^{n_n-1} (n_n - K)N_{fK}} \tag{3.220}$$

Equations (3.218)–(3.220) allow the description of porous cluster formation in a system having unlimited (infinite) size.

Figure 3.12 presents the behavior of N_{fne} for different n_e. The calculations are given for $n_n = 6$ (cubical situation: each cell has 6 neighbors, 2 on each coordinate in three-dimensional space) and low surface tension.

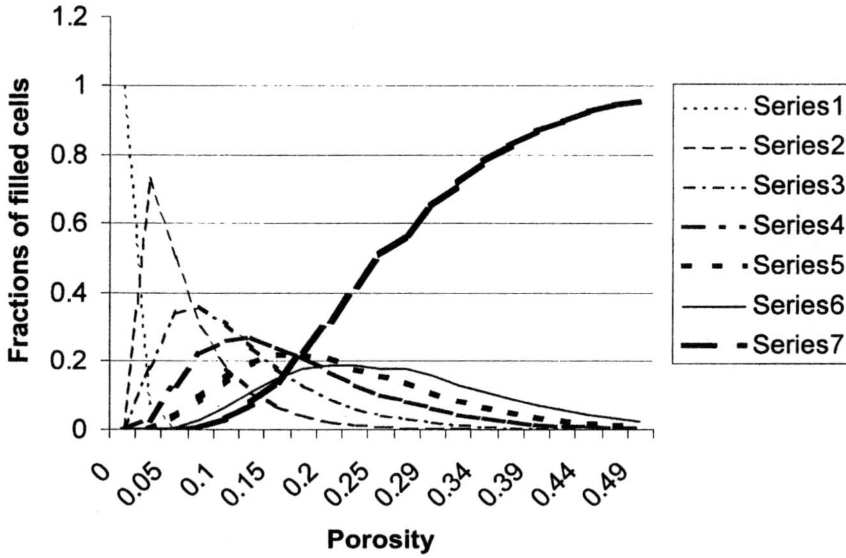

FIG. 3.12 Distribution of filled cells in the number of their empty neighbors during process of thermodynamically-limited pore formation in unlimited volume: Series 1: filled cells having no empty neighbor; Series 2: filled cells having (each) single empty neighbor; Series 3: filled cells having (each) two empty neighbors; Series 4: filled cells having (each) three empty neighbors; Series 5: filled cells having (each) four empty neighbors; Series 6: filled cells having (each) five empty neighbors; Series 7: filled cells having (each) six empty neighbors.

As follows from Fig. 3.12, the beginning of pore formation immediately reduces the number of filled cells having no empty neighbors almost to zero—of course, not because the loss of filled cells but mostly because of the appearance of filled cells of other kinds. The number of filled cells having empty neighbors only (isolated filled cells held inside the solid structure only by corners, $n_e = 6$) slowly increases monotonically. Other curves get maximum and then slowly decrease.

Figure 3.13 presents the internal surface area as a function of the volume, for the same conditions.

As follows from Fig. 3.13, the initial part of the curve corresponds to relatively low rising of the internal surface—that is because the low porosity corresponds mostly to filled cells having one or two empty neighbors.

More detailed analysis of structural properties of microporous clusters formed randomly under thermodynamic conditions in unlimited volume is given in Chapter 5.

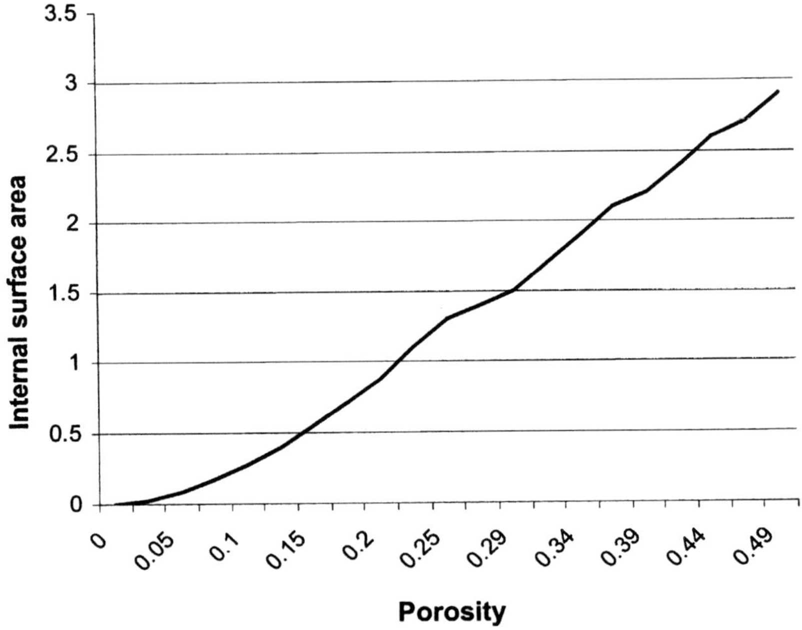

FIG. 3.13 Increase of internal surface area with the porosity.

E. Conjugation of Discrete and Continuous Approaches

The discrete and the continuous approaches are well similar but not identical. Let us mention some factors of their divergence resulting from the main assumptions:

Discrete approach:

Allows the description of pore formation on all stages.
Contains no unknown parameter.
Applicable mostly to equilibrium formation of pore.
Problematic for analytical studies.

Continuous approach:

Provides analytical solution in a simple form.
Allows calculations of thermodynamic functions and related measurable characteristics.
Contains three unknown parameters (α, ε_{min}, ε_{max}).
Allows the description of pore formation mostly after the beginning stage is over.

In any case, these two different approaches should provide the same physical results. Therefore, we may formulate the following conditions of their conjugation:

1. Equal values of calculated volume
2. Equal values of calculated energy
3. Equal values of calculated entropy

The third condition is realized, in most situations, automatically. Let us write two other conditions:

For the volume

$$V_0 = V_c + V_p = \sum_{k=0}^{n_n} N_{fk} + Vp = V_c + Q_0 \int_{\varepsilon_{\min}}^{\varepsilon_{\max}} f(\varepsilon)\, d\varepsilon \tag{3.221}$$

where V_0 is the total volume, V_c is the volume of the continuous phase, V_p is the volume of pores. As follows from Eq. (3.221), the conjugation of the discrete and the continuous solutions for volume provides no problem: one of them brings the information about the continuous phase; the other, about the pores.

For the energy:

$$E_\Sigma = Q_0 \int_{\varepsilon_{\min}}^{\varepsilon_{\max}} \varepsilon f(\varepsilon)\, d\varepsilon = \sigma \sum_{k=1}^{n_n} kN_{fk} \tag{3.222}$$

Let us assume $\varepsilon_{\min} = 0$, $\varepsilon_{\max} = \infty$. The first of these assumptions seems well lawful (that is just the free energy of plate surface), while the physical correctness of the second assumption will be analyzed in Chapter 5. These assumptions allow us to simplify Eqs. (3.221) and (3.222). In the steady-state situation, we obtain

$$V_p = \alpha Q_0 \tag{3.223}$$

$$E_\Sigma = Q_0 \alpha^2 \tag{3.224}$$

Thus, Eqs. (3.222)–(3.224) allow us to evaluate Q_0 and α from the conjugation conditions. If the porous cluster grows, the changes of Q_0 and α are found from the following equations resulting from Eqs. (3.222)–(3.224):

$$d\alpha = \frac{dE_\Sigma - \alpha dV_p}{\alpha Q_0} \tag{3.225}$$

$$dQ_0 = \frac{2Vp}{\alpha} - \frac{dE_\Sigma}{\alpha^2} \tag{3.226}$$

F. Nucleus Formation

As it was shown above, the process of the synthesis of active carbon out of carbon matrix needs formation of a nucleus of the solid carbon phase. In many cases, the organic phase is viscous liquid (high-content carbon increases chances for the successful synthesis of active carbon), but it can be sometimes amorphous solid.

This nucleus is located at the borderline "hot gas – organic phase" (see Fig. 3.14).

At the interface, hot gas causes the increase of the temperature of the organic surface; as a sequence, volatile components move to the gas phase and get burned. One of products of their combustion is carbon that is not compatible with the gas phase (because of the gravity factor), or organics, and is preferable trapped by the solid nucleus for cohesion reasons. That is the cause for nucleus growth. On the other hand, because of eventual factors (eventual increase of the temperature above the nucleus, fluctuation increase of the local concentration of oxygen, appearance of carbon-compatible components in the organic phase) the nucleus has a chance to be suddenly destroyed and dissolved.

The carbon nucleus interacts with both neighbor phases: receives carbon particles from both them, but also looses carbon because of burning

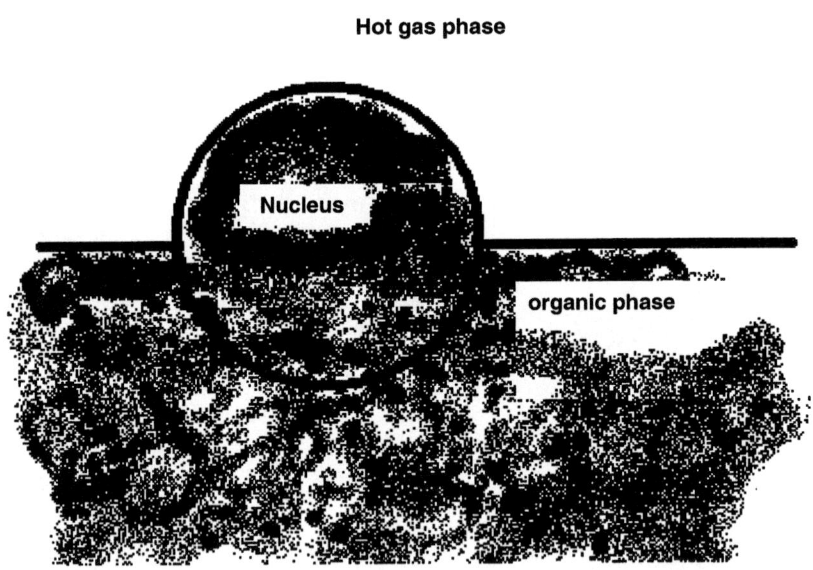

FIG. 3.14 Formation of solid porous nucleus at the frontier hot gas–organic phase.

in the gas phase or dissolving in the organic phase. Under high temperatures, carbon from the nucleus may interact with hydrocarbons in the organic phase, while such interactions are, in most cases, negligible.

In the energetic aspect, the nucleus growth increases the entropy of the system and reduces the interface tension energy between the gas and organic phases (due to reducing the surface area of the interface), increases the internal energy of the system (because of micropore formation), and provides the interface tension energy between "nucleus-gas" and "nucleus-organics" phases. The nucleus certainly contain void(s), but it seems incorrect to talk about its pore distribution: maybe the pore is single, maybe there are some pores, but, in any case, their number is not sufficient for the statistical equilibrium. However, the porosity itself exists and influences the statistical equilibrium of the nucleus with its neighbor phases. The unstable structure of the nucleus becomes the base for the future formation of all kinds of pores allowed by thermodynamics.

The considered system is obviously metastable: depending on eventual factors (fluctuations of intensive parameters), the nucleus may increase or decrease in size, and even get be lost. As follows from the above-presented method of thermodynamic cycles, we are allowed to write the thermodynamic functions as if the system was in equilibrium.

Let us assume that the considered system is under constant pressure and temperature. We assume also that the form of the nucleus does not change with time. As follows from the definition of potential Gibbs [Eq. (3.6)], its resulting value in such conditions is determined only by surface phenomena and mass transfer. To avoid complications in derivations, we assume that the form of the nucleus is close to spherical.

The conditions of the metastability are given as follows:

Thermodynamic metastability of the nucleus: Potential Gibbs has maximum, and its derivative by the nucleus size is zero.

Structural equilibrium: The interior parameters of the nucleus are stable against variations of energy distribution function of the nucleus.

The condition of the structural stability is indispensable, because otherwise the structural stability of the future pores would appear from another source, and the physical sense of the nucleus would be lost.

The equation for potential Gibbs is given by

$$\begin{aligned}\Delta G = \sigma_{sg} A_n \phi_{sg} + \sigma_{so} A_n (1 - \phi_{sg}) - \pi r^2 \sigma_{og} \\ + R_g T[\xi \ln \xi + (1 - \xi) \ln (1 - \xi)]\end{aligned} \tag{3.227}$$

where σ_{so} is the surface tension at the frontier solid carbon nucleus—organics, A_n is the surface area of the nucleus, ϕ_{sg} is the part of the nucleus contacting with the gas phase, σ_{sg} is the surface tension at the frontier solid carbon nucleus–gas, r is the radius of the nucleus, ξ is the current porosity of the nucleus, and σ_{og} is the surface tension at the frontier organics–gas. Let us note that three first terms in the right part of Eq. (3.227) correspond to surface phenomena related to the nucleus formation, and the rest term, the pore entropy contribution. Let us notice that in comparison with the situation of steady-state formation described by Eqs. (3.57)–(3.62), the condition of maximum of entropy is absolutely not applicable (because the system is not in steady state), while the condition of balance of substance stays valid:

$$\nu_\Sigma = \frac{4\pi\rho_c r^3}{1-\xi} \tag{3.228a}$$

$$m_n = (1-\xi)\rho_c(4\pi r^3) \tag{3.228b}$$

where ρ_c is the molar density of the continuous phase in the nucleus (mol/m^3), and m_n is the mass of the nucleus. The conditions of the statistical equilibrium are written as

$$\delta\nu_\Sigma = 0 \tag{3.229}$$

$$\delta\Delta G = 0 \tag{3.230}$$

Now, we obtain from Eqs. (3.227)–(3.230):

$$\delta r_n = \frac{r_n\delta\xi}{3(1-\xi)} \tag{3.231}$$

$$\begin{aligned} 8\pi[\sigma_{sg}\phi_{sg} + \sigma_{so}(1-\phi_{sg})]r_n\delta r_n - 2\pi r_n\sigma_{og}\delta_{rn} \\ + R_g T[\delta \ln\xi - \ln(1-\xi)\delta\xi] = 0 \end{aligned} \tag{3.232}$$

That results in the following condition of metastability of the carbon nucleus:

$$\begin{aligned} 8\pi[\sigma_{sg}\phi_{sg} + \sigma_{so}(1-\phi_{sg})]r_n^2 - 2\pi r_n^2\sigma_{og} \\ = 3(1-\xi)R_g T[\ln\xi - \ln(1-\xi)] \end{aligned} \tag{3.233}$$

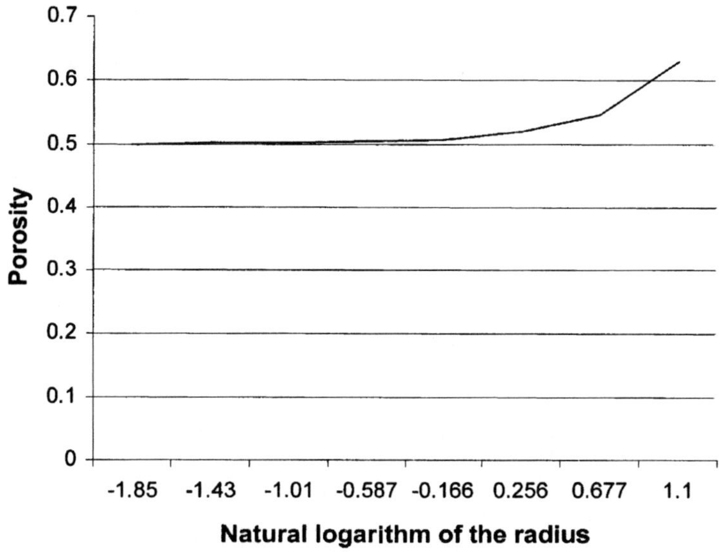

FIG. 3.15 Change of porosity of nucleus at the moment of formation: interactions between the nucleus and condensed phase are weaker than between the condensed phase and the gas.

Equation (3.233) allows us to evaluate the porosity of carbon nucleus before this begins its irreversible growth.

Figure 3.15 presents the relationship between the radius of the nucleus and its porosity.

Of course, the situation presented by Fig. 3.16 is much less typical and can be neglected.

As follows from Eq. (3.233) and Fig. 3.15, the nucleus radius influences the energy distribution functions for different nuclei. In a system containing a number of carbon nuclei, pores having the same energy will preferably form in large nucleus. On the other hand, if a nucleus is single for the system, its radius is not important, because inside it the distribution functions are normalized. The minimal porosity is about 50%, because of a high contribution of entropy factor to the free energy of the system for such values of porosity. The further irreversible growth of the nucleus is accompanied by capturing carbon particles, preferably to the voids, that certainly means to the decrease of the porosity.

It is important that the tendency of the change of porosity depends much on the interactions of the nucleus with both phases.

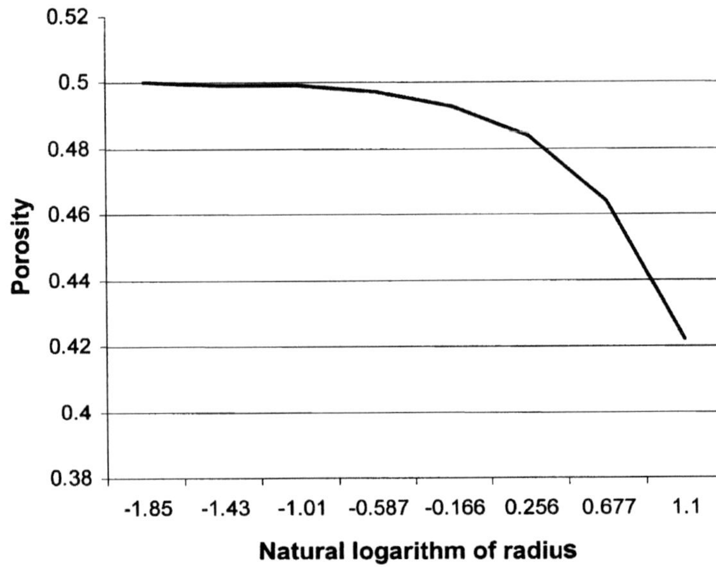

FIG. 3.16 Change of porosity of nucleus at the moment of formation: interactions between the nucleus and condensed phase are stronger than between the condensed phase and the gas.

Let us notify that this model is applicable, in principle, not only to the synthesis of active carbon but also to the preparation of active porous silicon (item 4).

V. CONCLUSIONS

1. In the light of the principal results obtained in two previous chapters, experimental studies of micropores, though extremely important in the practical aspect, are not sufficient for the characterization of micropores, and their results need theoretical treatment.
2. Among existing fundamental concepts applicable to condensed phase and interface, statistical thermodynamics is preferable, because of its capability to provide most relevant results, due to taking into account the irreversibility of processes related to formation of ensembles of micropores.
3. Among the methods used by statistical thermodynamics, the approach of maximal entropy in steady-state systems is definitely preferable, though it has several shortcomings, due

to its capability to treat the infinite number of sources of randomness.

4. Though the enormous variety of existing processes for synthesis of microporous materials, the list of theoretical tools needed for their description is short and comprises:
   a. Linear thermodynamic modeling of the growth of solid phase
   b. Thermodynamic modeling of the formation of solid nucleus in amorphous and polymeric organic phase
   c. Thermodynamic modeling of polymerization and aggregation.
5. For steady-state growth of microporous solid phase, the pore energy distribution is given by the exponential decrease of the distribution function with pore energy (Gibbs distribution). For non-steady state, pore-energy distribution has a more complex form and changes with time.
6. The requirement of the conjugation of local and substantial considerations of microporous material formation leads to differential equations, independent parameters such as time, local temperature, and pore energy.
7. Linear (chain) polymerization far from and close to equilibrium is effectively characterized by the evolution of thermodynamic potentials with time.
8. Among the existing models of branched polymerization, statistical polymer method is preferable, due to its capability to describe polymerization far from and close to equilibrium.
9. Micropore formation in polymerization processes is due to reactionable groups resting after the formation of "monomer–monomer" bonds in macromolecules. The conjugation between energy and size distribution functions allows, in principle, the estimation of size distribution function from the measured energy distribution function for micropores.
10. Multicomponent polymeric systems not only for branched but also for chain structures are described due to the statistical approach based on statistical polymer method.
11. Regarding fractals in homogeneous microporous media, one has to notify that their formation would be more correctly considered as transformation of fractals having the same symmetry. Fractal transformation is due to the self-organization of microporous structures, allowing the increase of their entropy, and can be compared to other known self-organization processes. The model of fractal transformation by self-organization is similar to some other existing models of self-organization, comprising models of

the beginning of the Universe by self-organization of hyper-cosmos and beginning of life on early earth.

12. In situations when the fractal formation (transformation) is so fast that its rate is comparable to this of the relaxation of micropore energy distribution, the resulting micropore energy distribution is seriously influenced by porous fractal growth.
13. Whereas the discrete model of micropore formation provides some results analogous to these obtained in the continuous model, the conjugation of both models allows the complete and close description of the pore formation process.
14. The beginning of the growth of a nucleus of porous solid phase (carbon) at the frontier between the organic phase (source of combustible volatiles) and the hot gas (where the volatiles are partly burned, with delivering carbon particles) is described as a metastable system. The condition of statistical equilibrium applied to such metastable system provides the equation for the evaluation of the porosity of the porous nucleus. The porosity is high, because the entropy factor influences much the free energy of the metastable nucleus.

REFERENCES

1. Landau, L.D.; Lifshitz, E.M. *Quantum Mechanics: Non-relativistic Theory*. 3rd ed.; Pergamon: Oxford, 1977; 673.
2. Landau, L.D.; Akhiezer, A.I.; Lifshitz, E.M. *General Physics: Mechanics and Molecular Physics*. Translated by Sykes, J.B., Petford, A.D. and Petford, C.L.; Pergamon: Oxford, 1967.
3. Prigogine, I.; Defay, R. *Chemical Thermodynamics*. Translated by Everett, D.H.; Longman: London, 1973.
4. Prigogine, I. *The End of Certainty: Time, Chaos, and the New Laws of Nature* (In collaboration with Stengers, I.); Free Press: New York, 1996.
5. Prigogine, I. *Introduction to Thermodynamics of Irreversible Processes*. 3rd ed. (no publisher): New York, 1967.
6. Romm, F. Critical size of a bubble in a superheated fluid and problems in its modeling. In: *Encyclopedia of Interface Science*, Hubbard, A., ed.; Marcel Dekker: New York, 2002.
7. Romm, F.A.; Trusov, V.P. Calculation of the chemical composition and other extensive parameters of the phases in a nonequilibrium multicomponent gas–liquid system with random fluxes. Russ. J. Phys. Chem. (USA), **1989**, *63*(6), 1595–1598.
8. Iler, R.K. *The Chemistry of Silica: Solubility, Polymerization, Colloid and Surface Properties, and Biochemistry*; Wiley-Interscience: New York, 1979.

9. Yehaskel, A. *Activated Carbon: Manufacture and Regeneration*; Noyes Data Corp.: Park Ridge, NJ, 1978.
10. Romm, F. Thermodynamic description of formation of microporous adsorbents (linear steady-state approximation). J. Coll. Interface Sci. **1996**, *179*(1), 1–11.
11. Romm, F. Thermodynamic description of formation of microporous adsorbents (non-steady-state approximation). J. Coll. Interface Sci. **1996**, *179*(1), 12–19.
12. Merris, R. *Graph Theory*. Wiley: New York, 2001.
13. Flory, P.J. *Principles of Polymer Chemistry*; Cornell University Press: New York, 1953.
14. Odian, G. *Principles of Polymerization*; 2nd ed.; Wiley-Interscience: New York, 1981.
15. Freire, J.J. Conformational properties of branched polymers: Theory and simulations. Adv. Polym. Sci. **1999**, *143*(Branched Polymers II): 35–112.
16. Gujrati, P.D. A binary mixture of monodisperse polymers of fixed architectures, and the critical and the theta states. J. Chem. Phys. **1998**, *108*(12), 5104–5121.
17. Batman, R.; Chhajer, M.; Gujrati, P.D. New statistical mechanical treatment of systems near surfaces. V. Incompressible blend of interacting polydisperse linear polymers. J. Chem. Phys. **2001**, *115*(10), 4890–4903.
18. Dusek, K. Network formation in curing of epoxy resins. Adv. Polym. Sci. **1986**, *78*(Epoxy Resins Compos. 3), 1–59.
19. Gujrati, P.D.; Bowman, D. Interplay between gelation and phase separation in tree polymers, and the calculation of macroscopic loop density in the postgel regime. J. Chem. Phys. **1999**, *111*(17), 8151–8164.
20. Bilgen, D.; Aykac, M.; Gujrati, P.D. Multicomponent system with polydisperse species. J. Chem. Phys. **1997**, *107*(21), 9101–9104.
21. Gujrati, P.D. Geometrical description of phase transitions in terms of diagrams and their growth function. Phys. Rev. E: Statistical Physics, Plasmas, Fluids, and Related Interdisciplinary Topics **1995**, *51*(2), 957–974.
22. Gujrati, P.D. Thermal and percolative transitions and the need for independent symmetry breaking in branched polymers on a Bethe lattice. J. Chem. Phys. **1993**, *98*(2), 1613–1634.
23. Chhajer, M.; Gujrati, P.D. New statistical mechanical treatment of systems near surfaces. IV. Surface and surface-induced capillary transitions in polymer solutions. J. Chem. Phys. **1998**, *109*(24), 11018–11026.
24. Kuchanov, S.I. *Methods of Kinetic Calculations in Polymer Chemistry*. (*Metody Kineticheskikh Raschetov v Khimii Polimerov*); Khimiya: Moscow, USSR, 1978; 367 pp. (Russian).
25. Romm, F. Derivation of the Equations for Isotherm Curves of Adsorption on Microporous Gel Materials. Langmuir **1996**, *12*(14), 3490–3497.
26. Romm, F. Thermodynamics of microporous material formation. In: *Surfactant Science Series, Interfacial Forces and Fields: Theory and Applications* (Monographic series, Hsu, J.-P. ed.), Chapter 2; Marcel Dekker: New York, 1999, pp. 35–80.

27. Romm, F. Evaluation of the weight distribution of polymers from reversible polymerization, using a statistical polymer method. J. Phys. Chem. **1994**, *98*(22), 5765–5767.
28. Romm, F. Elaboration of the model of branched cross-linked macromolecular structures and the preparation of new materials following theoretically obtained results. Pigm. Res. Technol. **2000**, *29*(6), 350–355.
29. Romm, F. Microporous materials: modeling of internal structure. In: *Encyclopedia of Interface Science*, Hubbard, A., ed.; Marcel Dekker: New York, 2002.
30. Romm, F. Modeling of random formation of microporous material following thermodynamic limitations. J. Coll. Interface Sci. **1999**, *213*(2), 322–328.
31. Romm, F. Modeling of internal structure of randomly formed microporous material of unlimited volume, following thermodynamic limitations. J. Coll. Interface Sci. **2000**, *227*, 525–530.

4
Characteristics and Properties of Microporous Structures

I. ADSORPTION AND DESORPTION IN MICROPOROUS MEDIA

A. Microporous Adsorbent from the Synthesis to the Uses

Most existing processes for synthesis of microporous adsorbents comprise, as the final stage, a treatment of the obtained porous materials under high temperatures. When the synthesis is achieved, the fresh micropores are not empty but contain some gas or vapor–gas mixture consisting of volatile components delivered from the solid phase during the synthesis and/or products of their oxidation. When the adsorbent is cooled, the micropores adsorb also the gas from the environment (probably air):

$$\mathrm{Ads} + \mathrm{G} \rightarrow \mathrm{Ads} \cdot \mathrm{G} \tag{4.1}$$

When the adsorbent is, after all, in contact with the gas, adsorption of which is needed, the interaction between the gas and the adsorbent consists not in the entering of the gas into voids inside the adsorbent but in the replacement of the previously adsorbed volatile components with the new adsorbate:

$$\mathrm{Ads} \cdot \mathrm{G} + \mathrm{G}^* \rightarrow \mathrm{Ads} \cdot \mathrm{G}^* + \mathrm{G} \tag{4.2}$$

The kinetics of reaction (4.2) is given by the following equation:

$$\frac{dC_{\mathrm{AdsG}}}{dt} = k_{-\mathrm{GG}^*} C_{\mathrm{AdsG}^*} P_{\mathrm{G}} - k_{+\mathrm{GG}^*} C_{\mathrm{AdsG}} P_{\mathrm{G}^*} \tag{4.3}$$

where k_{+GG*} and k_{-GG*} are constants of the rate of the direct and the reverse reactions (4.2), C_i is the concentration of ith component in the solid phase, and P_j is the partial pressure of jth volatile component in the gaseous phase.

Now, let us consider some options in the technique of adsorption performance:

1. The experiment is carried out in a closed box, and the delivered component G remains in the system; then the adsorption process continues until the equilibrium:

$$k_{-GG^*} C_{AdsG^*} P_G = k_{+GG^*} C_{AdsG} P_{G*} \tag{4.4}$$

   while the measured weight of the adsorbate brings the wrong information about the adsorptive properties of the initial adsorbent. If the adsorptive characteristics of G and G* are comparable and the molecular weight of G is equal or more than that of G* ($m_G \geq m_{G*}$), the researcher measures even "negative adsorption."
2. The experiment is carried out by injecting gas G* into the adsorbent carrier, while the gaseous products are removed away. Then, $P_G \approx 0$, and the solution of Eq. (4.3) is

$$\begin{aligned} C_{AdsG} &= C_{(AdsG)0} \exp(-k_{+GG*} P_{G*} t) \\ &= C_{(AdsG)0} \exp(-k_{+GG*} t) \end{aligned} \tag{4.5}$$

   where $C_{(AdsG)0}$ is the initial concentration of the adsorbent carrying the initial adsorbate. Then the resulting adsorbate consists only of AdsG*, but the loss in the weight of G should be taken into account.
3. Operations described above by item 2 are preceded by removal of G: before injecting G*, the adsorbent is heated on purpose to remove G as it is possible then cooled in vacuum. Then the resulting concentration of G is minor—but still not zero.
4. The cycle "heating adsorbent with removal of adsorbate, cooling in vacuum, and injecting G*″ is repeated several times, then the measured change of the weight is

$$\delta W_{AdsG^*} \approx \rho_{Ads} \nu_{Ads} [C_{AdsG^*}(T) - C_{AdsG^*}(T_0)] \tag{4.6}$$

   where ν_{Ads} is the amount of the adsorbent, ρ_{Ads} is its density, T_0 is the temperature to which the adsorbent is heated before being cooled, and T is the temperature under which G* is injected. If w_n is the value of δW_{AdsG^*} after n^{th} iteration, the condition of the correctness of the performance of the measurements is given by the following equation:

$$w_n = w_{n-1} \pm \delta w \tag{4.7}$$

where δw is the error of the measurements of the weight.

Unfortunately, as it was noted in Chapter 2, condition (4.7) for micropores can never be satisfied because of hysteresis, therefore, the correct organization of the measurements of adsorption is extremely complicated.

As an additional option, one might suggest the isolation of the adsorbent immediately after the achievement of its synthesis, but also such technique does not allow avoiding errors.

In several situations, for example, when the adsorbent is used for gas separation, it can be sufficient to measure the composition of the gas phase, first of all P_{G^*}.

In the further analysis of adsorption phenomena, we will assume that the experimental measurements are organized enough correctly and all errors are related only to hysteresis phenomena, as that was described in Chapter 2.

B. Reversible Adsorption and Desorption

The reversibility of the performance of adsorption–desorption processes means that

The processes are carried out very slowly, and the current partial pressure of G* in Eq. (4.2) differs a little only from that over the adsorbent (i.e., for adsorption it is a little *more* and for desorption, a little *less*).

The temperature is controlled (probably specified), and the only factor influencing the change of the weight of the product is the adsorption–desorption of G*, due to the different (while almost equal) pressures of G* just over the adsorbent and far above.

C. Fundamental Physical Causes of Adsorption

The main driving force for adsorption is the reduction of free energy of pores accumulated on the stage of pore formation. Adsorption process is always exothermal and always accompanied by the reduction of the entropy of the system "pore + adsorbate." In any solid or liquid, the particles at the surface are subject to unbalanced forces of attraction normal to the surface plane (if applicable). In discusses about the fundamentals of adsorption, it may be useful in several cases to distinguish between physical adsorption involving only relatively weak intermolecular forces (due to the free energy of pores, especially micropores) and chemisorption involving essentially the formation of a chemical bond between the adsorbate molecule and the surface of the adsorbent (simply said, a chemical reaction between the pore wall and adsorbate molecules). For chemisorption, the reduction of the pore free energy is not the only driving force: the second driving force is the heat effect of the chemical reaction of chemisorption.

Although the above distinction is conceptually useful, many real cases are intermediate between two those options, and it is not always possible to categorize a particular system unequivocally [1,2].

Physical adsorption can be distinguished from chemisorption according to one or more of the following criteria:

Physisorption does not involve the sharing or transfer of electrons, which always maintains the "individuality" (meaning no chemical changes) of the interacting species. The interactions are well reversible (except for micropores), enabling desorption to occur at the same temperature, although the process may be limited by the diffusion factor. Chemisorption involves chemical bonding and is definitely irreversible even in macropores.

Physisorption is not as specific as chemisorption: physically adsorbed molecules may move inside the pores, whereas chemically adsorbed molecules are immobilized at pore walls.

The absolute-value heat effect (the change of enthalpy in the sorption process) for physisorption is (except for micropores) much shorter than for chemisorption.

Hysteresis effects for adsorption–desorption cycle are observed for physisorption/only in micro- and mesopores/and for chemisorption (always, independent of the adsorbing structure).

In principle, chemisorption is typical for situations when the chemical reaction between the pore wall material and the adsorbate molecules is possible in normal conditions, without any regard to internal surface in the solid reagent. A very specific situation is found when the process of adsorption is accompanied by a chemical reaction, in which the adsorbate experiences chemical transformation without visible changes for the adsorbent (heterogeneous catalysis). As it was notified in Chapters 1 and 2, micropores allow reactions that would be just impossible in other conditions. In the case of heterogeneous catalysis, the driving force of the process is initially the reduction of the free energy of pores, but then it is the reduction of the chemical potential of the entire system, while the products do not usually remain inside the pores but leave them. Thus, heterogeneous catalysis can be considered as a very specific and complex form of chemisorption in micropores.

The divergence between these two mechanisms of adsorption is illustrated by Table 4.1.

As follows from Table 4.1, physisorption in micropores has intermediate features between physisorption in macropores on plate surface and chemisorption.

Physisorption in nonpolar solids is normally attributed to forces of interactions between the solid surface and the adsorbate molecules. Those forces are similar to the van der Waals forces (attraction–repulsion) between molecules. The forces of attraction involving electrons and nuclei of all components of the system have electrostatic origin and are termed *dispersion forces.* These exist in all kinds of matter, independently on its interior structure. The nature of the dispersion forces was first recognized in the 1930s by London.

The dipole–dipole interaction energy (for attraction—negative, of course) can be estimated from the following equation [2]:

$$E_D = -\frac{\mathrm{Const_{DD}}}{\mathrm{Dis}^6} \tag{4.8}$$

where E_D is the dispersion energy, $\mathrm{Const_{DD}}$ is a constant of dipole–dipole attraction, and Dis is the distance between the interacting particles.

In addition to dipole–dipole interaction, other possible dispersion interactions contributing physisorption include dipole–quadrupole and quadrupole–quadrupole interactions. If these are incuded into the expression for the interaction energy, it is written as follows:

$$E_D = -\frac{\mathrm{Const_{DD}}}{\mathrm{Dis}^6} - \frac{\mathrm{Const_{DQ}}}{\mathrm{Dis}^8} - \frac{\mathrm{Const_{QQ}}}{\mathrm{Dis}^{10}} \tag{4.9}$$

where $\mathrm{Const_{DQ}}$ and $\mathrm{Const_{QQ}}$ are constants of dipole–quadrupole and quadrupole–quadrupole interactions, respectively. Quadrupole interactions involve symmetrical molecules with atoms of different electronegativity like CO_2. The quadrupole moment of such molecules leads to their interactions with polar surfaces. Taking into account that the contribution of dipole–quadrupole and quadrupole–quadrupole interactions decreases fast with the distance, the decisive role is, in most situations, of dipole–dipole interaction, and dipole–quadrupole and quadrupole–quadrupole interactions can be neglected.

TABLE 4.1 Comparison of Physisorption to Chemisorption

Factor of comparison	Physisorption in macropores and plate surface	Physisorption in micropores	Chemisorption	Heterogeneous catalysis in micropores
Driving force for the adsorption process	Reduction of free energy of pores	Reduction of free energy of pores	Both chemical reaction and reduction of free energy of pores	First reduction of free energy of pores, then the chemical reaction
Reversibility	Reversible	Problematic reversibility	Irreversible	Irreversible
Absolute value of heat effect (kJ/mol)	<20	<80	<400	No limit
Motion of adsorbed molecules	Easily possible	Constrained	Impossible	Constrained
Hysteresis of adsorption–desorption	Negligible	Moderate	Strong	Strong

When an adsorbate molecule approaches close to a solid surface enough to allow interpenetration of the electron clouds, the repulsion factor arises; its contribution can be represented semiempirically by the following expression:

$$E_R = \frac{\text{Const}_\text{R}}{\text{Dis}^{12}} \tag{4.10}$$

where E_R is the repulsion energy (positive value, of course!) and Const_R is a constant of repulsion. Thus, the total potential energy of van der Waals interactions comprises two terms: dipole–dipole attraction (if dipole–quadrupole and quadrupole–quadrupole interactions are neglected) and repulsion:

$$E_{\sum} = -\frac{\text{Const}_\text{DD}}{\text{Dis}^6} + \frac{\text{Const}_R}{\text{Dis}^{12}} \tag{4.11}$$

D. The Polanyi Theory of Adsorption Potential

Among numerous theories of pore–adsorbate interaction considered in Ref. 3, we notice first of all the Polanyi theory of adsorption potential. This theory became the methodological base for volume-filling micropore theory (VFMT) largely employed in the adsorption science.

According to Polanyi, a quantification of the adsorption potential is required. It is defined as the amount of work needed for moving a molecule from a pore to infinity. The adsorption space (pore volume) is assumed to be composed of many equipotential adsorption energy surfaces varying from a maximum in the finest pores to zero in the bulk solution. Hence, for a given adsorbent, the equipotential surface should be the same for all adsorbates. Consequently, a plot of adsorbate volume vs. adsorption potential should result in a single characteristic curve applicable to all adsorbates on that particular adsorbent. Thus, this theory suggests that an adsorption isotherm could be predicted for any compound or any adsorbent for which a characteristic adsorption curve has been obtained [2].

In addition to adsorption isotherms, the adsorption equilibrium relationships are the isostere and the isobar. The isostere is, in general, a plot of the ln P vs. $1/T$ at a constant amount of vapor adsorbed. The slope of the isostere corresponds to the heat of adsorption [2].

E. Classification of Adsorption Isotherms

The majority of physisorption isotherms may be grouped into six types. In most situations, at sufficiently low surface coverage, the isotherm reduces to a linear form often compared to Henry's law region. If adsorbent is heterogeneous, this linear region may fall below the lowest experimentally measurable pressure. The principal types of isotherms are presented on Figs. 4.1–4.6.

The reversible type 1 isotherm (Langmuir isotherm) is concave to the argument P/P_0 axis and approaches a limiting value as $P/P_0 \rightarrow 1$. For micropores, the curve has a very sharp rising at the low pressure. Isotherms of such type are usual for homogeneous microporous solids, external surfaces of which are negligible (e.g., active carbon or molecular sieves). For such adsorbents, the adsorption process is determined mostly by the microporous substructure, and other factors can be neglected [2].

Langmuir isotherm is the most important for adsorption science but not alone. Among other curve types, very important is multilayer adsorption isotherm curve (Fig. 4.2).

The reversible type 2 isotherm is the normal form of isotherm obtained for nonporous or microporous adsorbent and represents unrestricted mono-multilayer adsorption. The beginning of the quasilinear middle section of the isotherm corresponds the monolayer completed and the multilayer adsorption beginning [2].

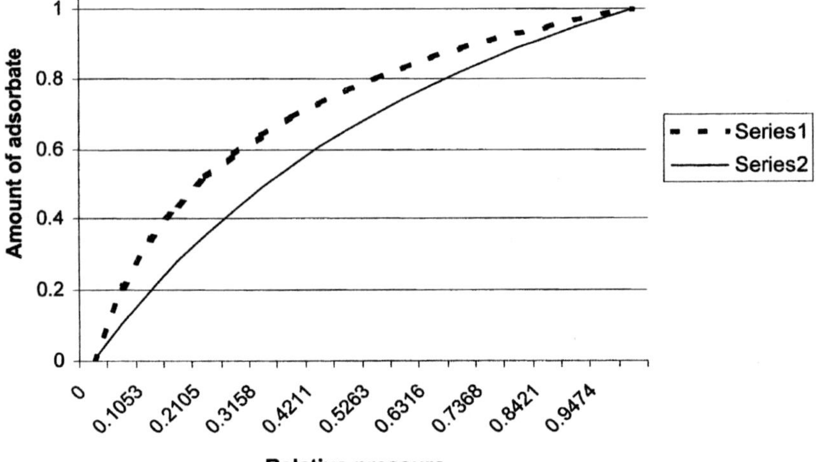

FIG. 4.1 Adsorption isotherm type 1 (Langmuir) for gas on homogeneous adsorbent. Series 1: microporous adsorbent; Series 2: external surface.

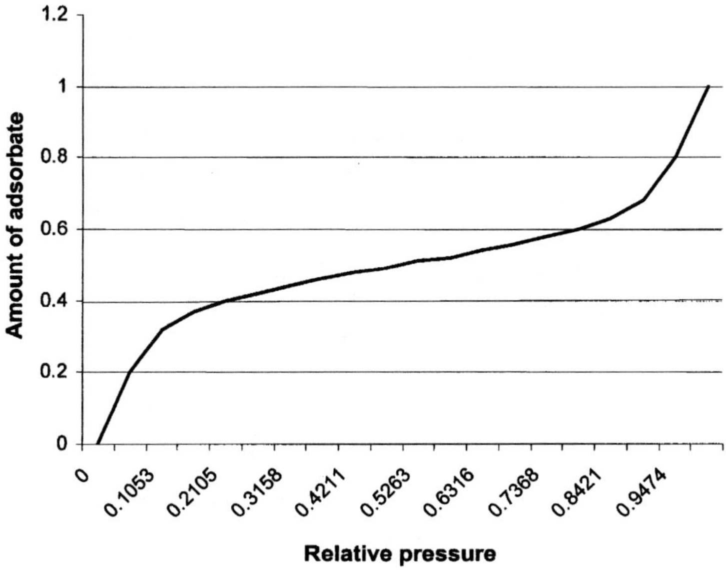

FIG. 4.2 Adsorption isotherm type 2 (BET) on homogeneous adsorbent.

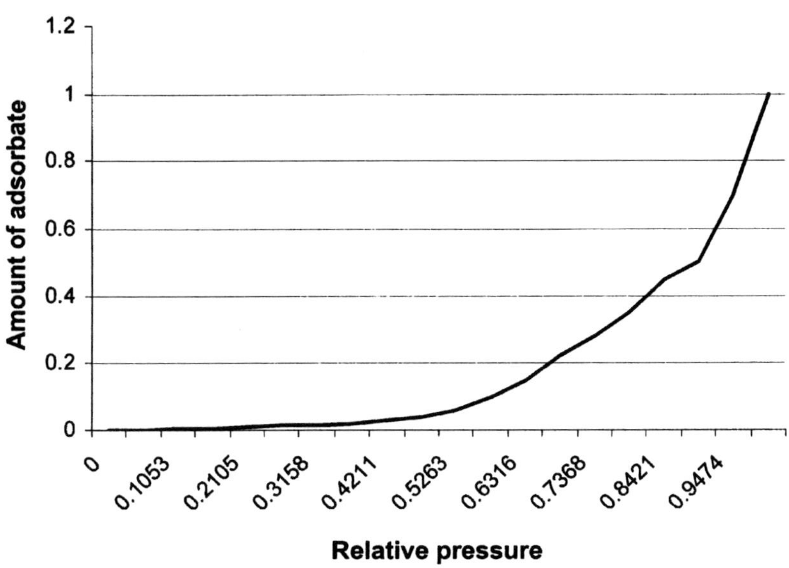

FIG. 4.3 Type 3 of adsorption isotherm curves.

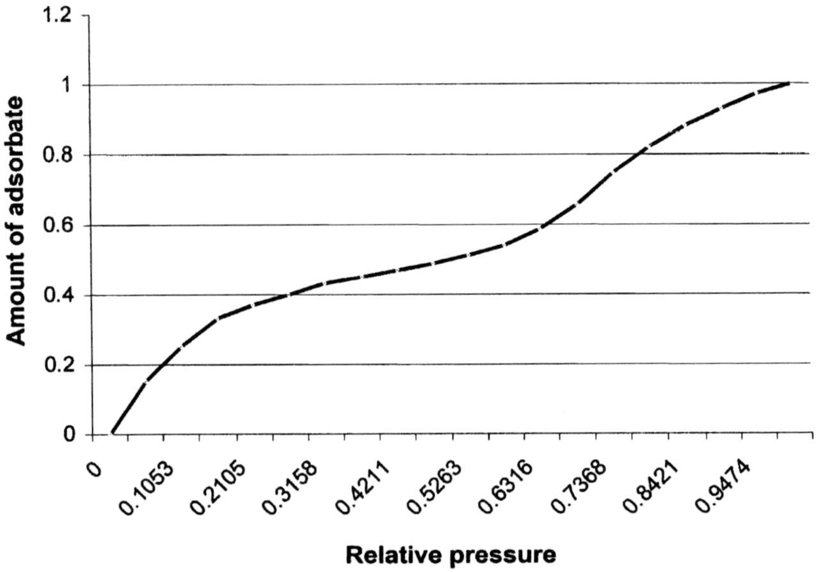

FIG. 4.4 Adsorption isotherm curve–type 4.

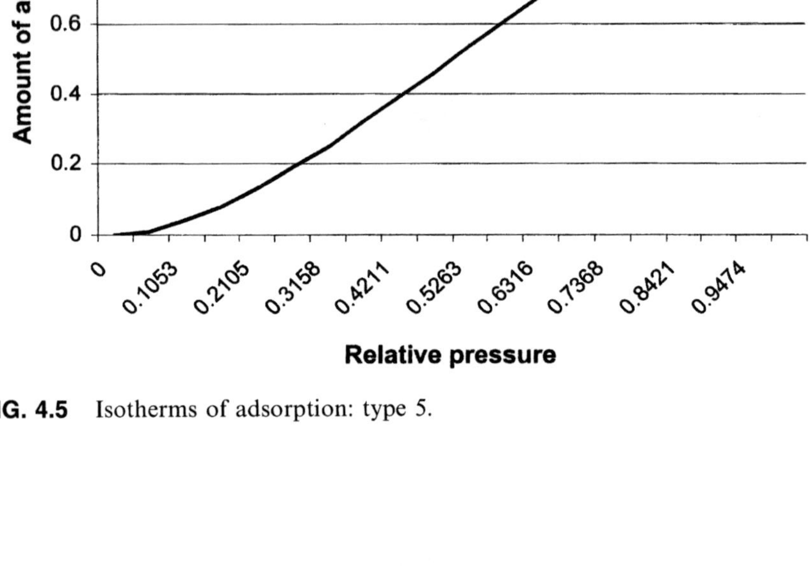

FIG. 4.5 Isotherms of adsorption: type 5.

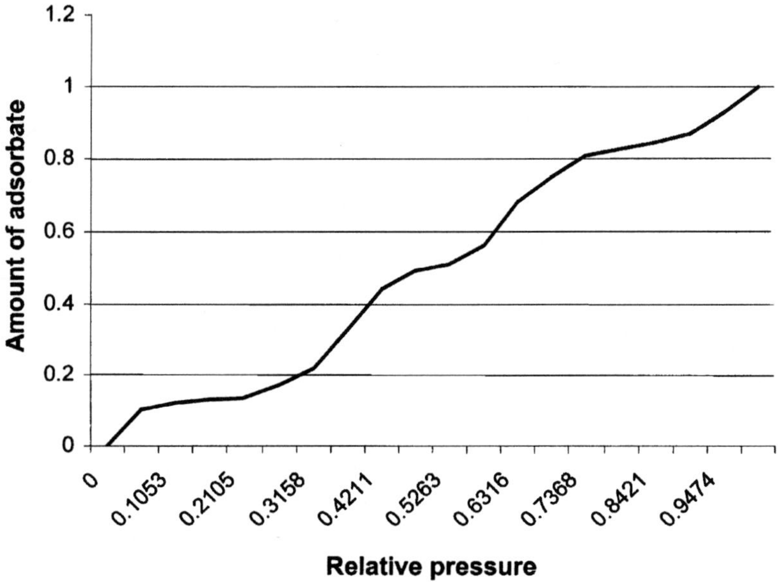

FIG. 4.6 Type 6 of adsorption isotherms.

Types 1 and 2 are found in condition that the adsorbate is compatible with the material of pore walls. Otherwise, under low pressures the amount of adsorbate is low (see Fig. 4.3).

The reversible type 3 isotherm is typical for such a specific situation we considered in Chapters 1 and 2: adsorption of vapors (e.g., water vapor) on hydrophobic adsorbents like active carbon. When $P/P_0 \rightarrow 1$, the vapor condensation in the micropores starts.

In the case when the adsorbate is compatible with the pore wall material, another form is obtained (see Fig. 4.4).

According to Ref. 2, the type 4 isotherms are irreversible and correspond to hysteresis loops described in Chapter 2. The form of these isotherms is usually attributed to capillary condensation in the mesopores. The limiting uptake for such isotherms is $P/P_0 \gg 1$. Other possible explanations for such effects include destruction of some pores because of rising pressure and, as result, increase of the internal volume or chemisortion-related effects. This form of adsorption isotherms is obtained for many of mesoporous industrial adsorbents, comprising silica and alumina gels adsorbing water [2–4].

The form of adsorbent isotherm curve for weak interactions adsorbent–adsorbate is given on Fig. 4.5.

According to Ref. 2, the type 5 isotherm is related to the type 3 isotherm in a specific situation when the adsorbent–adsorbate interaction is weak [2].

And the last example in this series: multilayer adsorption in micropores (see Fig. 4.6).

The type 6 isotherm represents stepwise multilayer adsorption on a uniform nonporous surface. The step height represents the monolayer capacity for each adsorbed layer and, in the simplest case, remains approximately constant for two or more adsorbed layers. Such curves are obtained with some noble gases on graphitized carbon blacks under temperatures of liquid nitrogen [2].

F. Models of Adsorption in Equilibrium

1. Henry's Law

The simplest adsorption isotherm, in which the amount adsorbed varies directly with the gas pressure, is very similar to Henry's law:

$$Q_{\rm ads} = K_{\rm ads} C_e \tag{4.12}$$

where $Q_{\rm ads}$ is the amount of the adsorbate, $K_{\rm ads}$ is the isotherm constant, and C_e is the equilibrium concentration in the fluid volume. Isotherms by Eq. (4.12) are typical for situations when the amount of the fluid over the adsorbent is much less than the adsorptive capacity of the adsorbent and the adsorbed layer is extremely diluted [2,5].

2. Freundlich Isotherm

The equilibrium relationships in adsorption are sometimes described by a Freundlich relationship. This is valid for adsorptive processes with no change in the configuration of the molecules in the adsorbed state. Freundlich isotherm has the following form:

$$Q_{\rm ads} = K_{\rm ads}(C_e)^{1/n_{fr}} \tag{4.13}$$

where n_{fr} is an empiric constant depending on the nature of the adsorbent and the adsorbate. Freundlich isotherm seems valid for adsorption on heterogeneous surfaces described by the discrete model (see Chapter 3), each class of cells obeying the Langmuir equation (see below). According to Freundlich isotherm, the amount adsorbed increases infinitely with the increase of the concentration or the partial pressure of the fluid to adsorb. Of course, such isotherm is not sufficient when coverage becomes enough high. At low concentrations, Freundlich isotherm does not reduce to the linear isotherm. Thus, Freundlich isotherm represents experimental results for moderate concentrations of the fluid to adsorb [2].

3. Langmuir Isotherm

This is applicable first of all to macroporous adsorbents, also to some microporous materials. This isotherm is applicable to both physisorption and chemisorption and derived under the following assumptions:

The adsorbed gas behaves ideally in the gas phase.
The adsorbate forms a monomolecular layer only.
The adsorptive surface is homogeneous.
Interactions between adsorbate molecules are negligible.
The motion of adsorbate molecules around the surface is impossible or negligible.

In accordance to the above assumptions, the equilibrium between the gas and the solid phases is described by the following equation [compare to Eq. (4.1)]:

$$\mathrm{Ads} + \mathrm{G} \Longleftrightarrow \mathrm{Ads} \cdot \mathrm{G} \tag{4.14}$$

Equation (4.14) leads to the condition of equilibrium:

$$K_{\mathrm{ads}} = \frac{C_{\mathrm{AdsG}}}{C_{\mathrm{Ads}} C_{\mathrm{G}}} = \frac{Q_{\mathrm{AdsG}}}{Q_{\mathrm{Ads}} C_{\mathrm{G}}} \tag{4.15}$$

where C_i is the concentration of ith component in equilibrium and Q_i is the amount of ith component in the solid phase.

The condition of the balance of the adsorbate is given by the following equation:

$$Q_{\mathrm{Ads}} + Q_{\mathrm{AdsG}} = Q_{\mathrm{Ads,\,0}} \tag{4.16}$$

where $Q_{\mathrm{Ads,0}}$ is the initial amount of the adsorbate. Equations (4.15) and (4.16) bring

$$Q_{\mathrm{AdsG}} = \frac{C_{\mathrm{G}} K_{\mathrm{ads}} Q_{\mathrm{Ads,\,0}}}{K_{\mathrm{ads}}\, C_{\mathrm{G}} + 1} \tag{4.17a}$$

Or

$$\theta_{\mathrm{AdsG}} = \frac{C_{\mathrm{G}}\, K_{\mathrm{ads}}}{K_{\mathrm{ads}}\, C_{\mathrm{G}} + 1} \tag{4.17b}$$

where $\theta_{AdsG} = Q_{AdsG}/Q_{Ads,0}$ and the graphical representation of Eq. (4.17a) is exactly Fig. 4.1. When $C_G \to \infty$, $Q_{AdsG} \to Q_{Ads,0}$. For low values of C_G, $Q_{AdsG} \approx C_G K_{ads} Q_{Ads,0}$ (Henry's law).

Langmuir isotherm is basic for the science of adsorption.

4. BET Equation

In 1938, Brunauer, Emmett, and Teller showed how to extend the Langmuir's approach to the multilayer adsorption, that is BET equation. The basic assumptions for BET are as follows:

Each molecule in the first adsorbed layer is considered to provide one site for the second and subsequent layers.

The molecules in the layers behave essentially as saturated liquid, and interactions between different layers do not depend on the interactions between the down layer and the adsorbent.

The above assumptions allow us to write the equations of reactions of multilayer formation:

$$\text{Ads} + \text{G} \Longleftrightarrow \text{Ads} \cdot \text{G} \tag{4.18a}$$

$$\text{Ads} \cdot \text{G} + \text{G} \Longleftrightarrow \text{Ads} \cdot \text{G}_2 \tag{4.18b}$$

$$\text{Ads} \cdot \text{G}_\text{k} + \text{G} \Longleftrightarrow \text{Ads} \cdot \text{G}_{k+1} \tag{4.18c}$$

Equations (4.18) correspond to the following equations for the condition of equilibrium:

$$K_{ads'} = \frac{C_{AdsG}}{C_{Ads} C_G} = \frac{Q_{AdsG}}{Q_{Ads} C_G} \tag{4.19a}$$

$$K_{ads''} = \frac{C_{AdsG2}}{C_{AdsG} C_G} = \frac{Q_{AdsG2}}{Q_{AdsG2} C_G} \tag{4.19b}$$

$$K_{ads} = \frac{C_{AdsG(k+1)}}{C_{AdsGk} C_G} = \frac{Q_{AdsG(k+1)}}{Q_{AdsGk} C_G} \tag{4.19c}$$

As for Langmuir isotherm, Eqs. (4.19) with the balance condition (4.16) are transformed to

$$\frac{Q_{AdsG} = C_G K_{ads} Q_{Ads,0} / K_{ads} C_G + 1 - C_G}{(1 - C_G)} \tag{4.20}$$

or

$$Q_{\mathrm{AdsG}} = (P/P_s)K_{\mathrm{ads}}Q_{\mathrm{Ads},0}/(K_{\mathrm{ads}}P/P_s + 1 - P/P_s)/(1 - P/P_s) \qquad (4.21)$$

where the physical sense of all terms is the same as in Eq. (4.17).

Let us notify that BET for low pressures is reduced to a Langmuir-like form.

G. Models of the Structure of Adsorbate

On the internal pore surface, an adsorbed molecule may hop along the surface when it attains sufficient activation energy and an adjacent adsorption site becomes available. Describing the porosity reduction in random structures, P. B. Visshher and J. E. Cates suggested a model for computing the motion of a solid–liquid interface by means of Lagrangian grid with constant-stress periodic boundary conditions [6].

H. Volume-Filling Micropore Theory

For micropores, derivations of adsorption isotherms need a comprehension about the internal structure of the adsorbent. The most widespread concept regarding this problem is volume-filling micropore theory [VFMT; sometimes also called *theory of volume-filling micropore* (TVFM)]. VFMT is based on the assumption that the adsorptive capacity of micropores is limited not by their surface but by their volume. As we noted in the previous chapters, this assumption is based on the recognized experimental data on properties of micropores. In this aspect, VFMT seems correct in principle. The adsorption is considered as the result of the adsorbate molecule entering a pore between the walls, as described according to the Polanyi concept above, regarding energetics of inside the empty space in solids. The accomplished work is proportional to $R_gT(\ln(P/P_s))$. A convenient starting point is the method of integral transforms, relating an overall isotherm θ_t to a local isotherm θ_l and the pore energy distribution [7]:

$$\theta_t(T, P) = \int_0^{\infty} \theta_l(T, P, \varepsilon)\chi_c(\varepsilon)\, d\varepsilon \qquad (4.22)$$

where $\chi_c(\varepsilon)$ gives a distribution of adsorption energies. If a variational technique called the *condensation approximation* (CA) is applied to the local isotherm, Eq. (4.22) is reduced to [7]

$$\theta_t(T, P) = \int_0^{\infty} \chi_c(\varepsilon)\, d\varepsilon \Longleftrightarrow \chi_c(\varepsilon) = -\frac{\partial\theta_t}{\partial\varepsilon} \qquad (4.23)$$

If θ_l is the Langmuir isotherm and gaussian pore-size distribution assumed, an overall isotherm is of the Dubinin–Radushkevich–Kaganer type [7]:

$$\theta_\Sigma = \mathrm{Exp}\left[-\mathrm{B_a}\frac{\mathrm{T}\ln(P/P_0)}{\beta_a}\right]^2 \tag{4.24}$$

$$\varepsilon = \varepsilon_0 - R_g T \ln(P/P_0) \tag{4.25}$$

where B_a is a semiempirical parameter depending on the adsorbent and β_a is the affinity coefficient according to CA.

Figure 4.7 below presents different profiles of adsorption isotherm curves obtained for different values of $b_a = B_a/\beta_a^2$.

As follows from Fig. 4.7, the choice of b_a is very important for the form of the resulting curve: Series 3 absolutely differs from Series 1! Let us also note that the resulting curve (especially its beginning at low pressures) does not always correspond to the Langmuir-like isotherm given by Fig. 4.1.

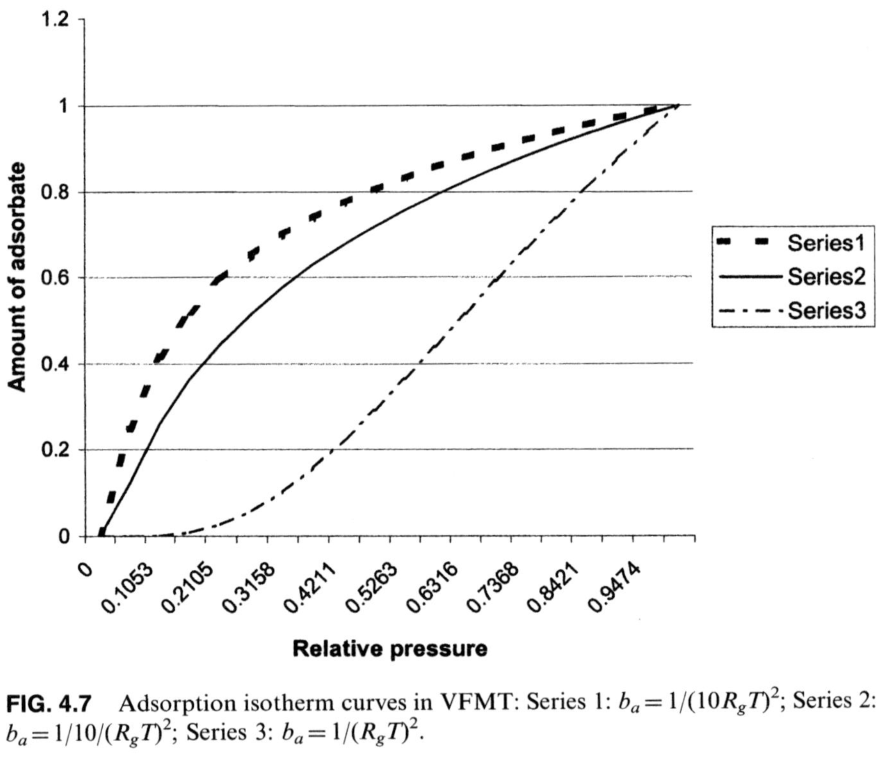

FIG. 4.7 Adsorption isotherm curves in VFMT: Series 1: $b_a = 1/(10R_gT)^2$; Series 2: $b_a = 1/10/(R_gT)^2$; Series 3: $b_a = 1/(R_gT)^2$.

Aiming to generalize VFMT, Dubinin and Astakhov replaced the gaussian distribution by "quasigaussian" and put forward a more general equation [8]:

$$\theta_\Sigma = \exp\left\{-B_a\left[\frac{T\ln(P/P_0)}{\beta_a}\right]^{n_e}\right\} \tag{4.26}$$

where n_e is an empirical parameter (in quasi–gaussian form, n_e is instead 2). For active carbons, n_e changes from 1 to 3 [8].

Aiming to apply VFMT to heterogeneous adsorbents, Stoeckli proposed a model taking into account their heterogeneity. Stoeckli assumed that a real adsorbent contains elements that can be separately described by the general equations of VFMT, but the semiempirical parameters change, depending on the adsorbent material. Stoeckli presented the solution for a heterogeneous material as the following integral [9–11]:

$$\theta_\Sigma = \int_0^\infty \theta_h(B_a)\exp(-B_a Y_\beta)\,dB_a \tag{4.27a}$$

where $\theta_h(B_a)$ is the local homogeneous solution for parameter B_a and

$$Y_\beta = \left(\frac{T\ \ln(P/P_0)}{\beta_a}\right)^2 \tag{4.27 b}$$

For particular solutions, θ_h is found from Eq. (4.22). Authors of VFMT pretend that Eq. (4.24) is applicable to all the possible homogeneous adsorbents.

McEnaney used VFMT for the estimation of sizes of micropores, based on the assumption that the decrease in the characteristic energy ε_0 in Eq. (4.25) is correlated with increases in micropore size determined by molecular probes Q_m, using an empirical equation [12]:

$$Q_m = 4.691\exp(-0.0666\varepsilon_0) \tag{4.28}$$

Stoeckli and Houriet showed that the parameters B_a and Q_{Ads0} change considerably for different active carbons [13].

VFMT was applied to description of the adsorption of vapors onto microporous adsorbents with both homogeneous and heterogeneous structures. One can take into account the association of the adsorbate molecules inside the pores for ordinary surfaces. The theoretical affinity coefficient β_a can be calculated by different methods, comprising molar volume, molar parachor or electronic polarization [14].

It was shown that taking into account the factor of heterogeneity is indispensable in the case of strongly activated carbons [14].

VFMT was used also for the solution of kinetic problems, such as diffusivity-limited adsorption on micropores [14]. VFMT allowed researchers to find that the micropores could be treated as slots between the graphitic planes of the microcrystallites [14].

The relation between Dubinin–Radushkevich parameters and thermodynamic functions, especially the enthalpy of immersion, also was studied [14]. It was shown that the correlation exists at the limits of the estimated experimental error.

An attempt to relate the Dubinin–Radushkevich equation to BET was carried out by Tsunoda [15–17]: the author suggested that the pre-exponential coefficient B_a in Eq. (4.26) can be related with the BET parameter K_{ads} in a linear form:

$$B_a = -a_d \ln K_{\text{ads}} + b_d \qquad (4.29)$$

where a_d and b_d are constant coefficients. Tsunoda suggested Eq. (4.29) for active carbons but did not mention what is the average dispersion of Eq. (4.29), particularly in comparison with the experimental error. The principal error of his suggestions seems to consist in establishing a formal correlation between BET (derived for surfaces) and Dubinin–Radushkevich equation designed for micropores. Such suggestion seems too doubtful without a serious theoretical research.

I. Analysis of Assumptions in VFMT

Now, following the previous review of VFMT, let us analyze the assumptions accepted in VFMT.

1. The main assumption in VFMT is the suggestion of adsorbate molecules reacting with pore volume—not pore wall, not pore surface, but the pore volume. This assumption correlates with all known experimental data about micropores, and there is no reason to make it in doubt.
2. The second assumption in VFMT is the condensation approximation given by Eq. (4.25). This assumption seems very problematic, because it is not clear how one takes into account the specificity of interactions between adsorbents and different gases and/or vapors. A pore, energy of which is found sufficient to adsorb fluids, is assumed to be filled entirely, without any intermediate

state. What about fluids having different compatibility with the pore wall material? This question stays without answer.

3. Another very problematic assumption in VFMT is suggestions about structural properties of adsorbents. These suggestions are not based on any analysis of structural characteristics determined by synthesis process, just some suppositions, and nothing more. Why gauss distribution of pores in size? Why the same distribution but in energy?
4. The fourth assumption in VFMT is more specific: Stoeckli's suggestion of Gibbs-like energy distribution of structural fragments in heterogeneous adsorbent. Though this assumption seems more logical and correlating with some of results obtained in Chapter 3, that contradicts the second assumption in VFMT (gauss distribution of micropores in size or energy). Though Stoeckli's suggestion seems more particular, that is very important, because most of industrial adsorbents are heterogeneous.

Thus, we have mentioned four principal assumptions in VFMT, among which three seem very doubtful. Is it possible to improve the logical structure of VFMT? This question is analyzed below.

J. Theory of Reversible Adsorption of Simple Gases by Micropore Volume (Langmuir-like Model)

Let us consider a very slow process of gas adsorption by a homogeneous microporous adsorbent, structural properties of which are determined by its preparation conditions and estimated by methods described in Chapter 3. Let us make the following assumptions about the pore–adsorbate interactions:

The process of micropore–gas interaction is close to equilibrium.
The adsorbed gas behaves ideally in the gas phase.
The adsorption in micropore is limited by pore volume only.
Interactions between adsorbate molecules are negligible.

These assumptions are very similar to those formulated for the derivation of Langmuir's isotherm.

In accordance to the above assumptions, the equilibrium between the gas G and the pore having the initial energy ε per unit of internal substance and containing adsorbate is described by the following equation [compare to Eq. (4.14)]:

$$\text{Pore}(\varepsilon) + \text{G} \longleftrightarrow \text{Pore}(\varepsilon) \cdot G \qquad (4.30)$$

Equation (4.30) leads to the following condition of equilibrium:

$$K_{\mathrm{ads}}(\varepsilon) = \frac{C_{\mathrm{Pore}}(\varepsilon)_{\mathrm{G}}}{C_{\mathrm{Pore}}(\varepsilon)P_{\mathrm{G}}} \tag{4.31}$$

where P_{G} is the partial pressure of G out pores, $C_{\mathrm{Pore}}(\varepsilon)$ is the volume part of the pore still able to adsorb gas G, and $C_{\mathrm{Pore}}(\varepsilon)_{\mathrm{G}}$ is the volume part of the pore already occupied by the gas. The values $C_{\mathrm{Pore}}(\varepsilon)_{\mathrm{G}}$ and $C_{\mathrm{Pore}}(\varepsilon)$ can be, in principle, found from the following equations:

$$C_{\mathrm{Pore}}(\varepsilon) = 1 - \frac{\theta_{\mathrm{G,c}}(\varepsilon)}{\theta_{\mathrm{G,max}}(\varepsilon)} \tag{4.32}$$

$$C_{\mathrm{Pore}}(\varepsilon)\mathrm{G} = \frac{\theta_{\mathrm{G,c}}(\varepsilon)}{\theta_{\mathrm{G,max}}(\varepsilon)} \tag{4.33}$$

where $\theta_{\mathrm{G,max}}(\varepsilon)$ is the maximal capacity of the micropore to adsorb the gas and $\theta_{\mathrm{G,c}}(\varepsilon)$ is the current amount of G in the micropore. Since the adsorption process reduces the free energy of pores, the value of $K_{\mathrm{ads}}(\varepsilon)$ increases exponentially with ε:

$$K_{\mathrm{ads}}(\varepsilon) = K_{\mathrm{ads}}(\varepsilon_0)\exp\frac{\gamma_{\mathrm{p}}\varepsilon}{R_g T} \tag{4.34}$$

where $K_{\mathrm{ads}}(\varepsilon_0)$ and γ_p are semiempirical parameters depending on both pore wall material and the adsorbate. Hence, the amount of the adsorbate in a micropore characterized by energy ε is found from Eq. (4.31):

$$\theta_{\mathrm{Pore}}(\varepsilon)_{\mathrm{G}} = \frac{K_{\mathrm{ads}}(\varepsilon)P_{\mathrm{G}}}{1 + K_{\mathrm{ads}}(\varepsilon)P_{\mathrm{G}}} \tag{4.35}$$

and the resulting equation for the adsorption isotherm is given by the following equation:

$$\theta_{\Sigma}(P_G) = \varsigma P_G \int_{\varepsilon_{\min}}^{\varepsilon_{\max}} \frac{f(\varepsilon)K_{\mathrm{ads}}(\varepsilon)}{1 + K_{\mathrm{ads}}(\varepsilon)P_G}\, d\varepsilon \tag{4.36}$$

where the sense of parameters ξ, $f(\varepsilon)$, and Q_0 is the same as in Chapter 3.

Now, let us analyze Eqs. (4.34)–(4.36).

1. Equations (4.34)–(4.36) contains two semiempirical parameters ($K_{\mathrm{ads}}(\varepsilon_0)$ and γ_p) depending not only on the adsorbent but also the adsorbate and some parameters ($\varepsilon_{\min}$, $\varepsilon_{\max}$, etc.) related to the

adsorbent preparation conditions. Since preparation-related parameters can be evaluated from the synthesis conditions, as it was described in Chapter 3, Eqs. (4.34)–(4.36) should be considered as two-parametric (containing two fitted parameters).

2. When $P_G \to 0$, obviously $\theta_\Sigma(P_G) \to 0$.
3. When $P_G \to \infty$, $\theta_\Sigma(P_G) \to \mathrm{CONST} \sim V_{\mathrm{por}}$ (V_{por} is the total volume of micropores in the treated sample).

Figure 4.8 presents adsorption isotherm curves obtained on the base of eqs. (4.34)–(4.36) for different values of γ_p.

As follows from Fig. 4.8, also Langmuir-like approach brings curves very sensitive to the value of the exponential term, but this does not influence the form of the entire curve (compare to Fig. 4.7). This result demonstrates one advantage of the proposed model: sensitivity to the exponential term is quantitative but not qualitative.

It is interesting to compare the adsorption isotherm curves obtained by the different methods. Figure 4.9 presents the curves obtained in accordance to Dubinin's VFMT (Eqs. (4.22)–(4.25)), the above-presented Langmuir-like approach [Eqs. (4.34)–(4.36)], and an intermediate curve, building of

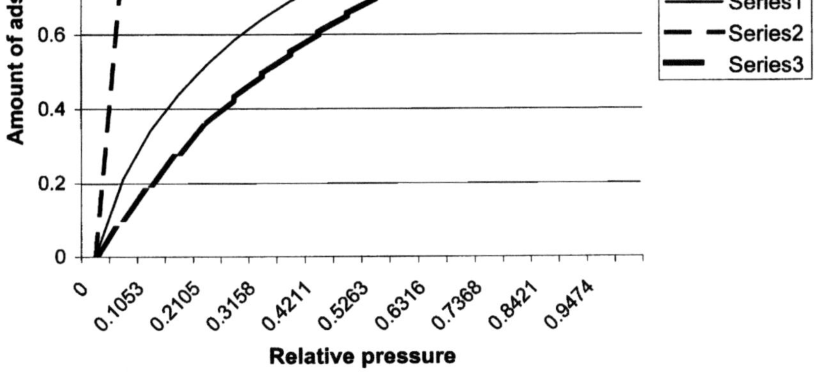

FIG. 4.8 Langmuir-like approach for micropores, adsorption isotherm curves: Series 1: $\gamma_p = 1$; Series 2: $\gamma_p = 10$; Series 3: $\gamma_p = 0.1$.

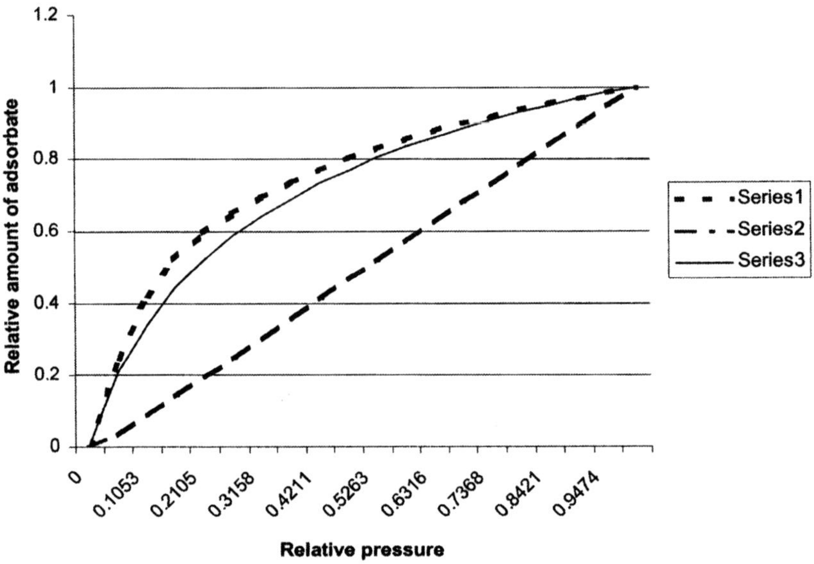

FIG. 4.9 Forms of Dubinin, Langmuir-like, and mixed (intermediary) adsorption isotherm curves: Series 1: Isotherm adsorption curve built according to VFMT; Series 2: Mixed (intermediary) curve built on the base of the combination of condensation approximation with Gibbs energy distribution; Series 3: Langmuir-like curve.

which combines condensation approximation from VFMT with Gibbs energy distribution.

As follows from Fig. 4.9, the intermediary solution (CA + Gibbs distribution) provides definitely wrong results. Both other curves are very similar. Hence, the main advantage of Langmuir-like building of the adsorption isotherm curve for micropores consists not in a "good" form but in the minimal number of empirical parameters to fit.

For homogeneous adsorbents, both models need two fitted parameters. However:

In VFMT, these are empirical parameters, physical sense of which is problematic and values of which can be found only from experiments, while in Langmuir-like model the fitted parameters are semiempirical and can be, in principle, found not only from experiments but also from quantum calculations.

For heterogeneous adsorbents, VFMT needs much more fitted parameters (from 4 to 6), they all are empirical, while the Langmuir-like model still needs two semiempirical parameters (because the heterogeneous situation also is modeled by methods of thermodynamics of nonequilibrium [18]).

Equations (4.34)–(4.36) were applied by the author to experimental isotherms of adsorption obtained on very different adsorbents: active carbons prepared in steady state [14], active carbons prepared in non-steady state [19], and silica and alumina gels [20]. Structural parameters of the adsorbents were estimated in accordance with the recommendations presented in Chapter 3. The comparison of the experimental data with the theoretical curves allowed the following conclusions:

The general form of the theoretical curve found by Eqs. (4.34)–(4.36) is always quasi-Langmuir: very similar to the traditional Langmuir isotherm, but the rising at $P_G \rightarrow 0$ is much more sharp than in the traditional case.

The error (evaluated by least square method) was always very short — on the level of the experimental error.

It was very interesting to notice that the increase of the number of experimental points sometimes reduced the calculated error of theoretical curves, which might be interpreted as an improvement of properties of the theoretical curve with the improvement of the experimental data.

K. Desorption in Equilibrium

Since desorption is the process of reversing adsorption, the hypothetical desorption in equilibrium should just repeat the paths of adsorption isotherm curves without any changes (but in the inverse direction). Although, in real practice, as it was noted in Chapter 2, it is absolutely impossible to perform reversible adsorption–desorption on microporous adsorbents.

L. Thermoporometry

Thermoporometry is one of experimental methods for studies of microporosity. This method, the basis of which consists in the measurement of the vapor pressure and/or amounts of water vapor delivered from micropores under various temperatures, was analyzed in Chapter 2. Now, we can suggest a theoretical base for this method.

Since water evaporation from microporous adsorbents can be approximately considered as desoption close to equilibrium, Eqs. (4.34)–(4.36) are applicable. However, contrary to the adsorption isotherm case, thermoporometry uses the varied temperature.

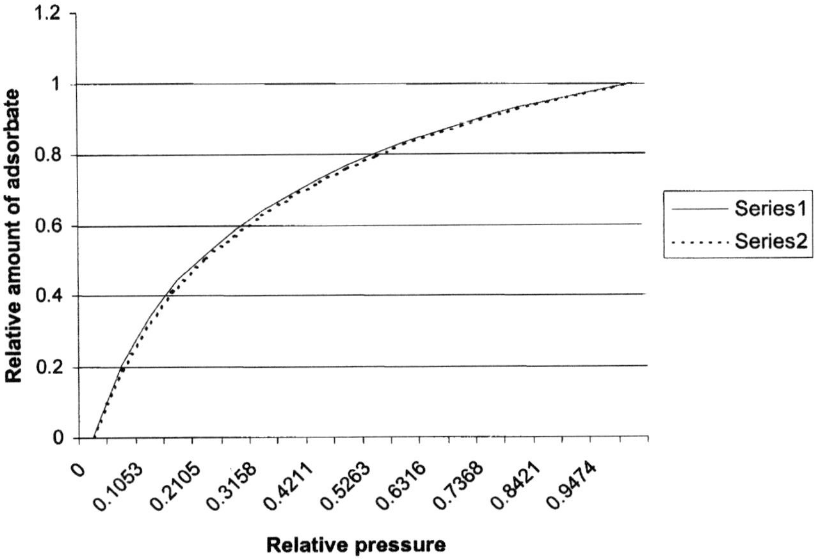

FIG. 4.10 Relative amount of adsorbate vs. relative pressure under varied temperatures. Series 1: Temperature 300 K; Series 2: Temperature 389 K.

Figure 4.10 presents thermoporometric curves for different temperatures: 300 K and 389 K. Since the total amount of adsorbate under the relative pressure = 1 depends much on the temperature, both curves were normalized (per the amount of adsorbate under $P/P_0 = 1$, that is, the relative amount of adsorbate) on purpose of detecting the sensitivity of the curve form to the temperature.

Table 4.2 illustrates the comparison between VFMT and Langmuir-like models.

M. Adsorption and Desorption in Nonequilibrium

All real processes of adsorption and desorption are irreversible and realized in nonequilibrium. The following factors cause nonequilibrium effects for adsorbent–fluid interactions [21]:

External diffusion: Limited velocity transfer of fluid molecules from the bulk fluid phase through a stagnant boundary layer surrounding each adsorbent particle to the external surface of the solid adsorbent – or back, in the case of desorption (limitation by fluid diffusion).

Internal diffusion: Limited velocity transfer of adsorbate molecules inside the adsorbent structure from fragments rich in the adsorbate

TABLE 4.2 Comparison Between VFMT and Langmuir-like Models

Criterion for comparison	VFMT	Langmuir-like
Suggested mechanism for pore–adsorbate interaction	Condensation approximation	Local Langmuir in volume
Suggested structure of micropores	Gauss distribution in size or energy	Determined by conditions of micropore formation
Accounting re-structuring (heterogeneity)	Gibbs distribution	Determined by conditions of microporous medium formation
Minimal number of fitted parameters (for homogeneity)	Two	Two
Number of fitted parameters for heterogeneity	Six	Two
Physical sense of fitted parameters	Empirical	Semiempirical
Principal accordance with experimental curves	Close	Identical

to fragments poor in the adsorbate, by the migration of the adsorbate molecules from the relatively small external surface of the adsorbent to the surfaces of the pores within each particle and/or by the diffusion of the adsorbate molecules through the pores of the particles—or back, for desorption (limitation by adsorbate diffusion).

Destruction and other irreversible changes of the adsorbent structure.

Eventual closing of narrow micropores with molecules of the adsorbate (or products of chemisorptive or catalytic reactions).

The problem of limitation by adsorbate diffusion will be considered below, in the part regarding phenomena of percolation and permeability in microporous materials. Phenomena concerning changes of the solid structure will be considered in Chapter 5.

Thus, the only thing we will touch here is the limitation by fluid diffusion, regarding the appearance of hysteresis loop.

The problem of mass transfer in fluids is widely studied in physical chemistry [2,21].

Let us consider the following scheme of fluid-diffusion-limited adsorption (see Fig. 4.11).

The adsorbed component G (its concentration C_G in the gas phase) comes slowly from bulk fluid to the solid–fluid interface and then penetrates (very fast) into micropores and becomes adsorbate. The concentration of G

FIG. 4.11 Fluid-diffusion-limited adsorption–desorption.

at the interface is quasistationary ($C_{G,int}$) and in the bulk is constant ($C_{G,bulk}$). Of course, $C_{G,int}<C_{G,bulk}$. According to second Fick's law, such a quasi-steady-state situation is described by the following equation:

$$\Delta C_G = 0 \tag{4.37}$$

where ΔC_G means the sum of the second derivatives of C_G by the spatial coordinates. Assuming one-dimensional mass transfer, we simplify Eq. (4.37):

$$\frac{d^2 C_G}{dl^2} = 0 \tag{4.38}$$

where l is the distance from the adsorbent surface. The boundary conditions are $C_G(0) = C_{G,int}$, $C_G(L_b) = C_{G,bulk}$, L_b being the distance from the adsorbent surface to the zone of the fluid, where the local concentration is approximately equal $C_{G,bulk}$. The solution of Eq. (4.38) is linear:

$$C_G(l) = C_{G,int} + \frac{l(C_{G,bulk} - C_{G,int})}{L_b} \tag{4.39}$$

The entropy increase rate is proportional $(C_{G,bulk} - C_{G,int})^2$. The rate of the transfer is found from the following equation [2]:

$$J_G = \gamma_G A_{Ads} \omega_{Ads} \frac{C_{G,bulk} - C_{G,int}}{\rho_b} \tag{4.40}$$

where J_G is the flux of fluid particles toward the adsorbent surface, γ_G is the external film mass transfer coefficient, A_{Ads} is the external surface of particle of the fluid, ω_{Ads} is the voidage between adsorbent granules, and ρ_b is the bulk density of particle of the fluid [2]. The external mass transfer coefficient for the fixed-bed adsorber can be evaluated from empiric equations derived by Wakao and Funazkri, Petrovic, and Thodos [2].

Now, if the adsorption process is replaced by desorption in the same regime, the bulk concentration of G is $C^*_{G,bulk}$ (respectively, $C_{G,int} > C^*_{G,bulk}$), and the equation for $C_G(l)$ is

$$C_G(l) = C_{G,int} + \frac{l(C^*_{bulk} - C_{G,int})}{L_b} \tag{4.41}$$

The measured isotherm of desorption will exhibit hysteresis: the distance of the desorption isotherm from the adsorption isotherm will be $(C_{bulk} - C^*_{bulk})$. Hence, the surface area of the hysteresis loop in such situation brings us information about the kinetic characteristics of the process of adsorption–desorption limited by fluid diffusion.

As we noted above, the limitation by fluid diffusion is not alone but one of possible factors of hysteresis appearance. How can we distinguish it from other factors? If we intensify the motion of the bulk fluid (e.g., by a mixer) and then the hysteresis loop seems to become more narrow, it is the sure symptom that the process is limited by fluid diffusion.

II. PERCOLATION AND PERMEABILITY

A. Description of Percolation and Permeability

General and experimental aspects of percolation and permeability were considered in Chapter 2. Now, let us analyze theoretical aspects of these phenomena: first of all, existing theoretical models of percolation and permeability.

First of all, note that the physical nature of percolation and permeability in microporous materials is similar to adsorption–desorption. In both situation, some fluid interacts with pores located in a microporous medium, and the physics of related interactions can be given by Eqs. (4.8)–(4.11). On the contrary to adsorption, the fluid in a permeable medium may leave the solid structure, if that is not forbidden by structural factors. Hence, percolation can be considered as a simultaneous combination of adsorption and desorption: the fluid is adsorbed but immediately desorbed—and, due to the gravity factor, leaves the pores.

Of course, it is possible that some components of the fluid are not allowed to penetrate through the porous medium, and their behavior can be described as adsorption without further desorption.

B. Existing Models of Percolation and Permeability

All existing models of percolation–permeability are based on the idea of connectedness. That means that the empty (porous) substructure inside the solid is continuous from "entrance" to "exit," and its "thickness" from the entrance to the exit is always sufficient to allow the particles of the fluid to pass through the medium.

All existing models of percolation and permeability can be methodologically divided into three groups:

1. Models applicable to both percolation and permeability: theoretical tools allowing the solution of both problems—estimation of percolation threshold and quantitative evaluation of permeability over the percolation threshold
2. Models applicable only to percolation
3. Models applicable only to permeability

In principle, any restrictions in the applicability of a model are the symptom of its shortcomings. It is obvious that a well-built model of permeability allowing the evaluation of fluid fluxes through a porous sample should show also the conditions, under which these fluxes equal zero, which is the percolation threshold by definition. On the other hand, a model, describing a porous material below and exactly at the percolation threshold, should be able to do that over this one, bringing information about the permeability. Nevertheless, the above division of percolation–permeability models is the fact we need to take into account.

The following approaches and techniques are used in studies of percolation and permeability:

1. Darcy Equation

Darcy equation is a phenomenological approach in the description of permeability, based on Onsager equations (see Chapter 3). Darcy equation is usually written in the following form:

$$J_n = \frac{\sum_k^{N_f} g_{kn} X_k}{\eta_n} \tag{4.42}$$

where k and n label the phases and the fluid fluxes, I_n is linked to the pressure gradient (forces) X_k, g_{kn} are the permeability coefficients, and η_n is the ith fluid viscosity. As in Onsager equations, coefficients g_{kn} are characterized by reciprocity ($g_{nk} = g_{kn}$) investigated by Flekkoy and Pride [22].

The main shortcoming of the Darcy equation consists in the non-clear physical sense of the coefficients g_{kn}, which makes impossible evaluations of percolation threshold. If their relationship with microscopic properties of the porous material is studied, it is not problem to generalize this approach onto percolation (an example of such expansion is given below).

Another shortcoming of Darcy equation consists in its invalidity in situations when both the exterior pressure and the fluxes have short values. In such situation, the linear approximation is wrong, and the system is described by Chapman–Enskog equation considered below.

2. Chapman–Enskog Equation

In a specific case of a porous material interacting with little amounts of gas (so little that the interaction can be considered as physisorption), the rate of fluid current from the surface to the interior parts of the porous medium is generally controlled by transport within the pore network. Intraparticular transport is consider as a diffusive process, kinetic data of which correlate in terms of a diffusivity defined in accordance with Fick's first law of diffusion. Intraparticular diffusion may occur by several different mechanisms, comprising ordinary diffusion, molecular diffusion, Knudsen diffusion, or surface diffusion. The concrete mechanism depends on pore-size and pore-energy distribution, the fluid concentration inside the porous medium, etc. For a dense fluid inside macropores, the mechanism corresponds to the molecular diffusion. If the fluid is a gas phase with low density and/or the fluid is found in mesopores, the molecules collide with pore walls much more frequently than with each other, which is the mechanism of Knudsen diffusion. Molecules found of the surface may evidence a considerable mobility. The transport by the motion of molecules on a surface is known as *surface diffusion* [2]. The Chapman–Enskog equation for the effective molecular diffusivity in a gas–solid system is written in the following form:

$$D_m = \frac{[B_{ce} T^{3/2} (1/M_1 + 1/M_2)^{1/2}]}{(P a_{12}^2 \nu_{ce})} \tag{4.43a}$$

$$B_{ce} = 10^{-4}\left[10.85 - \frac{2.5}{(1/M_1 + 1/M_2)^{1/2}}\right] \tag{4.43b}$$

where D_m is the molecular diffusivity, M_1 and M_2 are the molecular masses, P is the total pressure, $a_{12} = (a_1 + a_2)/2$ is the collision diameter from the Lennard-Jones potential (A), ν_{ce} is a function of $[\varepsilon_{ce}/(k_b T)]$, and $\varepsilon_{ce} = (\varepsilon_1 \varepsilon_2)$ is a Lennard–Jones force constant.

As Darcy equation, Chapman–Enskog equation does not allow the estimation of percolation threshold. Methodologically, these equations complete one other: for short pressures Chapman–Enskog equation works, for moderate and high pressure Darcy equation is valid.

3. Monte Carlo Simulations

This approach is based on the assumption that the motion of particles in the system is random, and there is no relationship between the current state of the system and its prehistory [23]. Monte Carlo modeling of fluid motion means that each element of the fluid flow can eventually penetrate into each of available fragments of the solid medium with the probability determined by some assumptions taken by the researcher, and then the appearance of percolation path(s) is confirmed or denied. Monte Carlo can be employed also for modeling of microporous structure formation by a casual mechanism, this aspect is considered in Chapter 5. In many cases, Monte Carlo simulations are performed on computer and do not provide analytical solutions. In principle, Monte Carlo approach is applicable to both percolations and permeability problems, but the absence of analytical solution reduces its usefulness.

4. Infinite Cluster (Fractal) Approach

This approach is based on the fractal concept, some aspects of which were already considered in Chapters 1 and 3. A percolation cluster is assumed to consist of porous clusters described as fractals of finite size or infinite. The main idea of the infinite cluster approach is that percolation is due to the formation of an infinite connected porous structure. This model was found useful in characterizing many types of disordered systems (not only solids), e.g., gels, fractures, dispersed ionic conductors and mixed alkali conductors, forest fires, and epidemics [24]. The site percolation in the infinite percolation cluster model is defined by choosing each site of the lattice to be occupied with a probability Pr (or, respectively, to be found empty with probability $(1 - \mathrm{Pr})$). A similar definition exists for bond percolation. There exists a critical value of Pr, $\mathrm{Pr_c}$, below which only finite-size clusters exist and above which an infinite cluster is generated. For a triangular lattice, $\mathrm{Pr_c} = 1/2$. For a square lattice, it was found numerically that $\mathrm{Pr_c} = 0.592745$ [24]. The percolation transition at $\mathrm{Pr_c}$ is described by the probability $\mathrm{Pr_{inf}}$

that a site in the lattice belongs to the cluster. Below $\mathrm{Pr_c}$, $\mathrm{Pr_{inf}}=0$, whereas above $\mathrm{Pr_c}$, $\mathrm{Pr_{inf}}$ increases with Pr [24]:

$$\mathrm{Pr_{inf}} \sim (\mathrm{Pr} - \mathrm{Pr_c})^{b_c} \tag{4.44}$$

where b_c is a parameter.

The diameter of the clusters below $\mathrm{Pr_c}$ is characterized by the correlation length λ, which can be understood as the mean distance between two sites belonging to the same cluster. When $\mathrm{Pr} \to \mathrm{Pr_c}$, $\lambda \to (\mathrm{Pr_c} - \mathrm{Pr})^{-\nu_c}$. The parameters b_c and ν_c are universal and depend only on the fractal dimensionality d_f (this is a main characteristic of the dimensionality of fractals; this question will be considered in Chapter 5 in more details). The infinite cluster for $\mathrm{Pr} > \mathrm{Pr_c}$ can be modeled as a regular lattice of fractal unit cells. Each fractal unit cell is of linear size $\lambda(\mathrm{Pr})$ and is similar to the fractal lattices modeling percolation at $\mathrm{Pr}<\mathrm{Pr_c}$ [24]. Diffusion in a percolation structure can be generated, in the infinite cluster model, by performing a random walk which can step only on the occupied sites belonging to the clusters (note that this detail attributes to infinite cluster models some features of Monte Carlo). The boundary unoccupied sites serve as reflecting sites. For $\mathrm{Pr}<\mathrm{Pr_c}$, the random walk is constrained to stay on the finite clusters and therefore its mean square displacement will increase with time until it reaches a plateau value characteristic of the linear size of the cluster, on which the random walk started. For $\mathrm{Pr} > \mathrm{Pr_c}$, for a random walk that started on the infinite cluster, its mean square displacement will increase with time [24].

The infinite cluster approach is applicable to both percolation and permeability problems: the percolation threshold is found from the value of $\mathrm{Pr_c}$, and fluid current passing through the porous structure is related to $\mathrm{Pr_{inf}}$ in Eq. (4.44).

Let us observe some shortcomings of the infinite-size cluster approach:

As all fractal models, the infinite-size cluster approach does not take into account the specificity of pores. Are they ultramicropores, supermicropores, mesopores—the result will be the same. Even the difference between open, semiopen, and closed pores is not taken into account anyway, though the predomination of semiopen or closed pores will obviously restrict any percolation.

Since this model requires the infinite size for the percolation cluster, the resulting value of parameters of percolation threshold will be overestimated. For big samples this overestimation can be negligible, but for thin films Eq. (4.44) will provide absurd overestimation.

In the estimation of permeability, Eq. (4.44) provides the problem of taking into account the specificity of different fluids.

5. Pore-Size Distribution Approach

This approach is based on the obvious assumption that pore size determines the permeability of porous structure, hence, pore-size distribution (PSD) determines the total percolation–permeability properties of the system [25–30].

We can notify two principal shortcomings in PSD method:

As we wrote in Chapter 1, the sense of pore size is extremely unclear in the case of micropores; in any case, micropore size cannot be directly measured or calculated from experimental data.

Fluid–solid interactions in porous fractals are determined not only by the geometrical factor (pore size) but also the compatibility between the fluid and pore wall material.

6. "Pore-Core" Model

This approach is empirical. The main idea consists in the estimation of the relationship between different measurable parameters, e.g., percolation and permeability characteristics and Hg-intrusion results [31,32].

In principle, the relationship between all measurable parameters of porous structures is doubtless (see more details on this subject in Chapter 1), but the empirical representation of this relationship has all shortcomings of empirical methods. In principle, the researcher using empirical approach can never be sure what happens in the zone of properties not yet measured, cannot evaluate the errors of the approximation, etc.

7. Finite-Size Cluster Approach

The finite-cluster approach in modeling percolation and permeability is based on the assumption that experimental research deals not with infinite porous samples but some pieces having limited size (in the case of films this size can be very short) and corresponds to the situation given on Fig. 1.1 (see Chapter 1).

The finite-size cluster approach is based on the following assumptions [33]:

The primary sample of microporous material (not for tests!) is prepared in the form of a very big cube—so big that border effects are negligible. In such case, the sample may be considered as homogeneous system, and the Law of Large Numbers is applicable;

The sample for percolation–tortuosity tests (secondary, or tested sample) is a small cube (size L) eventually taken from the primary sample.

As follows from results presented in Chapter 3, micropore formation in the primary sample leads to the appearance of porous clusters of varied volume and surface, but all they have the same geometrical form as the primary sample. Hence, they are cubes, though their fractal dimensionalities differ. Each cluster is characterized by volume V and one of the following parameters: fractal dimensionality d_f, surface area A, or size D. Clusters are distributed in V and $d_f/A/D$, the distribution function being $\psi(V,d_f) = \psi(V,A) = \psi(V,D)$. The correlation between these parameters is given by [33]

$$V = D^{d_f}, \qquad A = 2d_f D^{(d_f - 1)} \tag{4.45}$$

Percolation effect is observed for the tested sample if two following conditions are completed:

The tested sample having size L is totally or partly taken from a microporous structure containing cluster with the size $D > L$.

The distance from the geometrical center of the tested sample to each point belonging to the border of the mentioned cluster is at least $L/2$.

According to the criteria proposed above, certain (nonzero) percolation in the tested sample is always possible, because all methods of microporous material preparation may eventually cause formation of big clusters that may eventually become zones from which the tested sample are taken. Moreover, percolation cluster does not need large volume; it is enough if its dimensionality is short (close 1 as possible). The probability of eventually taking the tested sample from a cluster with characteristics (V, D) is

$$P(V,D) = \left(1 - \frac{L}{D}\right)\psi(V,D) \tag{4.46}$$

For a heterogeneous structure, the total probability of appearance of percolation for the tested sample is given by [33]

$$P_\Sigma = \int_{d_{f,\,\min}}^{d_{f,\,\max}} \int_{V_{\min}}^{V_{\max}} Q_0\left[1 - \frac{L}{D(V,d_f)}\right]\psi(V,d_f)\,dV\,d(d_f) \tag{4.47a}$$

where Q_0 is the normalizing coefficient, $d_{f\min}$ and $d_{f\max}$, $V_{\min}$ and $V_{\max}$ are minimum and maximum values of fractal dimensionality and cluster volume, respectively. For a homogeneous structure, $d_{f,\min} = d_{f,\max}$. One may also assume $V_{\min} = L^{d_{f,\min}}$, $V_{\max} = \infty$ [33].

Let us note that, as follows from Eqs. (4.46) and (4.47a), depending on the conditions of preparation of microporous material determining the value of ψ, percolation can be sometimes gained in samples having low porosity.

The value of the normalizing coefficient Q_0 is found from the following condition:

$$\int_{d_{f,\min}}^{d_{f,\max}} \int_{V_{\min}}^{V_{\max}} Q_0 \psi(V, d_f)\, dV\, d(d_f) = 1 \tag{4.47b}$$

The value ψ is found from energy distribution of pores, as it is done in Chapter 3.

Let us consider clusters enclosed in several fragments of structure of the primary sample. These fragments contain particles of the continuous solid phase, which may be very near the enclosed clusters. In principle, inside such a fragment containing "exterior" cluster one may find some other fragments containing "interior" clusters of shorter size. We note that in such situations external clusters may eventually have less volume (because of shorter dimensionality) than internal ones.

The total values of the volume and the surface area of the porous phase are found from [33]

$$V_\Sigma = \int_{d_{f,\min}}^{d_{f,\max}} \int_{V_{\min}}^{V_{\max}} Q_0 V \psi(V, d_f)\, dV\, d(d_f) \tag{4.48a}$$

$$A_\Sigma = \int_{d_{f,\min}}^{d_{f,\max}} \int_{V_{\min}}^{V_{\max}} Q_0 A \psi(V, d_f)\, dV\, d(d_f) \tag{4.48b}$$

In the light of the considered approach, the sense of such notions as percolation threshold and permeability also changes. Percolation threshold should be considered as such minimum of probability of percolation [that is found from Eq. (4.47)] that below this threshold percolation is neglected, according to the practical interests of the researcher. Permeability is estimated from Eq. (4.47), while tortuosity is found from the equation [33]

$$\tau = \frac{D^3}{(VLP_\Sigma)} \tag{4.49}$$

Now, let us analyze shortcomings of the finite-size cluster model. In comparison with the infinite-size cluster model, the finite-size cluster model solves the problem of short size and can be, in principle, applied to films. Although, the problem of taking into account the specificity of micropores remains. Also the problem of different fluids (that should influence the value of P_Σ when this is interpreted as permeability) stays without solution.

8. Random Trajectory Model for Percolation (Slab Symmetry)

As the finite-size cluster method can be methodologically considered as a development of the infinite-size cluster approach, the random trajectory approach can be considered as a development of the finite-size cluster method. As in this one, the random trajectory method is based on the idea of the finite path needed for the percolation effect. However, the random trajectory method considers not fractals but pores themselves—moreover, the solid particles belonging to the continuous phase. In this sense, the random trajectory method should be considered as a discrete model of porous solid structure.

The random trajectory method considers the solid structure, in the simplest situation, as a slab having thickness L, with the microporous structure having porosity ξ and divided into cells: filled cells belonging to the continuous phase or empty cells (voids) belonging to the porous medium (that can be compared to Pore-Core). As it was shown in Chapter 3, filled cells are distributed in the number of empty neighbors, and this distribution can be found from differential equations containing the formation-related parameters like surface tension σ and temperature T.

Similar equations can be written for the analogous distribution of empty cells. Since interface and entropy effects have a symmetric form against both filled and empty cells, the number of voids having $(n_n - m)$ empty neighbors at porosity ξ_1 ($\xi_1 = V_2/V_\Sigma$, where V_Σ is the total volume of the system) is equal to the number of filled cells having m empty neighbors at porosity $\xi_2 = \xi = (1 - \xi_1)$. Hence, we obtain the corresponding distribution of voids $\theta_m(\xi)$, m being the number of empty neighbors; θ_m is equal to the fraction of voids having (each) m empty neighbors: $\theta_m(\xi) = \nu_{(n-m)}(1 - \xi)$. The change of the distribution of voids in the number of empty neighbors is given by the following equations:

$$W_r(m) = Q_0 m \exp\left[\frac{-\sigma(n_n - m)}{R_g T}\right] \tag{4.50}$$

$$d\theta_m(V_{\text{pore}}) = W_r(m)\,dV_{\text{pore}} + \frac{\psi\,dV_{\text{pore}}[\theta_{m-1}(n_n - m + 1) - \theta_m(n_n - m)]}{Z_\Sigma} \tag{4.51}$$

$$Z_\Sigma = \sum_{k=0}^{n_n-1}(n_n - k)\theta_k$$

$$\psi = \sum_{k=0}^{n_n-1}(n_n - k)\theta_k \frac{\exp[-\sigma(n_n - k)/(R_g T)]}{Y_\Sigma}$$

$$Y_\Sigma = \sum_{k=0}^{n_n-1}\theta_k \exp\left[\frac{-\sigma(n_n - k)}{(R_g T)}\right]$$

$$d\theta_0(V_{\text{pore}}) = \frac{dV_{\text{pore}}\psi\theta_0 n_n}{\sum_{k=0}^{n_n-1}(n_n - k)\theta_k} \tag{4.52}$$

$$d\theta_{nn}(V_{\text{pore}}) = -W_r\,dV_{\text{pore}} + \frac{dV_{\text{pore}}}{\sum_{k=0}^{n_n-1}(n_n - k)\theta_k} \tag{4.53}$$

We assume that the orientation of empty neighbors for voids is eventual, meaning all orientations being possible with the same probability, while the percolating flow prefers the orientation "down." If this one is impossible, the flow prefers a horizontal orientation, and only in the case where the only open orientation is "up," the flow gets up. If the flow has "choice" between different trajectories, it "prefers" the shorter way (shorter in comparison with alternative trajectories).

The motion of the percolating flow is only through the voids having at least two empty neighbors. Let us characterize each void by its level h, the distance from the upper size of the considered cube; for the upper size, $h = 0$, and for the bottom, $h = L$. For neighbor cells on the same vertical, h differs by 1.

There are three principal options for the motion of the flow from a void located on level h into a void on a close level h' [34]:

1. $h' = h + 1$, the flow gets down, and its further motion has two options only: down or horizontally (obviously, because the upper orientation is already closed by the incoming flow). The option down is obtained if the considered void on level h' has an empty neighbor on level $h'' = h' + 1$, this probability is equal to $(m - 1)/(n_n - 1)$, where n_n is the coordination number (for a three-dimensional cube, $n_n = 6$, for a two-dimensional cube, $n_n = 4$).

Respectively, the probability of the horizontal motion is $(n_n - m)/(n_n - 1)$.

2. $h' = h$, the flow gets horizontal trajectory, and its further motion has three options:
   a. Down, with the probability $(m-1)/(n_n-1)$.
   b. Horizontally, with the probability $(n_n - m)/(n_n - 1)$ in condition that $m > 2$.
   c. Up, only if $m = 2$ and the orientation down is closed.
3. $h' = h - 1$, the flow gets up, and its further motion has two options only: up or horizontally. The option up is obtained only if $m = 2$ and the only open orientation is up; otherwise, that is a horizontal motion.

Now, let us consider a percolation way comprising l_+ acts of motion down, l acts of motion up (certainly, $l_+ - l_- = \mathrm{L}$), and l_0 acts of horizontal motion. Some examples of percolation paths are given on Fig. 4.12.

Let us note that the sequence of the acts of motion up, down, and horizontally has no importance. The total length of such percolation way is $l^* = l_+ + l_- + l_0$, and the percolation system gets two factors of freedom: l_0 and l^*.

The probability of an act of motion down is [34]

$$p_+ = \varsigma \sum_{m=2}^{n_n} \theta_m \frac{m-1}{n_n - 1} \tag{4.54}$$

The total probability of the walk l_+ down is $(p_+)^{l+}$.

The probability of each act of motion up is

$$p = \frac{\xi\theta_2}{n_n - 1} \tag{4.55}$$

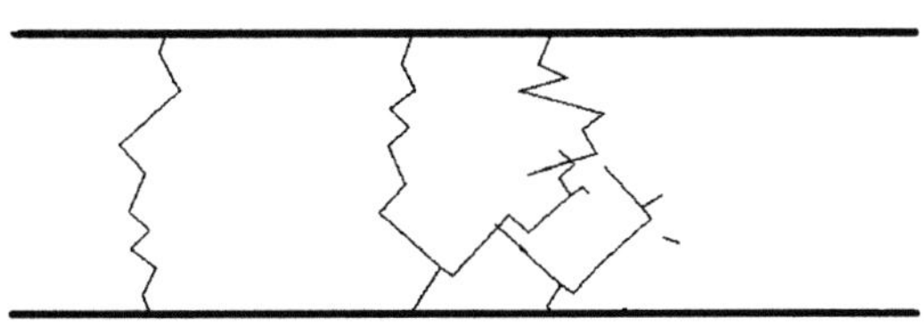

FIG. 4.12 Variants of percolation paths.

The coefficient $1/(n_n - 1)$ is written in Eq. (4.55), because only the $1/(n_n - 1)$-th part of the situation $m = 2$ allows the orientation up. The total probability of the walk l_- down is (p_-).

The geometrically averaged probability of each act of horizontal motion is [34]

$$p_0 = \frac{\xi(l_+ - l)\, S_1 + (l_0 - l)(S_2 + S_3) + (l - S_4)}{(l^* - 2)} \qquad (4.56)$$

$$S_1 = \sum_{m=2}^{n_n} \theta_m \frac{(n_n - m)}{(n_n - 1)}$$

$$S_2 = \sum_{m=3}^{n_n} \theta_m \frac{(n_n - m)}{(n_n - 1)}$$

$$S_3 = \frac{\theta_2}{n_n - 1}$$

$$S_4 = \frac{\theta_2/(n_n - 2)}{n_n - 1}$$

We note that p_0 depends on l^* and l_0, because horizontal orientation has much more options for realization than vertical. The total probability of horizontal walk l_0 is $p_0^{l_0}$.

The probability of a percolation trajectory having parameters (l^*, l_0) is found from the following equation [34]:

$$P(l^*, l_0) = p_+^{l_+} p_0^{l_0} p_-^{l_-} \qquad (4.57)$$

The total probability of percolation is found as the sum over all possible percolation trajectories [34]:

$$P_\Sigma = \sum_{l_1=L}^{l_{\max}} \sum_{l_2=0}^{l_1-L} P(l_1, l_2) \qquad (4.58)$$

where l_1 and l_2 have the sense of l^* and l^0, respectively, and $l_{\max}$ is the maximal possible walk corresponding to all voids forming the same percolation way, found as follows:

$$l_{\max} = \xi \frac{V_0}{(w_0)^{2/3}} \qquad (4.59)$$

where $\omega_0 = 1$ is the volume of a cell.

Comparing the obtained solution to the finite-size cluster model, we note that different percolation trajectories may belong to different clusters. The condition of the conjugation of the obtained solution with the cluster model is given as follows:

$$P_{\Sigma 1} = P_{\Sigma 2} \tag{4.60}$$

$$P_{\Sigma 1} = \sum_{l_1=L}^{l_{\max}} \sum_{l_2=0}^{l_1-L} P(l_1, l_2) \tag{4.61}$$

$$P_{\Sigma 2} = \int_{f_{\min}}^{f_{\max}} \int_{V_{\min}}^{V_{\max}} Q_0 \left(1 - \frac{L}{D(V,f)}\right) \Psi(V,f)\, dV\, df \tag{4.62}$$

where all values regarding the finite-size clusters have the same sense as in the previously considered model.

Thus, the random trajectory model allows the prediction of percolation threshold from the distribution of empty cells in the number of empty neighbors. This model takes into account the finite size of the real porous structure and closing or semiopening of pores. However, the above model is not still appropriate for estimations of permeability.

9. Random Trajectory Model for Permeability (Slab Symmetry)

This version of the random trajectory model assumes that all percolation trajectories (for which $l_+ = l_- + L$) contribute passing through of the fluid flow. The total capability of random trajectories to allow the penetration of the fluid from the head to the bottom of the sample determines the permeability properties of the solid structure. For hydraulic reasons, the shapes of the paths are not important for the fluid penetration rate.

Thus, in the random trajectory model for permeability, this is considered as the result of the motion of a fluid along percolation paths. The principal difference between both versions of the random trajectory model (percolation and permeability versions) consists in the absence of preferable orientations: for hydraulic reasons, all available orientations have the same probability to get the fluid flow. Each percolation path begins at the head of the considered sample and finishes at its bottom. Each percolation path is characterized by the values of l_Σ and l_0 and the sequence of steps down, horizontally, and up. At the same pair (l_Σ, l_0), the variety of percolation paths is numerically characterized by the number of possible combinations. This is found as

$$g(l_\Sigma, l_0) = C_{l_\Sigma}^{l_+} C_{l_\Sigma - l_+}^{l_-} \tag{4.63}$$

where $C_{l_\Sigma}^{l_+}$ is the number of possible placements of l_+ steps down among l_Σ steps totally and $C_{l_\Sigma - l_+}^{l_-}$ is the number of possible placements of l_- steps up among $l_\Sigma - l_+$ steps remaining after steps down.

For a chosen trajectory characterized by the pair (l_Σ, l_0), the probability of its availability for percolation is determined by the following factors [35]:

All cells belonging to the same percolation path must be empty (probability of such event is ς^{l_Σ}).

Each of these cells needs at least two empty neighbors (probability of such event is $[(1-\theta_0)\ (1-\theta_1)]$, where θ_μ is the number of voids having (each) μ empty neighbors.

The values of θ_0 and θ_1 are found from the same equations as written for the random trajectory model for percolation.

The total intensity of the flow penetrating through the sample per unit of surface area is [35]

$$\Pi_f = \sum_{l_\Sigma = L}^{l_{\max}} \sum_{l_0 = l_{\min}}^{l_\Sigma - L} g(l_\Sigma, l_0)\pi(l_\Sigma, l_0) \tag{4.64}$$

$$\pi(l_\Sigma, l_0) = \varsigma^{l_\Sigma}(1-\theta_0)^{l_\Sigma}(1-\theta_1)^{l_\Sigma} A_{\theta_1} A_{\theta_2} \tag{4.65}$$

$$A_{\theta_1} = \left[\sum_{m=2}^{n_n} \frac{\theta_m m(m-1)}{n_n}\right]^{l_+ + l_-} \tag{4.66}$$

$$A_{\theta_2} = \left[\sum_{m=2}^{n_n} \frac{\theta_m m(n_n-2)}{n_n}\right]^{l_0} \tag{4.67}$$

The resistance of (l_Σ, l_0)-th trajectory to the flow is determined not by its shape, not by the sequence of steps, but only by the length of the trajectory. The pressure needed for the flow passing through is

$$\delta P = \eta l_\Sigma \omega(l_\Sigma) \tag{4.68}$$

where $\omega(l_\Sigma)$ is the linear velocity of the flow inside a path having the total length l_Σ, and η is the friction coefficient. The amount of the fluid penetrating through the sample is [35]

$$\Theta_f = \sum_{l_\Sigma=L}^{l_{max}} \sum_{l_0=l_{min}}^{l_\Sigma-L} \omega(l_\Sigma) g(l_\Sigma, l_0) \pi(l_\Sigma, l_0) \tag{4.69}$$

From Eqs. (4.68) and (4.69) we obtain

$$\Theta_f = \sum_{l_\Sigma=L}^{l_{max}} \sum_{l_0=l_{min}}^{l_\Sigma-L} \frac{\delta P g(l_\Sigma, l_0) \pi(l_\Sigma, l_0)}{(\eta l_\Sigma)} \tag{4.70}$$

Equation (4.70) can be considered as Darcy equation for microporous medium [compare to Eq. (4.42)].

An analysis performed in Ref. 35 on the base of Eq. (4.70) showed that nonzero permeability is found even for very low porosity, but there is a very narrow interval of values of the porosity, in which the value of Π_f gets a very sharp rising. This regime corresponds to Π_f getting value up to 1 and can be called *visible percolation threshold* (VPT), because it is the parameter really available for experimental measurements. The notion of VPT is commendable for the characterization of permeability properties.

It was shown in Ref. 35 that VPT decreases with rising of the coordination number n_n, because of the increase of the number of parameters of freedom in the system, and Π_f increases.

Also, the influence of connectedness [numerical definition of which is given in Chapter 2] was studied in Ref. 35. It was shown that the increase of connectedness first reduces much VPT, but then has almost no influence.

The influence of thickness also was analyzed in Ref. 35. It was shown that, though VPT always increases with thickness (that is normal), the derivative of this increase is negative.

Thus, the random trajectory approach allows the solution of all problems related to percolation and permeability. The principal comparison of various methods of characterizing percolation and permeability is illustrated by Table 4.3.

Thus, the random trajectory method is definitely commendable for the solution of traditional problems related to percolation and permeability.

TABLE 4.3 Applicability of Existing Methods of Modeling Percolation and Permeability

Method or model	Sphere of applicability	Physical sense of the method	Limitations or shortcomings of the method
Darcy equation	Permeability only	Linear thermodynamics, Onsager equations	Applicable only over percolation threshold; applied pressure needs to be high enough
Chapman–Enskog equation	Permeability only	Linear thermodynamics of diffusion in multi-particle system	Applicable only over percolation threshold; applied pressure needs to be short enough
Monte-Carlo simulations	Both percolation and permeability	Chaos in particle motion	Pre-history of the system not taken into account
Infinite cluster approach	Both percolation and permeability	Infiniteness of the co-existing fractals	Not applicable to films
Finite-size cluster approach	Both percolation and permeability	Finite size of co-existing fractals	Specificity of micropores not taken into account
PSD	Both percolation and permeability	Pore-size distribution assumed	Difficulties in estimation of pore-size distribution for micropores
Pore-Core	Both percolation and permeability	Semiempirical correlations between measurable properties	Absence of fundamental solution
Random trajectory approach	Both percolation and permeability	Finiteness of percolation paths	Difficulties in studies of big solid samples

10. Random Trajectory Method for Modeling Permeability of System Having Sphere Symmetry (Heterogeneous Catalysis)

Now, let us use the random trajectory method for the solution of such practically important problem as the estimation of permeability of sphere. Such problem appears, particularly, in heterogeneous catalysis. The principal scheme of a catalytic particle is presented on Fig. 4.13.

The catalytic particle consists of two parts: the exterior sphere (porous carrier of the catalyst) and the interior sphere (porous catalytic nucleus). The exterior sphere (usually prepared of aluminium oxide or silica dioxide) aims to protect the catalytic nucleus (this can be prepared of precious materials, e.g., platinum, palladium) from eventual mechanical damage and chemical pollution. Catalytic particles are usually treated in the regime of fluidized bed (turbulent fluid flow coming from the bottom makes solid particles fly), they collide and make contact with walls, products of reaction, etc. If catalytic nucleus was unprotected, it would be destroyed and/or contaminated very fast.

The reagents from the exterior fluid flow penetrate into the carrier (exterior sphere having radius R) and pass through the pores of the exterior sphere into the catalytic nucleus (interior sphere with radius r), where the main reaction is performed. The products of the reaction are removed through the same pores of the carrier. Thus, the effectiveness of the catalytic process depends much on the permeability of the carrier.

As in the case of slab sample, the permeability is determined by percolation trajectories. However, in the considered situation, percolation

FIG. 4.13 Principal scheme of catalytic particle.

trajectories begin at the surface of the exterior sphere and finish at the surface of the interior sphere. Percolation path comprises l_+ steps toward the sphere center, l_- steps back (toward the exterior sphere surface), and l_0 steps in the plate perpendicular the radius vector, on which steps toward the center and back are measured. The motion perpendicular to the radius vector is characterized by two coordinates (see Fig. 4.14), as in the traditional consideration of three-dimensional space: steps parallel to the radius vector correspond to the direction Z, and perpendicular to the radius vector correspond to directions X, Y, and their combinations.

The percolation takes place, if

The balance of steps along the Z axis toward the center and back satisfies the condition

$$l_+ - l_- = R - r; \tag{4.71}$$

The balances of steps along X and Y axes in directions "plus" and "minus" allow getting the target (the catalytic nucleus) hence, their absolute values remain in the limit given by the interior radius r;

$$|l_{1+} - l_{1-}| \leq r, \qquad |l_{2+} - l_{2-}| \leq r \tag{4.72}$$

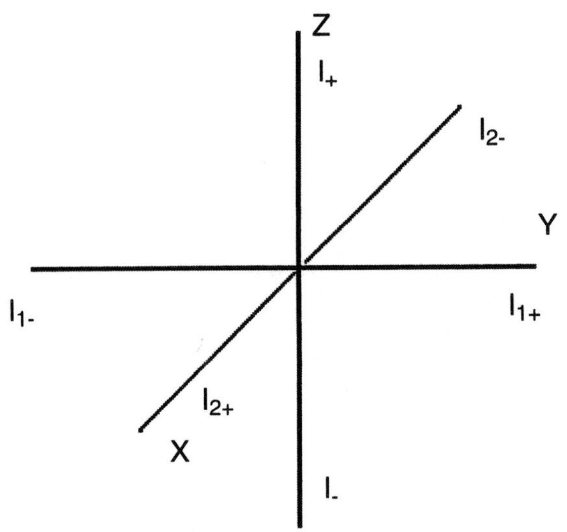

FIG. 4.14 Scheme of the choice of coordinates for the motion of flow through porous catalytic sphere.

Each percolation path is characterized by the group of five independent parameters: the total length (l_Σ), l_+, l_{1+}, l_{1-}, and l_{2+}. Their relationship is, obviously,

$$l_\Sigma = l_+ + l_- + \mathrm{l}_{1+} + l_{1-} + l_{2+} + l_{2-} \tag{4.73}$$

As in the case of slab, the sequence of steps is not important (while, in principle, steps perpendicular the radius vector are constrained by the border of the exterior sphere). At the same values of these parameters, the number of possible combinations of percolation paths is found from the following equation:

$$g(l_\Sigma, l_+, l_{1+}, l_{1-}, l_{2+}) = C_{l_\Sigma}^{l_+} C_{(l_\Sigma - l_+)}^{l_{1+}} C_{(l_\Sigma - l_+ - l_{1+})}^{l_{1-}} C_{(l_\Sigma - l_+ - l_{1+} - l_{1-})}^{l_{2+}} \tag{4.74}$$

where the value C_n^m has the sense (as for the slab symmetry) of the number of possible placements of m steps among n available cells. The maximal length available for percolation path is equal numerically to $l_{\text{max}+} = 4\pi(R^3 - r^3)/3$. The total intensity of the flow penetrating from the surface of the exterior sphere to the surface of the interior sphere is given by Eq. (4.64), and the total current getting to the catalytic nucleus is

$$J_\Pi = 4\pi R^2 \sum_{l_\Sigma = R - r}^{l_{\max}} \sum_{l_0 = 0}^{l_\Sigma - (R - r)} g(l_\Sigma, l_+, l_{1+}, l_{1-}, l_{2+}) \pi(l_\Sigma, l_0) \tag{4.75}$$

where $l_0 = l_\Sigma - l_+ - l_-$, as in the case of slab symmetry.

The results of calculations of various parameters related to the permeability of particle having a sphere symmetry are presented below on Figs. 4.15–4.16.

Figure 4.15 presents the appearance of VPT for a structure having sphere symmetry (like a catalytic particle).

As follows from Fig. 4.15, as in the case of slab symmetry, the permeability is very sensitive to the value of porosity, and a very sharp rising of the flow through the exterior sphere is found in a very narrow interval of values of porosity. As in the case of slab symmetry, this allows us to talk about visible percolation threshold (VPT) corresponding to the really measurable percolation. An experiment performed in conditions corresponding to these for Fig. 4.15 would bring us the result: percolation threshold corresponds to porosity about 0.617.

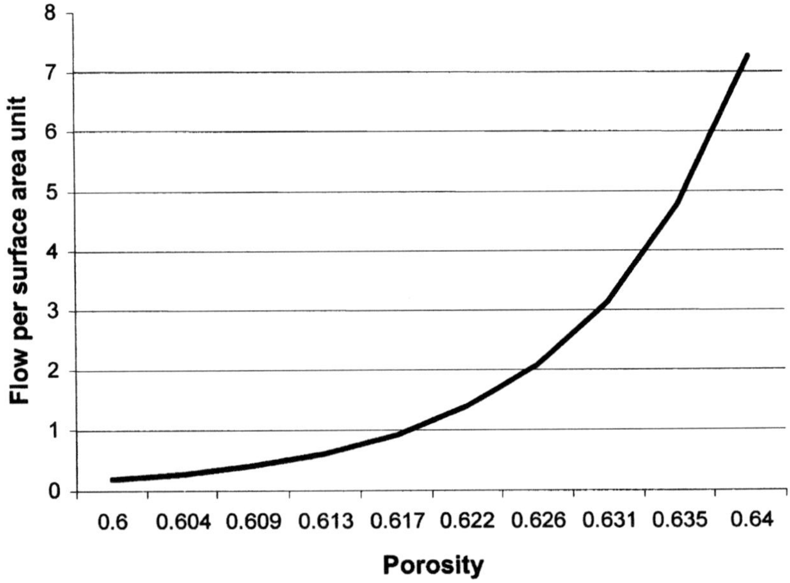

FIG. 4.15 Appearance of visible percolation threshold (VPT) for sphere symmetry, coordination number 6, $\varphi = 1$, $R = 15$, $r = 5$.

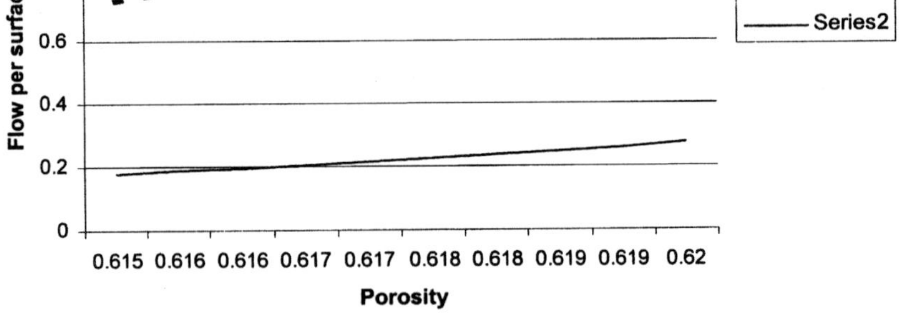

FIG. 4.16 Mass-transfer for heterogeneous catalysis: appearance of VPT for varied coordination number: Series 1: $n_n = 6$, Series 2: $n_n = 4$.

In the further analysis, we will characterize all parameters related to percolation threshold by their relationship with VPT, meaning the corresponding porosity, for which the flow per *exterior* surface unit is equal 1 (on all graphs, the text *flow per surface area unit* means the exterior surface).

Figure 4.16 presents the relationship between the permeability of sphere and the coordination number of particles (n_n); $n_n = 4$ corresponds to the plate situation (two-dimensional case) and $n_n = 6$, to the space (three-dimensional situation).

Figure 4.16 brings us a trivial result: the permeability depends much on the coordination number. This fact is doubtless, because the coordination number influences the degrees of freedom of the system: larger is the coordination number, more the flow has options for its motion, larger is the resulting flow through the medium. A similar result was obtained also for the slab symmetry.

Figure 4.17 presents the relationship between the permeability and the connectedness parameter φ.

As follows from Fig. 4.17, the connectedness parameter φ influences the value of VPT. Increase of the connectedness parameter reduces VPT

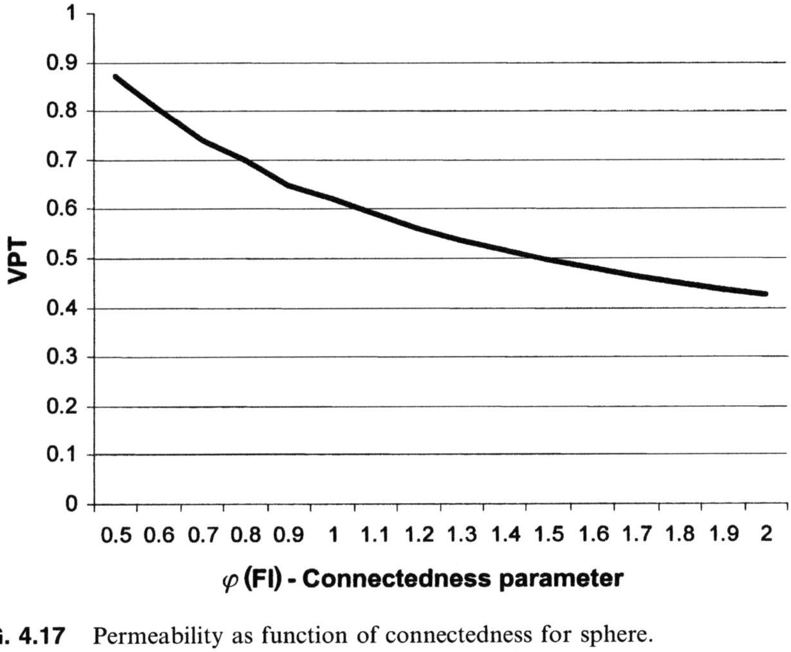

FIG. 4.17 Permeability as function of connectedness for sphere.

and, respectively, increases the permeability. In the physical point of view, this is because more regular structure (caused by high surface tension and corresponding to high values of the connectedness parameter) allows the easier penetration of the flow through the porous medium.

Very important, in the aspect of computing, is the question about the real value of l_{max} needed to be accounted in Eqs. (4.71)–(4.75). If the value is assumed very large, the calculation is very precious but takes extremely long time. If it is very short, the calculation is very fast, but the error of such calculation is too large. Figure 4.18 presents the calculated VPT vs. various values of (l_{max}/R).

As follows from Fig. 4.18, for $l_{max}/R \geq 2$, the error in the estimation of VPT becomes negligible. Therefore, all calculations of VPT presented above and below are made with the assumption that $l_{max}/R = 2$.

Now, let us analyze the influence of specific parameters of sphere onto the permeability. The sizes of the considered system can be characterized by a pair of parameters taken from the following list: r, R, (r/R). Let us choose the pair (r, R).

Figure 4.19 presents the relationship between the exterior radius (R) and VPT.

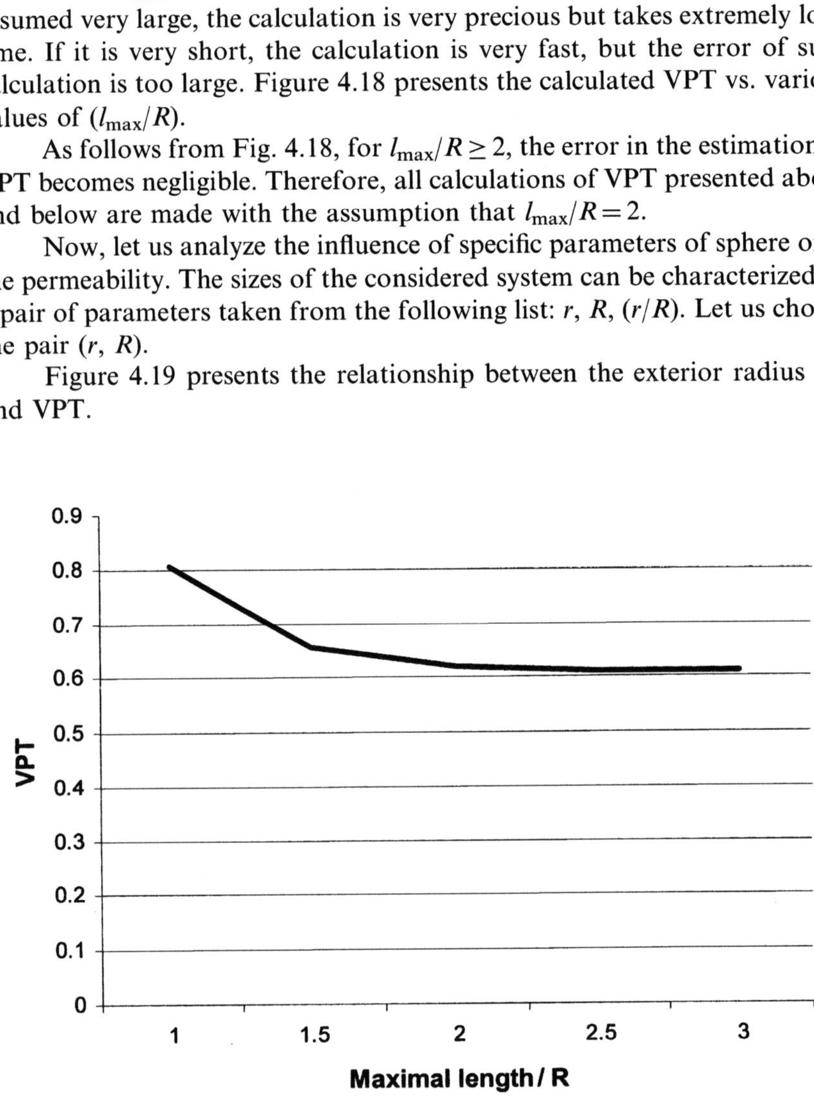

FIG. 4.18 Influence of the accounted length onto VPT calculated.

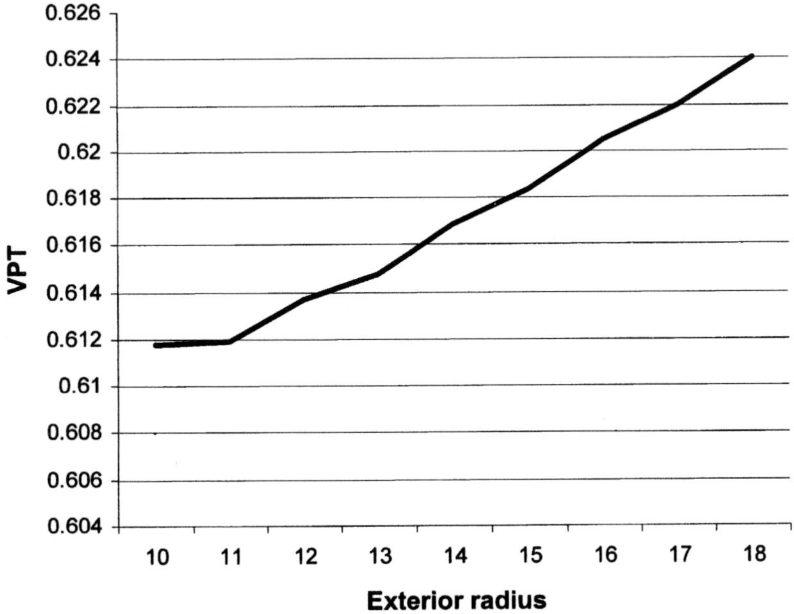

FIG. 4.19 Influence of the exterior radius on VPT.

As follows from Fig. 4.19, VPT increases with *R*. This result seems obvious, because the increase of *R* reduces the chance for a path beginning on the exterior surface to finish on the interior surface.

The influence of the interior radius *r* (radius of the catalytic nucleus) on VPT is given on Fig. 4.20.

As follows from Fig. 4.20, VPT decreases with *r*—for the same reason that VPT increases with *R*.

A parameter very important for practical needs is the total flow through the exterior surface found from Eq. (4.75), that is, the parameter presenting the technical merit of the catalytic system. As follows from Eq. (4.75), two factors compete: rising of *R* increases VPT, but also the total surface area increases!

Figure 4.21 presents the relationship between the total flow and the exterior radius. As follows from Fig. 4.21, the total flow passes through maximum with rising *R*. Why ? the total decrease of the flow when *R* is very large is obvious: because of the increase of VPT. Why the total flow increases to maximum for moderate values of *R*, that is, due to the increase of the exterior surface area: the large number of flows compensates the loss in the intensity of each of them.

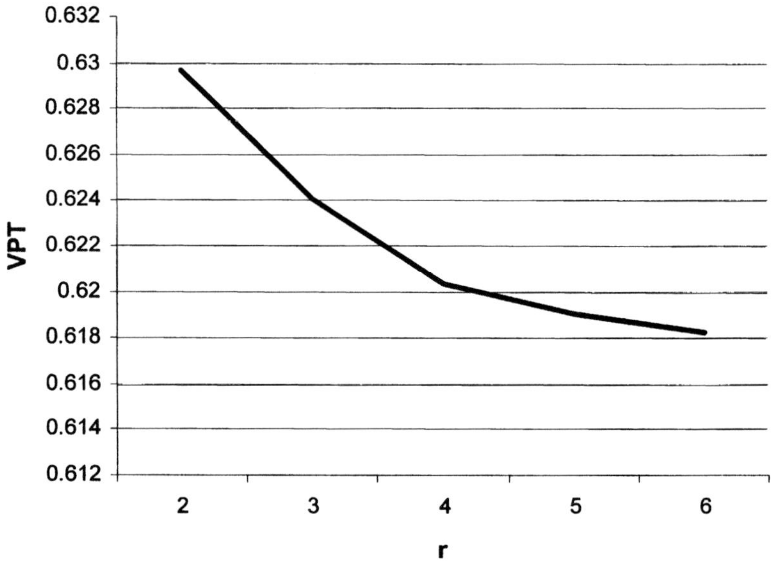

FIG. 4.20 Influence of the interior radius on VPT.

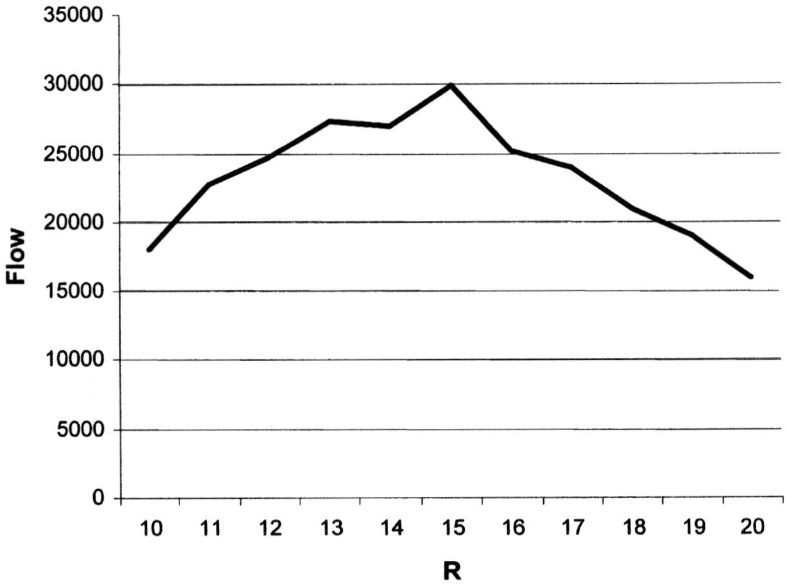

FIG. 4.21 Influence of the exterior radius on the total flow through the total exterior surface.

III. MECHANICAL PROPERTIES OF MICROPOROUS MATERIALS

A. Existing Theories of Mechanical Properties of Porous Solids

Existing theoretical methods for study of mechanical stability and resistance of materials are based on traditional theories of solid state. The employed approaches comprise

Statistical mechanics and thermodynamics [36–38]
Perturbation method [39,40]
Interface and colloid theory [41–43]
Phonon theory [44]
Cumulative damage theory [45]
Empirical models [46–48]

All these methods can be presented in analytical or numerical form. Some of them (e.g., Refs. 36–38, 41 and 42) are similar to the technique analyzed in Chapter 2. However, they do not account for the structural features of microporous materials. The fractal approach in the estimation of mechanical properties of solids was developed in Ref. 49.

In several cases, authors of models of mechanical properties of porous solids assume some geometry of microporous media. For example, Boccaccini et al. [43] elaborated a theoretical model for the prediction of Young's modulus of elasticity of porous materials, assuming a definite microstructural spheroidal model. The effective Young's modulus is given in Ref. 43 as a function of the volume fraction of closed porosity and the microstructural parameters: shape (axial ratio of the spheroidal pores) and orientation. According to Ref. 43, the microstructural parameters involved in the equation could be obtained from real microstructural data; then no fitting would be involved.

Since we already gave some criticisms of theories containing some assumptions of shape of micropores (see Chapters 1 and 3), we do not consider this model in detail.

As in the previous analysis, we find commendable to apply thermodynamics of nonequilibrium to the description of mechanical properties of porous materials. The principal advantages of thermodynamic approach to the solution of this problem consist in the following:

Thermodynamics allows taking into account the irreversibility of all processes related to deformations (if those are irreversible).

Thermodynamics considers all sequences of deformations as the motion of the treated system to a new thermodynamic equilibrium.

On the contrary to adsorption and permeability, mechanical properties of microporous structures are determined by all voids, comprising closed pores. This fact allows paying less attention to the microporous structure itself, and different solutions, particularly related to restructuring of porous media, become applicable.

B. Modeling of Mechanical Properties of Porous Solids, Using the Fractal Description

The fractal model of mechanical properties of solids is applicable to homogeneous randomly structured microporous materials. The fractal model assumes that the mechanical stability of solid structures is determined by elements of the continuous solid phase (fibers) [49]. The fiber substructure is considered as a fractal having dimensionality f_c. It is assumed that all fragments of the continuous solid phase (fibers) are fixed in the solid structure. Recall Chapter 1, we note that the considered material can be characterized, instead the pore distribution in energy or size, by an analogous distribution of fibers: energy distribution $F(E)$ or size distribution $\Phi(\rho)$, respectively.

From the considered material, a cubic sample is taken so carefully that the border effects can be neglected. The size of the sample is H; its volume is $V_0 = \mathrm{H}^3$. The sample is placed under a press performing pressure P on all sides. Because of the homogeneity of the sample, its deformation (h) is the same at all coordinates, and the mechanical resistance is characterized by the function $P(h)$. All changes in the sample are irreversible. At beginning of the process, $h = 0$, $P(h = 0) = 0$. The process is finished when all pores are destroyed: $h(\text{finish}) = h' = H\xi^{(1/3)}$, $P(h = h') = P'$, the pressure determined by the compressibility of the continuous solid phase. Every increase of the pressure causes destruction of several elements of the continuous solid phase. Such destruction depends on the thickness of these elements. Fiber having size ρ is destroyed into fractal cubes, each of which has dimensionality f_c and size ρ; such destruction needs energy:

$$E_d = Q_c \sigma_c \rho^{d_{fc}-1} \Phi(\rho) \tag{4.76}$$

where σ_c is the cohesion tension, Q_c is a normalizing coefficient, and d_{fc} is the fractal dimensionality of the fiber.

For the evaluation of mechanical stability and resistance of a microporous structure, the following subproblems need to be solved:

Correlation between different kinds of distribution functions found for the system

Thermodynamic functions of structural elements
Volume and energy balances regarding distribution functions

The process of formation of pores is related to the appearance of their walls belonging to fibers. The fibers form the continuous phase, the fractal with dimensionality f_c in a cubic initial volume $\omega_0 = h_0^3$, the porosity obtained is ξ_0, and the size of fiber is ρ. Such a system is described by the following equations:

$$\omega_p = \varsigma\omega_0 = r^{d_{f,p}} \quad (4.77)$$

$$\omega_c = (1 - \varsigma)\omega_0 = \rho^{d_{f,c}} \quad (4.78)$$

$$A_p = \varsigma_0 A_e + A_i = 2d_{f,p} r^{d_{f,p}-1} \quad (4.79)$$

$$A_c = (1 - \varsigma_0)A_e + A_i = 2d_{f,c}\, \rho^{d_{f,c}-1} \quad (4.80)$$

$$A_i = A_0\xi_0(1 - \xi_0) \quad (4.81)$$

where A_0 is the initial total surface area, d_{fp} is the fractal dimensionality of porous cluster, A_i is the internal surface area, A_c and A_p are surface areas of the fiber and the pore, respectively, $A_e = 6h_0^2$ is the exterior surface area of the initial cube.

The combinatorial entropy of the continuous phase is found from:

$$\Delta S = R_g\omega_0[-\ln(\xi_0) - \ln(1 - \xi_0)] \quad (4.82)$$

The free energy of the continuous phase is

$$\Delta G_{cp} = \sigma_c A_e - T\Delta S \quad (4.83)$$

The chemical potential of the continuous solid phase is

$$\mu_{cp} = \Delta G_{cp}/V_{cp} \quad (4.84)$$

where V_{cp} is the volume of the continuous solid (fibers) phase.

Thus, we get a sequence of parameters available for evaluation: $\varepsilon - r - \rho$. Hence, knowing the function $f(\varepsilon)$, we get all necessary forms of distributions.

1. Breaking-Up of Fibers

A fiber having characteristic size ρ has cohesive bond with the solid structure, the bond strength being $\sigma_c\rho^{(fc-1)}$. The volume of such fibers is proportional to their volume fraction $Q_c\Phi(\rho)\,d\rho$, while their number, respectively, is the same divided by the fiber volume: $Q_c\Phi(\rho)\,d\rho/\rho^{fc}$. Hence, the total strength of the cohesive bond of these fibers to the solid structure is $\sigma_c Q_c\Phi(\rho)\,d\rho/\rho$. The work performed by the exterior force destroying the fibers is found from

$$3P(h)(H-h)^2\,dh = Q_c\sigma_c\Phi(\rho)\,\frac{d\rho}{\rho} \tag{4.85}$$

On the other hand, the volume change is related to the reduction of the porous volume:

$$3(H-h)^2\,dh = q_p f(\varepsilon)\,d\varepsilon = Q_p\phi(r)\,dr \tag{4.86}$$

where ε and r are, respectively, the energy (per volume unit) and the size of the pores coexisting with the broken fibers and Q_p is a normalizing coefficient.

2. Size and Energy Distribution in the Continuous Solid Phase

In Chapter 3 we found the correlation between $f(\varepsilon)$ and the conditions of the preparation of microporous materials. Evaluations of other distributions (pores in size and fibers in energy and size) were not performed. However, it was shown in Ref. 50 that geometrical properties of porous and continuous phases are similar, and these substructures differ in the fractal dimensionality only. Considering them as cubes having different fractal dimensionalities (since fractals are not pores, we do not assume any shape of micropores), we may write the following equations:

$$V_p = \varsigma V_0 = L_p^{d_{f,p}} \tag{4.87}$$

$$V_c = (1-\varsigma)V_0 = L_c^{d_{f,c}} \tag{4.88}$$

$$A_p = 2d_{f,p}L_p^{d_{f,p}-1}, \qquad A_c = 2d_{f,c}\,L_c^{d_{f,c}-1} \tag{4.89}$$

where L_p and L_c are the formal linear sizes of the porous and continuous-solid clusters (fractals), respectively;

$$V_p = \int_{r_{\min}}^{r_{\max}} Q_p\phi(r)\,dr \tag{4.90}$$

$$V_c = \int_{R_{\min}}^{R_{\max}} Q_c \Phi(\rho)\, d\rho \tag{4.91}$$

$$U_p = \int_{r_{\min}}^{r_{\max}} Q_p \varepsilon(r) \phi(r)\, dr \tag{4.92}$$

$$U_c = \int_{R_{\min}}^{R_{\max}} Q_c E(\rho) \Phi(\rho)\, d\rho \tag{4.93}$$

On the other hand, the total pore volume and energy are found from the following equations, respectively,

$$V_p = \int_{\varepsilon_{\min}}^{\varepsilon_{\max}} q_e f(\varepsilon)\, d\varepsilon \tag{4.94}$$

$$U_p = \int_{\varepsilon_{\min}}^{\varepsilon_{\max}} q_e \varepsilon f(\varepsilon)\, d\varepsilon \tag{4.95}$$

There are two balances we can write for the system: those of volume and energy.

Volume balance is written as

$$dV = q_e f(\varepsilon)\, d\varepsilon \tag{4.96}$$

or:

$$dV = -\, Q_c \Phi(\rho)\, d\rho \tag{4.97}$$

where ρ is the size of a continuous solid substructure (fibers), q_e, Q_ρ is normalizing coefficients, and Φ is the function of distribution of fibers in size.

The energy balance of the system is given by

$$P(V)dV = Q_c \sigma_c \rho^{d_{f,c}-1} \Phi(\rho)\, d\rho \tag{4.98}$$

Equations (4.97) and (4.98) provide the correlation between V and P; hence, the system of equations (4.77)–(4.98) has the consistent solution and provides the information about such mechanical properties of the considered microporous system as mechanical stability and resistance.

Based on Eqs. (4.77)–(4.98), an example of calculations of mechanical resistance of a porous material vs. the deformation was presented in Ref. 49. It was found that the function resistance vs. deformation has a minimum. In the initial part of values of the arguments (high porosity) the resistance is high, due to the large fraction of micropores having small size, walls of which are not yet destroyed; in the end part of the values of the argument (very low porosity), the resistance rises sharply because there the fibers play much less important role than the resistance of the continuous phase entire (in which pores almost do not stay already).

C. Mechanical Properties of Polymers

The model presented above is applicable mostly to nonelastic porous solids, properties of which irreversibly change under an exterior tension loaded. That model is not valid for elastic microporous solids, first of all polymers, pore distribution in which may change because of the deformation caused by an exterior factor. Such a system is described by the statistical polymer method presented in Chapter 3.

1. Bonds in the Macromolecular Structure

The considered macromolecular system is characterized by the following interior bonds [51]:

Intermonomer bonds forming the macromolecules (not cross-linkage).

Cross-links: They influence the form and the characteristic size of macromolecules, in their energy of formation they are similar to intermonomer bonds, but their formation reduces the system entropy, and cross-links may get eventual tension because of the steric factor.

Weak intermacromolecular bonds: these form the structure from macromolecules.

Under an exterior pressure loaded, intermacromolecular bonds are destroyed first, this process being completely or partly irreversible. If the pressure loading is accompanied by heating, many cross-links are broken too, also this process is reversible. Intermonomer bonds and the rest of the cross-links (that eventually do not cause steric problems to their macromolecules) are destroyed under much higher pressures.

2. Structure of Macromolecular System

A macromolecular system with branching and without numerous cross-links may have a very high porosity. For several systems, that can be more 80% (e.g., silica gel [52]). If such a macromolecular mixture is treated under pressure, there are found two factors reducing the porosity: additional cross-link formation and interpenetration of macromolecules. Interpenetration is understood as such a form of interactions of macromolecules that results in finding monomeric units belonging to a macromolecule inside the volume limited by monomeric units belonging to another macromolecule. The physical sense of interpenetration is the occupation of voids inside several macromolecules by other macromolecules.

The condition of additional cross-link formation is given by the equations derived in Chapter 3 for cross-linking, while interpenetration of two or more (μ) macromolecules is possible on condition that

$$\mu < M = \frac{1}{1-\xi} \tag{4.99}$$

where ξ is the maximal available local porosity (this corresponds to a system containing infinite macromolecules without cross-links or interpenetration); $\xi = (Z^3 - Z^{df})/Z^{df}$, where $Z(N)$ is the effective size of statistical N-mer. Equations for the evaluation of $Z(N)$ will be derived in Chapter 5, d_f is the fractal dimensionality of statistical N-mer.

A macromolecular system acted by an exterior pressure P is described by the following equation:

$$P\,dV_{mm} = dE_{mm} \tag{4.100}$$

where dV_{mm} and dE_{mm} are the changes of the volume and the interior energy of the macromolecular system, respectively:

$$dV_{mm} = dV_{cl} + dV_{ip} \tag{4.101}$$

$$dE_{mm} = dE_{cl} + dE_{ip} \tag{4.102}$$

where the terms with indices *cl* and *ip* mean cross-linking and interpenetration, respectively. For many of systems, the cross-linkage does not significantly depend on pressure, and the deformation is determined mostly by interpenetrations.

The whole volume of the macromolecular system can be divided into M zones containing (each) μ interpenetrating macromolecules. For the zones in which interpenetration is not found, we assume (by definition) $\mu = 1$, whereas the maximal available value of μ is M [see Eq. (4.99)]. The value of M is estimated by the statistical polymer method, as it is described above.

The system is characterized by distribution function $F(\mu)$ presenting the fraction of monomeric units found in zones of μ-times interpenetration. The amount of monomeric units, the volume, the energy, and the entropy are estimated from the following equations:

$$\nu_\Sigma \sim \sum_{\mu=1}^{M} F(\mu) \tag{4.103}$$

$$V_\Sigma \sim \sum_{\mu=1}^{M} \frac{F(\mu)}{\mu} \tag{4.104}$$

$$E_\Sigma = \sum_{\mu=1}^{M} \varepsilon_w \mu F(\mu) \tag{4.105}$$

$$S_\Sigma = -R_g \sum_{\mu=1}^{M} [F(\mu) \ln F(\mu)] \tag{4.106}$$

where ε_w is the energy of a single weak bond. Let us assume that the porosity of macromolecules does not directly depend on μ (meaning that the interpenetration does not change the form of macromolecules). For a system close equilibrium, the entropy is in a maximum, therefore little variations of left terms in Eqs. (4.103)–(4.106) give

$$\sum_{\mu=1}^{M} \delta F(\mu) = 0 \tag{4.107}$$

$$\sum_{\mu=1}^{M} \frac{\delta F(\mu)}{\mu} = 0 \tag{4.108}$$

$$\sum_{\mu=1}^{M} \mu \delta F(\mu) = 0 \tag{4.109}$$

$$\sum_{\mu=1}^{M} \delta F(\mu)[\ln F(\mu) + 1] = 0 \tag{4.110}$$

From Eqs. (4.108)–(4.110) we obtain

$$\sum_{\mu=1}^{M} \delta F(\mu)[\ln F(\mu) + \gamma_1 + \mu\gamma_2 + \gamma_3/\mu] = 0 \tag{4.111}$$

where γ_1, γ_2, γ_3 are some coefficients. Eq. (4.111) can be rewritten, taking into account that these coefficients are related to the parameters of interpenetration:

$$F(\mu) = A_F \exp\left[\frac{\beta_v}{\mu} + \beta_e \varepsilon_w \mu\right] \tag{4.112}$$

Equations (4.108)–(4.112) provide the complete description of a macromolecular structure with interpenetration, while other effects (e.g., cross-linking) should be accounted for separately.

If interpenetration is accompanied by cross-linking, the contribution of each process to the total mechanical resistance is estimated from the following equations for volume, energy, and entropy:

$$\delta V_\Sigma = \delta V_{ip} + \delta V_{cl} \tag{4.113}$$

$$\delta E_\Sigma = \delta E_{ip} + \delta E_{cl} \tag{4.114}$$

$$\delta S_\Sigma = \delta S_{ip} + \delta S_{cl} = \text{maximum} \Longrightarrow \delta S_{ip} = \delta S_{cl} \tag{4.115}$$

Let us notice that the mechanical equilibrium for branched macromolecules is related to several cross-linking and interpenetration. When we talk about the interpenetration and cross-linking mechanisms for the deformation, we mean "additional" cross-linking or/and interpenetration. For deformations caused by compression (the current volume below the equilibrium volume), the number of cross-links is more than in equilibrium, while for decompression (the current volume above the equilibrium volume), the situation is contrary (reduction of the number of cross-links). The tendency to interpenetration, respectively, predominates under the compression, while decompression decreases the factor of interpenetration.

The above equations allow us to estimate mechanical characteristics of macromolecular systems. For example Fig. 4.22 presents the value of pressure needed for cross-linking of different macromolecules.

As follows from Fig. 4.22, the most contribution to the mechanical resistance of polymeric materials is due to low-degree macromolecules. Since cross-linking is a high-energy process (on the contrary to interpenetration, which is a low-energy process, as noted above), cross-linking takes place mostly under lower pressures. The pressure needed for the formation of one additional cross-link in a macromolecule depends much on its weight: the increase of the degree of polymerization leads to the reduction of the pressure needed. This result is in accordance with known facts, such as the steric difficulties for the formation of cross-link in short macromolecules.

Now, let us consider the aspect of the relationship between interpenetration and the needed pressure. Figure 4.23 presents the value of pressure needed for interpenetration of macromolecules.

As follows from Fig. 4.23, the contribution of interpenetration increases with the deformation. The curve pressure vs. deformation exhibits a slow increase of the derivative. Physically, that means that the interpenetration is nonlinear factor of deformation.

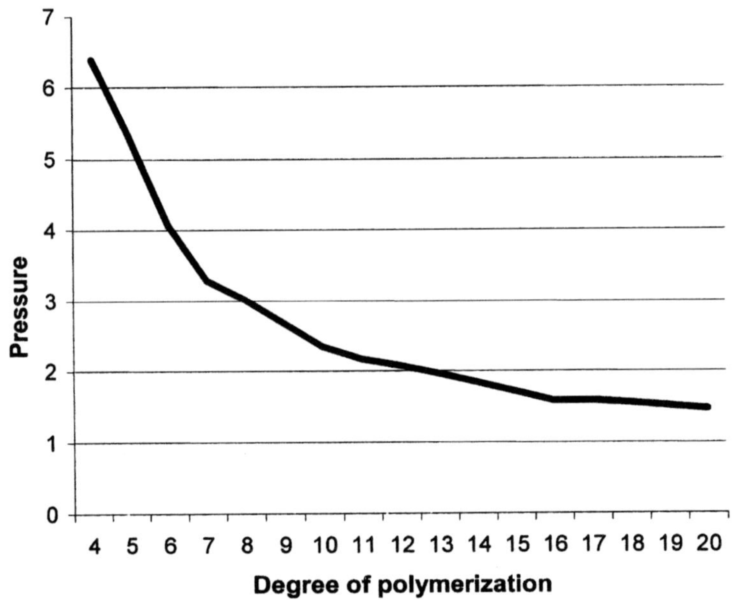

FIG. 4.22 Pressure needed for additional cross-linking in macromolecules (number of branching $m = 5$).

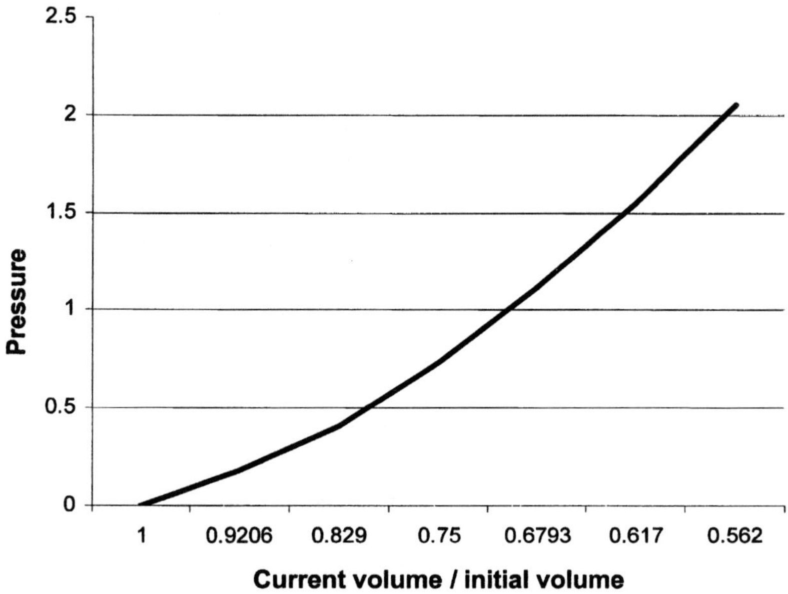

FIG. 4.23 Pressure needed for interpenetration of macromolecules.

We notify that all kinds of deformations of polymeric materials can be considered as declinations of cross-linkage and interpenetration from their equilibrium values.

IV. CONCLUSIONS

1. A principal problem in the preparation and the uses of adsorbents consists in the difficulties in avoiding the adsorption of undesirable fluids. In real industrial conditions, freshly prepared adsorbents may contain undesirable volatile components that reduce the adsorptive effectiveness of industrial adsorbents and may cause wrong results of experiments with the adsorbents.
2. The principal drawbacks of the existing volume-filling micropore theory (VFMT), assuming that adsorption in micropores is limited by their volume, include
   a. The condensation approximation assuming that filling of micropores is determined only by their energy
   b. Baseless assumptions about the energy distribution of micropores

3. The recently proposed Langmuir-like theory of filling of micropores, based on the assumption of Langmuir-like mechanism of pore–adsorbate interactions, neutralizes both above disadvantages of VFMT.
4. Both VFMT and Langmuir-like models suggest adsorption isotherm curves, the form of which is in accordance with experimental data. However, fitted parameters appearing in the Langmuir-like model are not empirical but semiempirical, on the contrary to VFMT. These parameters are more sensitive to properties of adsorbate. In the case of heterogeneous structures, the advantages of the Langmuir-like model become even more significant, because this does not need additional fitted parameters for taking into account the factor of heterogeneity (on the contrary to VFMT).
5. Among the existing models of percolation and permeability, the random trajectory approach is definitely preferable (for modeling both percolation and permeability), due to its ability to bring the relationship between the microscopic characteristics of a porous structure (distribution of solid particles in the number of their near neighbors) and its measurable properties.
6. The random trajectory method allows the solution of practically important problems, such as the evaluation of the permeability of catalyst carrier (in the simulation of mass transfer in the heterogeneous catalysis). All principal results obtained in the random trajectory model of catalyst carrier are in the accordance with known experimental facts.
7. Among existing theories of mechanical properties of porous materials, the following thermodynamic approaches are preferable:
   a. The fractal-based model of irreversible changes in porous material acted by exterior tension
   b. The statistical polymer-based model of reversible changes in polymeric porous material under exterior tension

REFERENCES

1. Gregg, S.J.; Sing, K.S.W. *Adsorption, Surface Area, and Porosity*. 2nd ed.; Academic Press: London, 1982.
2. Noll, K.E.; Gounaris, V.; Hou, W.-S. *Adsorption Technology* (*For Air and Water Pollution Control*); Lewis Publishers: Chelsea, 1991.
3. Rudzinski, W.; Everett, D.H. *Adsorption of Gases on Heterogeneous Surfaces*; Academic Press: London, 1992.

4. Heffelfinger, G.S.; Van Swol, F.; Gubbins, K.E. Adsorption hysteresis in narrow pores. J. Chem. Phys. **1988**, *89* (8), 5202–5205.
5. Bojan, M.J.; Van Slooten, R.; Steele, W. Computer simulation studies of the storage of methane in microporous carbons. Sep. Sci. Technol. **1992**, *27* (14), 1837–1856.
6. Visscher, P.B.; Cates, J.E. Simulation of porosity reduction in random structures. J. Mat. Res. **1990**, *5* (10), 2184–2196.
7. Dubinin, M.M. Physical feasibility of Brunauer's micropore analysis method. J. Coll. Interface Sci. **1974**, *46* (3), 351–356.
8. Dubinin, M.M.; Astakhov, V.A. Development of theories on the volume filling of micropores during the adsorption of gases and vapors by microporous adsorbents. 1. Carbon adsorbents. Izvestiya Akademii Nauk SSSR, Seriya Khimicheskaya **1971**, (1), 5–11.
9. Stoeckli, F.; Morel, D. On the physical meaning of parameters E0 and β in Dubinin's theory. Chimia **1980**, *34* (12), 502–503.
10. Huber, U.; Stoeckli, F.; Houriet, J.P. A generalization of the Dubinin–Radushkevich equation for the filling of heterogeneous micropore systems in strongly activated carbons. J. Coll. Interface Sci. **1978**, *67* (2), 195–203.
11. Stoeckli, F.; Ballerini, L. Evolution of microporosity during activation of carbon. Fuel **1991**, *70* (4), 557–559.
12. McEnaney, B. Estimation of the dimensions of micropores in active carbons using the Dubinin–Raduskevich equation. Carbon **1987**, *25* (1), 69–75.
13. Stoeckli, H.F.; Houriet, J. Ph. The Dubinin theory of micropore filling and the adsorption of simple molecules by active carbons over a large range of temperature. Carbon **1976**, *14* (5), 253–256.
14. Romm, F. Thermodynamic description of a microporous material formation (linear steady-state approximation). J. Coll. Interface Sci. **1996**, *179* (1), 1–11.
15. Tsunoda, R. Determination of micropore volumes of active carbon using Dubinin–Radushkevich plots. Bul. Chem. Soc. Japan **1977**, *50* (8), 2058–2062.
16. Tsunoda, R. Adsorption of water vapor on active carbons: estimation of pore width. J. Coll. Interface Sci. **1990**, *137* (2), 563–570.
17. Tsunoda, R. Relationships between adsorption parameters of the BET and the Dubinin–Radushkevich equations in nitrogen adsorption on active carbons. J. Coll. Interface Sci. **1987**, *117* (1), 291–292.
18. Romm, F. Thermodynamics of microporous material formation. In: *Surfactant Science Series, Interfacial Forces and Fields: Theory and Applications* (Monographic series, ed.: Hsu, J.-P.), Chapter 2; Marcel Dekker: New York, 1999, pp. 35–80.
19. Romm, F. Thermodynamic description of formation of a microporous material (non-steady-state approximation). J. Coll. Interface Sci. **1996**, *179* (1), 12–19.
20. Romm, F. Derivation of the equations for isotherm curves of adsorption on microporous gel materials. Langmuir **1996**, *12* (14), 3490–3497.
21. Rudzinski, W.; Steele, W.A.; Zgrablich, G. Equilibria and dynamics of gas adsorption on heterogeneous solid surfaces. *Stud Surf. Sci. Catal.*, 1997; 104. Elsevier: Amsterdam, Neth., 1997.

22. Pride, S.R.; Flekkoy, E.G. Two-phase flow through porous media in the fixed-contact-line regime. Geosciences Rennes, Universite de Rennes 1, Rennes, Fr. Phys. Rev. E: Statistical Physics, Plasmas, Fluids, and Related Interdisciplinary Topics **1999**, *60* (4-B), 4285–4299.
23. Kling, A.; Barao, F.; Nakagawa, M.; Tavora, L.; Vaz, P. (Eds.). *Advanced Monte Carlo for Radiation Physics, Particle Transports Simulation and Applications*. Proc. Conf. 2000, Lisbon, Portugal, 2001.
24. Cohen, R.; ben-Avraham, D.; Havlin, S. Percolation critical exponents in scale-free networks. Phys. Rev. E: Statistical, Nonlinear, and Soft Matter Physics **2002**, *66* (3-2A), 36113/1–36113/4.
25. Davies, Graham M.; Seaton, Nigel A. Predicting adsorption equilibrium using molecular simulation. AIChE J. **2000**, *46* (9), 1753–1768.
26. Heuchel, M.; Davies, G. M.; Buss, E.; Seaton, N. A. Adsorption of carbon dioxide and methane and their mixtures on an activated carbon: Simulation and experiment. Langmuir **1999**, *15* (25), 8695–8705.
27. Murray, K.L.; Seaton, N.A.; Day, M.A. Use of mercury intrusion data, combined with nitrogen adsorption measurements, as a probe of pore network connectivity. Langmuir **1999**, *15* (23), 8155–8160.
28. Davies, G.M.; Seaton, N.A. The effect of the choice of pore model on the characterization of the internal structure of microporous carbons using pore size distributions. Carbon **1998**, *36* (10), 1473–1490.
29. Boulton, K. L.; Lopez-Ramon, M. V.; Davies, G. M.; Seaton, N. A. Effect of assumed pore shape and interaction parameters on obtaining pore size distributions from adsorption measurements. Special Publication—Royal Society of Chemistry 1997, *213*(Characterisation of Porous Solids IV), 504–511.
30. Lopez-Ramon, M.V.; Jagiello, J.; Bandosz, T.J.; Seaton, N.A. Determination of the Pore Size Distribution and Network Connectivity in Microporous Solids by Adsorption Measurements and Monte Carlo Simulation. Langmuir **1997**, *13* (16), 4435–4445.
31. Schoelkopf, J.; Ridgway, C.J.; Gane, P.A.C.; Matthews, G.P.; Spielmann, D.C. Measurement and network modeling of liquid permeation into compacted mineral blocks. J. Coll. Interface Sci. **2000**, *227* (1), 119–131.
32. Gane, P.A.; Kettle, J.P.; Matthews, G.P.; Ridgway, C.J. Void space structure of compressible polymer spheres and consolidated calcium carbonate paper-coating formulations. Ind. Engng. Chem. Res. **1996**, *35* (5), 1753–1764.
33. Romm, F. Modeling of percolation–permeability in microporous systems. J. Coll. Interface Sci. **2000**, *232* (1), 121–125.
34. Romm, F. Random Trajectory Modeling of limited-volume percolation in a microporous structure. J. Coll. Interface Sci. **2001**, *240* (1), 368–371.
35. Romm, F. Evaluation of permeability of microporous media, using the modified random-trajectory approach. J. Coll. Interface Sci. **2002**, *250* (1), 191–195.
36. Kikuchi, R.; Chen, L.-Q. Theoretical investigation of the thermodynamic stability of nanoscale systems III. Thin film with an IPB. Nanostruct. Mat. **1995**, *5* (7/8), 745–754.

37. Henderson, J. R. Potential-distribution theorem. Mechanical stability and Kirkwood's integral equation. Mol. Phys. **1983**, *48* (4), 715–717.
38. Bessenrodt, R.; Schnittker, P. Applications of a nonlinear stability theory for the stationary states of isothermal networks. Zeitschrift fuer Physik **1974**, *268* (2), 217–224.
39. Blasbalg, D.A.; Salant, R.F. Numerical study of two-phase mechanical seal stability. Tribology Transactions **1995**, *38* (4), 791–800.
40. Reiser, B. On the lattice stability of metals. II. Two perturbation formalisms for the comparison of crystals with different lattices. Theoret. Chim. Acta **1976**, *43* (1), 37–43.
41. Mastrangelo, C.H.; Hsu, C.H. Mechanical stability and adhesion of microstructures under capillary forces. Part I: Basic theory. J. Microelectromech. Sys. **1993**, *2* (1), 33–43.
42. Pendle, T.D.; Gorton, A.D.T. The mechanical stability of natural rubber latexes. Rub. Chem. Technol. **1978**, *51* (5), 986–1005.
43. Boccaccini, A.R.; Ondracek, G.; Postel, O. Young's modulus of porous ceramics. Silic. Indus. **1994**, *59* (9–10), 295–299.
44. Moleko, L.K.; Glyde, H.R. Vibrational stability and melting. Phys. Rev. B: Condensed Matter and Materials Physics **1983**, *27* (10), 6019–6030.
45. Tateishi, T. Continuum theory of cumulative damage. Bull. JSME **1976**, *19* (135), 1007–1018.
46. Polyakov, V.V.; Golovin, A.V. Elastic characteristics of porous materials. Prikladnaya Mekhanika i Tekhnicheskaya Fizika **1993**, (5), 32–35.
47. Cooper, R. E.; Rowland, W.D.; Beasley, D. Survey of the effects of porosity on the elastic moduli and strength of brittle materials with particular reference to beryllium. U.K. At. Energy Auth., At. Weapons Res. Estab., Rep. **1971**, AWRE O *25*/71.
48. Ferrari, F.; Rossi, S.; Bonferoni, M. C.; Caramella, C. Rheological and mechanical properties of pharmaceutical gels. Part I: Nonmedicated systems. Boll. Chim. Farm. **2001**, *140* (5), 329–336.
49. Romm, F.; Figovsky, O. Estimation of mechanical stability and resistance of microporous materials prepared by pyrolysis. J. Solid State Chem. **2001**, *160* (1), 13–16.
50. Romm, F. Modeling of the internal structure of randomly formed microporous material of unlimited volume, following thermodynamic limitations. J. Coll. Interface Sci. **2000**, *227* (2), 525–530.
51. Romm, F.; Figovsky, L. Theoretical modeling of mechanical resistance and stability and related characteristics of polymeric systems with branching/cross-linking. J. Solid State Chem. **2002**, *164* (2), 237–245.
52. Iler, R.K. *The Chemistry of Silica : Solubility, Polymerization, Colloid and Surface Properties, and Biochemistry*. Wiley-Interscience: New York, 1979.

5
Models of Microporous Structure

I. CLASSIFICATION AND CHARACTERIZATION OF MICROPOROUS STRUCTURES

The general analysis of classifications of models of porosity was presented in Chapter 1. Now let us consider the classification of models of microporous structures.

According to these assumptions, the existing models of microporous structures can be classified according to their

Methodology: analytical, numerical, or empirical
Vision of microstructures: determinist or random (casual)
Accounting or not accounting quantum effects: continuous or discrete structure
Limited or unlimited number of counted elements of the porous structure (for discrete models)

In principle, all models of microporous structures are related to some comprehension of microporous medium formation. Let us consider some existing models of microporous structures in the aspect of their relationship with the processes of micropore formation.

A. Models Assuming Shape of Pores

Models of this kind are the simplest in physical chemistry of porosity. Researchers assume shape of pores and their size distribution that allow some estimations regarding structural characteristics of pores and measurable properties of porous materials. An example of such models is given by the Russel model of heat exchange in porous medium, assuming that pores are cubes, all of the same size, with solid walls of uniform thickness, then an equation for the thermal conductivity of porous medium is obtained [1]:

$$K_{\text{pore}} = K_{\text{cont}} \frac{B_1}{B_2} \tag{5.1a}$$

$$B_1 = \frac{K_{\text{dis}}}{K_{\text{cont}}} \xi^{2/3} + 1 - \xi^{2/3} \tag{5.1b}$$

$$B_2 = \frac{K_{\text{dis}}}{K_{\text{cont}}} (\xi^{2/3} - \xi) + 1 - \xi^{2/3} + \xi \tag{5.1c}$$

where K_{pore} is the thermal conductivity of porous medium, K_{cont} is the thermal conductivity of continuous phase, K_{dis} is the thermal conductivity of the dispersed phase, and ξ is the porosity.

Shape-assuming models of porous structures result from extreme difficulties in experimental and theoretical studies of pores, especially micropores. The general criticism of shape-assuming models of porosity, especially in the case of micropores, was given in Chapter 1, which remains valid also for Russel-like models of structure of porous medium. Physically, the only example when one may talk about some specifications of pore shape and size distribution is, probably, given by porous crystals (zeolites).

B. Casual Models: Discrete and Continuous

The main idea of the casual model consists in the absolute randomness of microporous structure. That is possible, of course, only if the pore formation process is also absolutely random: the previous history of the system does not influence its further evolution. Such random process of pore formation is described by

Monte Carlo simulations for discrete models (e.g., Ref. 2)
Gauss distribution in the case of the continuous models (as it was assumed by Dubinin and his colleagues in VFMT, see Chapter 4)

When casual models are valid? Only in systems without prehistory, first of all low-porosity structures, where each pore is so far from others that they "do not feel" one other. This corresponds to the initial stage of pore formation, when pore formation in a fragment of the structure cannot influence the further formation of voids just because these are too low number.

Practically, it is very difficult to cite examples of such structures used in industry. Low-porosity materials do not have most of the advantages of porous materials but have most of their disadvantages. Even methodologically, the initial stage of microporous structure formation is better described by the model of random formation with thermodynamic limitations considered below.

C. Equilibrium Thermodynamic Model

This model assumes that microporous structure is characterized by the following pore energy distribution [3]:

$$f(\varepsilon) \sim \exp\left(\frac{-\varepsilon}{R_g T}\right) \tag{5.2}$$

where, as always, ε is the energy of substance inside a micropore (per 1 mol of the substance), T is temperature (K), and R_g is gas constant; $f(\varepsilon)$ is the probability to find 1 mol of substance inside pores in the energetic state from ε to $d\varepsilon$, where $d\varepsilon$ is negligibly short. Though Eq. (5.1) is written for continuous microporous structure, the same is valid for discrete structure, while ε changes from $\varepsilon + \delta\varepsilon$, and this time, $\delta\varepsilon$ is not negligibly short.

Equilibrium thermodynamic model of microporous structure corresponds to the equilibrium process of micropore formation (its discrete version was considered in Chapter 3). Though the practical significance of the equilibrium thermodynamic model is minor (among real structures that is applicable only to some products of sedimentation), this model is very important methodologically, because it allows the conjugation of discrete and continuous models of pore formation and it allows the derivation of thermodynamic functions of micropores.

D. Thermodynamic (Nonequilibrium) Model

This model was developed in a series of publications [4–8], the main idea of which consisted in the theoretical study of the relationship between preparation conditions of microporous materials, their internal structure and measurable properties. Some aspects of the thermodynamic model were already considered in Chapter 3 (problem of microporous medium formation) and in Chapter 4 (problem of adsorptive, percolation–permeability, and mechanical properties of microporous media). The thermodynamic model is formulated for both continuous [4–7] and discrete representations of the microporous structure.

Methodologically, the thermodynamic model can be considered as the development of the equilibrium thermodynamic model of Cerofolini, while the equilibrium model did not allow any solution of problems considered in Chapters 3 and 4.

E. Casual Model with Thermodynamic Limitations

This model suggested a compromise between the casual and thermodynamic (equilibrium) models [9]. The main idea of this model, assuming that even random processes may happen only under several thermodynamic restrictions, was presented in Chapter 3. The main advantage of this model consists in the methodological conjugation between the casual and nonequilibrium thermodynamic models, allowing the continuous description of microporous medium formation "from zero" (when the process just starts) "to the end" (when the microporous material is ready).

F. Polymeric Thermodynamic Models

These models present the specific application of discrete equilibrium and nonequilibrium models to polymeric and aggregate structures. Among this kind of models, we notify the statistical polymer method designed especially for the description of microporous media formed in macromolecular structures without or with branching (maybe also cross-linking). The facilities of this method were demonstrated in Chapter 3 (theoretical analysis of various problems related to branched polymerization) and Chapter 4 (modeling of mechanical resistance of elastic macromolecular system).

Below, among all models of polymeric structures, we will deal only with the statistical polymer method.

G. Fractal Models

As it was notified in Chapter 3, fractal models of porosity describe not pores themselves but macroscopic sequences of their formation in conditions of homogeneity—so-called random fractals. Fractal models do not distinguish among open, semiopen, and closed pores, macro-, meso-, and micropores, continuality and discreteness of the microporous phase but treat all that "in general" (Table 5.1).

TABLE 5.1 Relationship Between Various Models of Microporous Structures and the Corresponding Processes of Micropore Formation, Methodology, and Sphere of Applicability

Model of structure	Main physical assumption	Corresponding process of formation of the structure	Main method(s) used in the model	Sphere of applicability
Shape-assuming models	Shape and size distribution of pores are specified	Pore formation in crystal	Analogy between pores and macroscopic objects	Zeolites?
Casual (discrete)	Randomness of the structure	Randomness of formation	Monte Carlo	Very low porosity
Casual (continuous)	Randomness of the structure	Randomness of formation	Gauss distribution	Very low porosity
Equilibrium thermodynamic (continuous)	Equilibrium in porous system	Pore formation in equilibrium	Gibbs energy distribution	Structures prepared in equilibrium
Equilibrium thermodynamic (discrete)	Equilibrium in porous system	Pore formation in equilibrium	Discrete Gibbs energy distribution	Structures prepared in equilibrium
Thermodynamic model (nonequilibrium), continuous	Applicability of thermodynamics of nonequilibrium	Pore formation in nonequilibrium	Equations of statistical thermodynamics of nonequilibrium for continuous structure	Noncrystalline structures

Thermodynamic model (nonequilibrium), discrete	Applicability of thermodynamics of nonequilibrium	Pore formation in nonequilibrium	Equations of statistical thermodynamics of nonequilibrium for discrete structure	Noncrystalline structures
Casual model with thermodynamic limitations	Random structure constrained by thermodynamics	Random formation under thermodynamic limitations	Monte Carlo constrained by thermodynamic equations	Low and moderate porosity
Models of polymeric structures	Discrete thermo-dynamic models formulated for macromolecules	Polymerization and aggregation	Models of polymers	Polymeric and aggregated structures
Fractal models	Formation of random fractals	Homogeneous pore formation	Geometry of space having dimen-sionality below 3	Homogeneous porous structures

H. Analysis of the Thermodynamic Model

The thermodynamic model of microporous structure is analytically based on the same equations that were derived in Chapter 3:

Gibbs distribution of pores in energy, for structures prepared in steady-state processes
Equations of the chemical model, when Gibbs distribution is not applicable or not sufficient

I. Gibbs Energy Distribution of Micropores

As it was shown in Chapter 3, steady-state synthesis of microporous material leads to Gibbs distribution of micropores in energy:

$$f(\varepsilon) = \exp\left(\frac{-\varepsilon}{\alpha}\right) \tag{5.3}$$

where $f(\varepsilon)$ is the amount of substance, contained in micropores and having energy from ε to $(\varepsilon + d\varepsilon)$ per 1 mol of the substance contained in these micropores, and α is a parameter related to the degree of nonequilibrium of the system. The value of ε changes from ε_{min} to ε_{max}. All additive (extensive) parameters of the microporous structure are evaluated from the following equation:

$$\Pi = Q_0 \int_{\varepsilon_{min}}^{\varepsilon_{max}} \Pi(\varepsilon) f(\varepsilon)\, d\varepsilon \tag{5.4}$$

where Q_0 is the normalizing coefficient, ε_{min} and ε_{max} are the limits of integration of ε, and $\Pi(\varepsilon)$ is the value of the parameter Π for micropores, the substance which has energy from ε to $(\varepsilon + d\varepsilon)$. In the cases when the parameter Π is just the amount of the substance in micropores $[\Pi(\varepsilon) = 1]$ or the internal energy of micropores $[\Pi(\varepsilon) = \varepsilon]$, we obtain from Eq. (5.4)

$$\nu_\Sigma = Q_0 \int_{\varepsilon_{min}}^{\varepsilon_{max}} f(\varepsilon)\, d\varepsilon \tag{5.5}$$

$$E_\Sigma = Q_0 \int_{\varepsilon_{min}}^{\varepsilon_{max}} \varepsilon f(\varepsilon)\, d\varepsilon \tag{5.6}$$

The evaluation of ν_Σ is not a problem, e.g., by direct experimental measurements or from material balance. The evaluation of E_Σ is much more difficult, and this parameter is, in most of situations, unknown and needing to be found anyway. Hence, the four equations (5.3)–(5.6) practically contain six unknown parameters and do not have consistent solution without additional derivations or assumptions.

The simplest assumption allowing us to reduce the number of unknown parameters in Eqs. (5.3)–(5.6) is the suggestion that $\varepsilon_{\min} = 0$. Physically, that means that the energy of the largest micropores (or narrow mesopores) is negligible in comparison with that of ultramicropores. Though the results obtained due to such assumption are inexact, this assumption allows in many cases a good representation of the physical image. In principle, one may suggest that

If it is important to evaluate $\varepsilon_{\min}$ as exactly as possible, then it can be done by quantum methods, the application of which to largest supermicropores should not be too difficult, because the effects of near walls for them are minimal.

If it is not important to evaluate $\varepsilon_{\min}$ as exactly as possible, the assumption $\varepsilon_{\min} = 0$ is not absurd.

Following the same logic, that seems attractive to assume approximately $\varepsilon_{\min} = \infty$. However, what will be the error caused by such an approximation?

Aiming to answer this question, let us present the parameter $\Pi(\varepsilon)$ for Eq. (5.4) as the following range:

$$\Pi(\epsilon) = \sum_{k=0}^{\infty} \pi_k \varepsilon^k \tag{5.7}$$

From Eqs. (5.4) and (5.7) we obtain

$$\Pi = Q_0 \sum_{k=0}^{\infty} \int_{\varepsilon_{\min}}^{\varepsilon_{\max}} \pi_k \varepsilon^k f(\varepsilon)\, d\varepsilon \tag{5.8}$$

For $k = 0$, the error is obviously zero (because there is no energetic divergence), hence, the error of the approximation $\varepsilon_{\min} = \infty$ is determined for steady-state regime product by the following integrals:

$$J_k = \int_{\varepsilon_{\max}}^{\infty} \varepsilon^k \exp\left(\frac{-\varepsilon}{\alpha}\right) d\varepsilon \tag{5.9}$$

where k changes from 1 to ∞. The real value of Π is

$$\Pi = Q_0 \sum_{k=0}^{\infty} \int_0^{\infty} \pi_k \varepsilon^k f(\varepsilon)\, d\varepsilon - \sum_{k=1}^{\infty} J_k \tag{5.10}$$

For $k = 1$, the integration is simple:

$$J_1 = \int_{\varepsilon_{\max}}^{\infty} \varepsilon \exp\left(\frac{-\varepsilon}{\alpha}\right) d\varepsilon = \alpha^2 (1 + \varepsilon_{\max}/\alpha) \exp(-\varepsilon_{\max}/\alpha) \tag{5.11}$$

For other J_k, one may prove by the method of the mathematical deduction:

$$J_k\left(\frac{\varepsilon_{\max}}{\alpha}\right) = -\left(\frac{-\varepsilon_{\max}}{\alpha}\right)^k \exp\left(\frac{-\varepsilon_{\max}}{\alpha}\right) - kJ_{k-1} \tag{5.12}$$

Let us note that not $\varepsilon_{\max}$ but $\varepsilon_{\max}/\alpha$ influences the value of the error. The same is true about $\varepsilon_{\max}$. Among all possible integrals J_k,

J_1 corresponds to linear effects.
All J_k for $k > 1$ correspond to nonlinear effects.

Let us call integrals J_k *error-integrals*. Figure 5.1 presents the various error-integrals J_k vs. $\varepsilon_{\max}/\alpha$.

As follows from Fig. 5.1, the linear factor of error is very short, hence, for large values of $\varepsilon_{\max}/\alpha$ the linear factor of error is negligible. The increase of k is accompanied with a very significant rising of the modulus of J_k; however, the sign of J_k is opposite to those of J_{k-1} and J_{k+1}: the linear error-integral is positive, the bilinear is negative, trilinear is again positive, etc. Hence, even for nonlinear situation error-integrals compensate one other. As that was easy to predict, the modulus of J_k drops with $\varepsilon_{\max}/\alpha$.

We are interested first of all in linear effects, because

In the experimental aspect, it is too difficult to determine and measure thermodynamic parameters in nonlinear systems (see the related analysis in Chapter 3).

In the aspect of microporosity, microporous structures get formation almost always in conditions of synthesis performed in linear systems (that was also considered in Chapter 3).

In the aspect of properties of porous media, such characteristics as adsorption–desorption, permeability–percolation, and mechanical resistance are studied mostly in linear conditions (see Chapter 4).

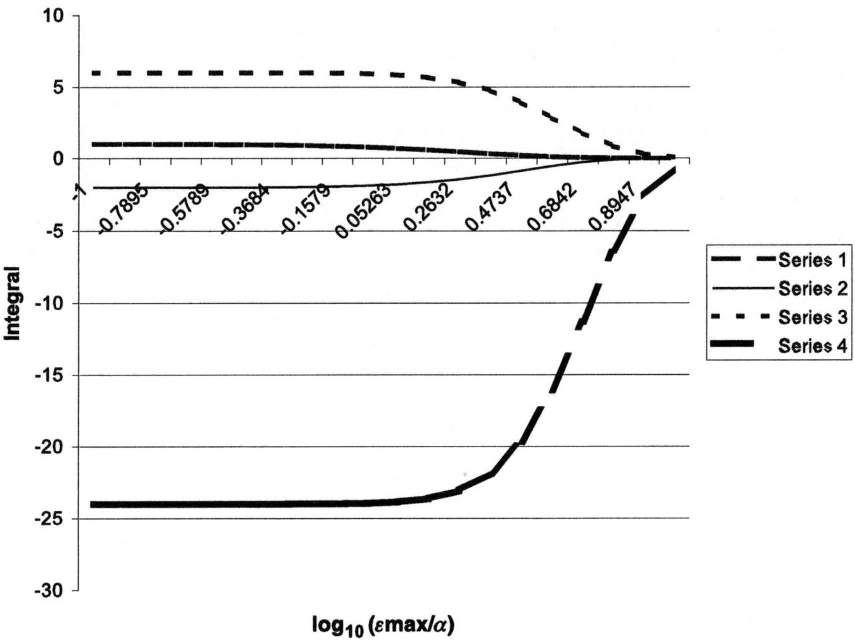

FIG. 5.1 Behavior of error-integrals $J_k(\varepsilon_{max}/\alpha)$: series 1: $k=1$; series 2: $k=2$; series 3: $k=3$; series 4: $k=4$.

In the applicative aspect, the existing forms of uses of microporous materials are effective only in conditions of linear systems.

Since, for the above reasons, we are interested first of all in studies of linear systems, among all integrals J_k we are interested mostly in J_1; Fig. 5.1 proves that values of J_1 are moderate, hence, the error of the approximation "$\varepsilon_{max} = \infty$" should not be large for most of problems interesting us, and we may use this approximation, taking into account that several errors, nevertheless, occur.

J. Thermodynamic Potentials of Micropores

The notions of thermodynamic potentials are conventionally applied to chemical substances, their mixtures, and derivatives. Is it correct to talk about thermodynamic potential of pores—specific two-phase systems, special properties of which are determined not by their chemical components but mostly by their exceptional structure? Since thermodynamics is applicable to various kinds of systems, comprising multiphase systems, there is nothing incorrect with defining thermodynamic functions for micropores. The additional arguments supporting such approach are as follows:

Pores are correctly characterized by such measurable parameters as pressure, temperature, volume, surface area, surface tension, and chemical composition.

Assemblages of pores are correctly characterized by such nonmeasurable parameters as entropy, internal energy, negentropy, and free energy Gibbs.

Pores participate in chemical reactions and behave very specifically in some of such reactions.

Thus, pores have all properties of chemical substances and, respectively, can be characterized by thermodynamic potentials, comprising chemical potential (or generalized chemical potential).

As in Chapter 3, we derive the chemical potential of micropore, using the method of thermodynamic cycles: though the process of pore formation is irreversible, we decompose it to some reversible stages that allow us to define and derive the chemical potential of micropore.

We define the *chemical potential of micropore* as the measure of the ability of micropore to react with various chemical substances, evaluated as the free energy of the system "micropore walls plus the content of micropore" under constant pressure, temperature, and other intensive parameters minus the chemical potentials of all components of the system "micropore walls plus the content of micropore" per 1 mol of the chemical substance(s) contained in the micropore.

For the derivation of the equation for the chemical potential of micropore, let us consider the hypothetical process of the equilibrium and reversible formation of micropore (just hypothetical! We notified in Chapter 3 that almost always the micropore formation process is nonequilibrium and irreversible!) under constant pressure and temperature:

$$\text{Solid} \Longleftrightarrow \text{Solid} \cdot \text{Pore}\,(\varepsilon, \text{T}) \tag{5.13}$$

The value of the chemical potential is given by the following:

$$\mu_{\text{pore}}(\varepsilon) = \mu^0_{\text{pore}}(\varepsilon) + R_g T \,\text{In}\, X(\varepsilon) \tag{5.14}$$

where T is the current temperature, $\mu^0_{\text{pore}}(\varepsilon)$ is the standard chemical potentials of the micropore with energy ε per 1 mol of contained substances, respectively, and $X(\varepsilon)$ is the volume fraction of pores having energy from ε to $\varepsilon + d\varepsilon$ per 1 mol of the contained substance(s). As we showed in Chapter 3, the equation for $X(\varepsilon)$ in steady state is given by

$$X(\varepsilon) = \xi \exp\left(\frac{-\varepsilon}{\alpha}\right) \tag{5.15}$$

where ξ is the porosity and α is a parameter related to the degree of non-equilibrium of the system. For equilibrium, $\alpha = R_g T_f$, where T_f is the temperature of pore formation in equilibrium by reaction (5.15), and Eq. (5.14) is transformed to the following:

$$\mu_{\text{pore}}(\varepsilon) = \mu^0_{\text{pore}}(\varepsilon) + R_g T \operatorname{In} \xi - \varepsilon \frac{T}{T_f} \tag{5.16}$$

In the physical sense of the standard potential, $\mu^0_{\text{pore}}(\varepsilon) + R_g T \operatorname{In} \xi = \varepsilon$, and Eq. (5.16) is transformed to

$$\mu_{\text{pore}}(\varepsilon, T) = \varepsilon\left(1 - \frac{T}{T_f}\right) \tag{5.17}$$

Obviously, for $T = T_f$, $\mu_{\text{pore}}(\varepsilon) = 0$, because the occurrence or the absence of the micropore has no thermodynamic advantage in equilibrium. For $T > T_f$, $\mu_{\text{pore}}(\varepsilon) < 0$, and this micropore loses the contained substance. For $T < T_f$, $\mu_{\text{pore}}(\varepsilon) > 0$, and the micropore gets the tendency to adsorb a fluid from exterior. That is the thermodynamic reason for the appearance of adsorptive properties for micropores under temperatures below the temperature of reversible formation.

For the free energy Gibbs, we obtain from Eq. (5.17) for the specified temperature T:

$$dG_{\text{pore}}(\varepsilon, T) = \varepsilon\left(1 - \frac{T}{T_f}\right) Q_0 \exp\left(\frac{-\varepsilon}{R_g T_f}\right) d\varepsilon \tag{5.18a}$$

For the assemblage of micropores, the value of the free energy Gibbs is obtained by the integration of Eq. (5.18a):

$$\Delta G_{\text{pore},\Sigma}(T) = \left(1 - \frac{T}{T_f}\right) E_\Sigma \tag{5.18b}$$

where E_Σ is the internal energy of the microporous substructure under the real preparation conditions. As follows from Eqs. (5.18a) and (5.18b), the same tendency notified above for the chemical potential of micropores is found also for the free energy Gibbs: for $T=T_f$, $dG_{\text{pore}}(\varepsilon, T=T_f)=0$ and $\Delta G_{\text{pore},\Sigma}$ $(T=T_f)=0$; for $T>T_f$, $dG_{\text{pore}}(\varepsilon,T=T_f)<0$ and $\Delta G_{\text{pore},\Sigma}$ $(T=T_f)<0$; for $T<T_f$, $dG_{\text{pore}}(\varepsilon,T=T_f)>0$ and $\Delta G_{\text{pore},\Sigma}(T=T_f)>0$. The physical sense of these results does not differ from that formulated for the chemical potential.

It is interesting to note that the derivatives of $dG_{\text{pore}}(\varepsilon,T)$ and $\Delta G_{\text{pore},\Sigma}(T)$ by temperature are negative, while "normally" the free energy Gibbs increases with temperature.

The entropy of micropores is found from the following universal equation:

$$dS_{\text{pore}}(\varepsilon,T) = \frac{-\partial G_{\text{pore}}(\varepsilon,T)}{\partial T}\frac{dT}{T} \tag{5.19a}$$

$$dS_{\text{pore},\Sigma}(T) = \frac{\partial G_{\text{pore}}(\varepsilon,T)}{\partial \mathrm{T}}\frac{dT}{T} = E_\Sigma \frac{dT}{TT_f} \tag{5.19b}$$

$$S_{\text{pore}}(\varepsilon,T) = S_{\text{pore}}(\varepsilon,T_f) + \varepsilon Q_0 \exp\left(\frac{-\varepsilon}{R_g T_f}\right)\frac{\ln(T/T_f)}{T_f} \tag{5.20a}$$

$$S_{\text{pore},\Sigma}(T) = S_{\text{pore},\Sigma}(T_f) + E_\Sigma \frac{\ln(T/T_f)}{T_f} \tag{5.20b}$$

hence, the entropy of micropores after the synthesis changes with temperature like that of nonporous solids! (See Table 5.2.)

K. Aging and Hysteresis of Micropores

The free energy of microporous system becomes the driving force for all processes related to micropores, first of all adsorption, desorption, aging,

TABLE 5.2 Comparison of Thermodynamic Properties of Nonporous Solids and Micropores

Criterion for comparison	For traditional chemicals	For micropores
Validity of thermodynamic functions	Valid	Valid
Main thermodynamic correlations	Valid	Valid
Chemical potential shows the reaction-ability	Right	Right
Potentials increase with temperature	Right	Wrong
Entropy increase with the temperature	Logarithmic	Logarithmic

and hysteresis. Energetic aspects of adsorption and desorption were considered in Chapter 4; now, let us consider aging and hysteresis related to the free energy of the structure. The free energy of these processes is estimated from Eq. (5.18b).

As it was mentioned in Chapter 2, the principal difference between aging and hysteresis consists the spontaneous appearance of aging, while hysteresis is related to exterior actions. In other aspects, hysteresis can be considered just as a form of aging.

Let us mention the following possible mechanisms for aging and hysteresis:

Interactions of micropores with the internal substance (contained in the micropores)
Microcracks and destruction of micropores
Reaction to eventual exterior mechanical, chemical, or energetic (e.g., electromagnetic) actions

In principle, interactions with the internal substance should not cause any fast changes in micropores, because this substance is the volatile product of the process of synthesis of microporous material (that is typically done under high temperature); however, this effect after a long time cannot be neglected. The high energy of micropores stimulates all kinds of possible chemical reactions, first of all endothermic ones, and narrow pores, energy of which is higher, are destroyed first. If the microporous material has a polymeric or aggregated structure, the polymerization and cross-linking processes are stimulated until the system gets to the equilibrium weight distribution of macromolecules and loses the free energy. The sequences of aging by chemical interactions comprise:

Pollution of micropores with products of such reactions
Stimulation of cracking processes

Microcracks take place in all known solid structures, mostly on their exterior surface; however, the high energy of micropores makes them especially sensitive to cracking. The main sequence of microcracking consists in such a destruction of micropores that the relief of their walls loses its sharpness (and, consequently, the energy is reduced). An example of such softening relief is given by Figs. 5.2 and 5.3.

Exterior effects destroy micropores in the same manner that interior chemical interactions and cracking, but their sequences appear much faster.

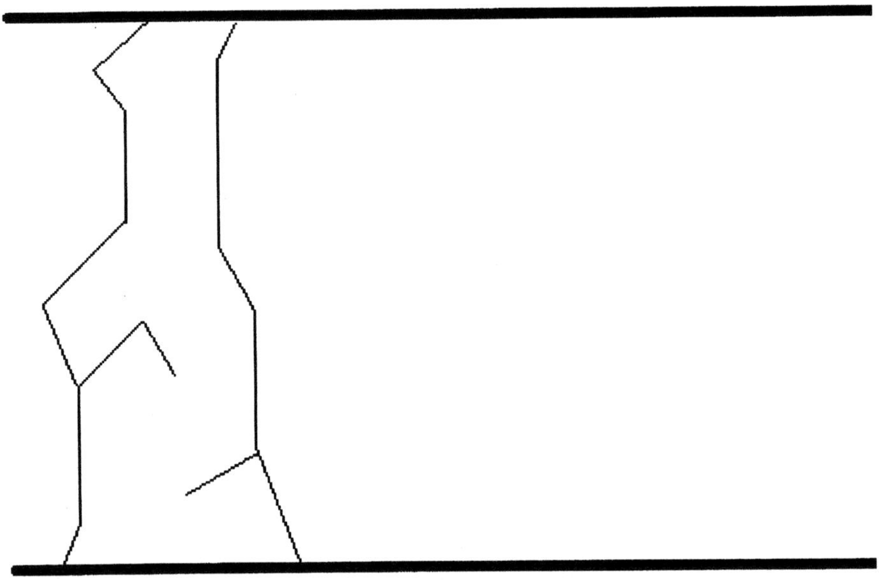

FIG. 5.2 Relief in micropore before softening.

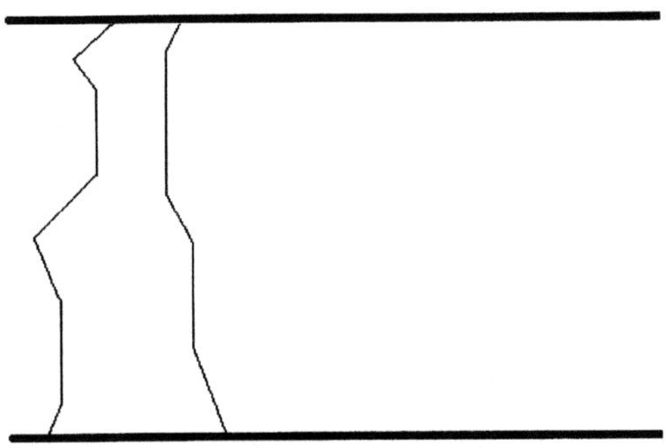

FIG. 5.3 Relief in micropore after softening.

Thus, we can suggest the mechanism for hysteresis during adsorption–desorption cycles:

During the stage of adsorption, the heat delivered due to the adsorptive interactions (especially in the case of chemisorption) intensifies all processes related to aging, especially cracking.

The adsorbed substance may eventually change the chemical composition of the near layers of the continuous solid phase, stimulating its cracking.

During the stage of desorption accomplished by heating, the received heat stimulates the acceleration of aging, and hysteresis appears.

Since aging can be approximately considered as a process in an insulated system, the loss of negentropy with time is estimated from Eq. (3.28), where $dV = 0$:

$$dU_0 = -\sigma_s dt \tag{5.21}$$

In the linear approximation, $\sigma_s = \gamma_0 U_0$, where the parameter γ_0 has the sense of the aging parameter. The solution of Eq. (5.21) has a simple exponential form:

$$U_0 = U_{0,in} \exp(-\gamma_0 t) \tag{5.22}$$

where $U_{0,\text{in}}$ is the initial value of the negentropy of the microporous system, corresponding to the fresh product and evaluated as described in Chapter 3.

On the other hand, since micropores have many features of traditional chemicals, aging can be interpreted as a chemical reaction, product of which is the nonporous medium. A destroyed micropore is not always transformed into a pore-free fragment of the structure, in most of cases it is transformed into another micropore having lower energy. As a result, the fraction of micropores having energy close to ε is not only reduced because of the destruction of these micropores but also increased due to the destruction of micropores having energy above ε, and the evolution of the microporous system entire can be considered as the competition of processes of micropore destruction and transformation. The resulting kinetic equation for aging of micropores having energy close to ε is written as

$$\frac{\partial X(t,\varepsilon)}{\partial t} = \int_{\varepsilon}^{\infty} k_a(\varepsilon,\varepsilon')X(\varepsilon')\, d\varepsilon' - \int_0^{\varepsilon} k_a(\varepsilon'',\varepsilon)X(\varepsilon)\, d\varepsilon'' \tag{5.23}$$

Analytical solution of Eq. (5.23) is too complicated, therefore, for an approximated estimation of tendencies related to aging, let us assume that each kind of micropore "gets aging" independently of others, and its aging rate is

$$\frac{\partial X(t,\varepsilon)}{\partial t} \approx Z_a \exp\left(\frac{-E_a}{R_g T}\right) X(t,\varepsilon) \tag{5.24}$$

where Z_a is the pre-exponential coefficient for aging rate and E_a is the activation energy for aging. Let us estimate the value of the aging parameter $[\partial \ln X(t,\varepsilon)/\partial t]$ in relation with various factors.

1. The pre-exponential coefficient. The relationship between the aging parameter and the pre-exponential coefficient in Eq. (5.24) is given on Fig. 5.4. As we can see from Fig. 5.4, the mentioned relationship is very simple—just logarithmic. However, we have to notify that even very low values of the coefficient cause some measurable aging effect.
2. Influence of the activation energy. The relationship between the logarithm of the aging parameter and the activation energy in Eq. (5.24) is given by Fig. 5.5. Also this time the correlation is simple. We note that large values of the activation energy over 45 kJ make the aging process slow enough. However, from general considerations we may suggest that the activation energy for very narrow pores is short; hence, their aging rate is not negligible.
3. Hysteresis. After all, we are interested in estimating the influence of hysteresis on the aging. Let us consider a situation when a microporous adsorbent is kept under 300 K and used for such an adsorption–desorption cycle that the temperature of the regeneration by desorption is 1000 K. Figure 5.6 presents the influence of the exploitation time onto aging. As follows from Fig. 5.6, the time of exploitation influences the aging: if that is short, one may

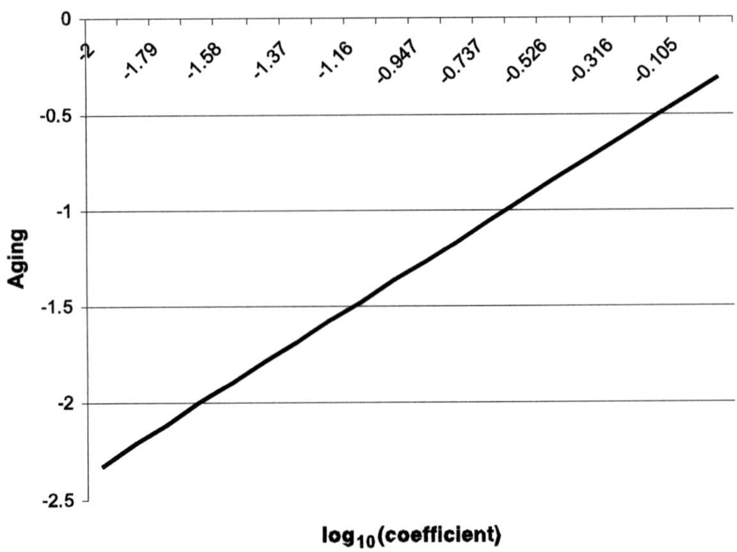

FIG. 5.4 Relationship between the aging parameter and the pre-exponential coefficient: activation energy of aging: $E_a = 30\,\text{kJ/mol}$.

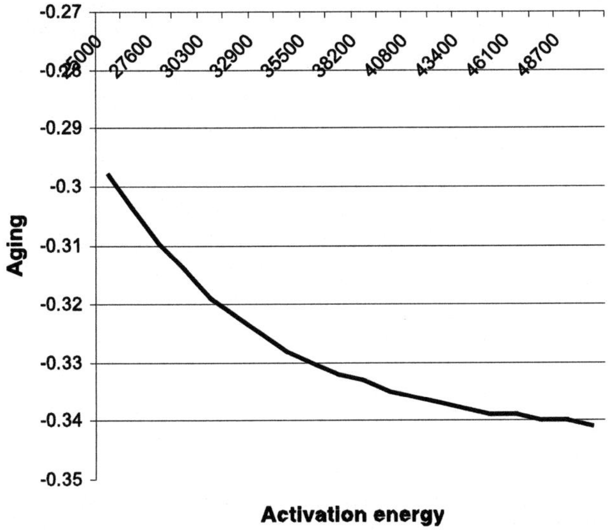

FIG. 5.5 Relationship between the aging parameter and the pre-exponential coefficient: constant of the rate of aging: $Z_a = 1$ Hz.

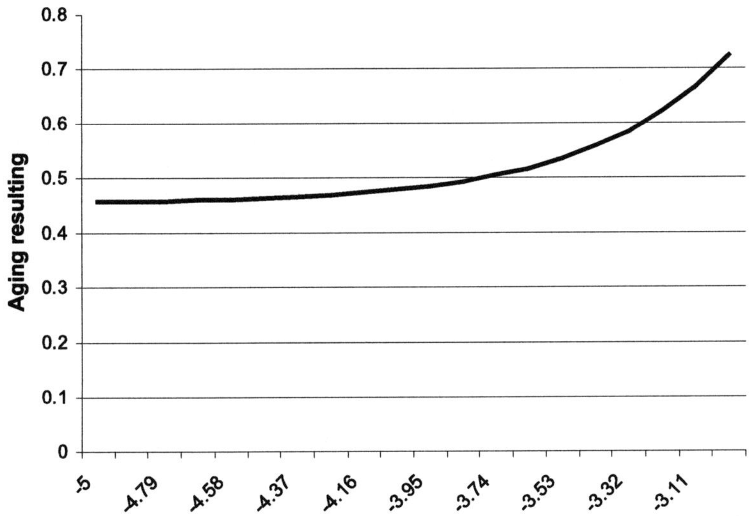

FIG. 5.6 Relationship between the time in use and aging of microporous adsorbent: temperature of desorption is 1000 K, constant of the rate of aging is $Z_a = 1$ Hz, and activation energy of aging is $E_a = 30$ kJ/mol.

neglect its influence, but the longer the exploitation is, the aging gets a very sharp rising.

II. ANALYSIS OF THE POLYMERIC MODEL

A. Characteristic Size of Macromolecules: Estimation by the Statistical Polymer Method

The polymer size estimation presented below is based on the structural considerations in the statistical polymer method described in [6,7,10].

Let us consider a branched non-cross-linked statistical polymer (see Fig. 5.7).

A monomer unit in the statistical polymer is chosen as a basis. All other monomer units can be divided into groups located on the same distances from the basis.

Such distance is expressed in monomer units. The group located on the distance of k monomer units from the basis is characterized (by definition) by $(k+2)$-th level. The basis corresponds to the first level, its neighbors form a second level, etc. Since cross-linkage is absent, units on kth level $(k>1)$ can be connected to

Units on $(k-1)$-th level
Units on $(k+1)$-th level

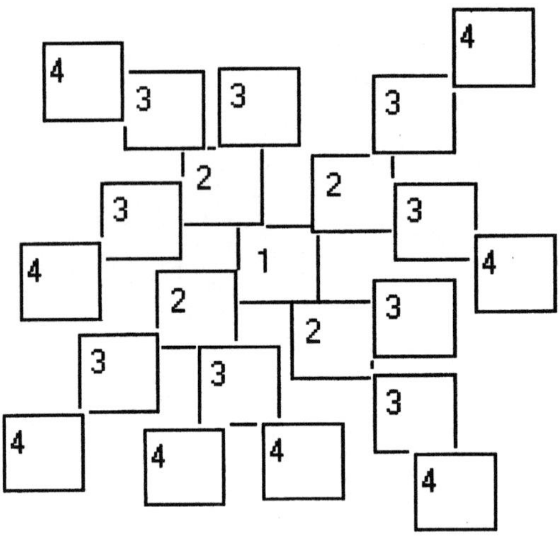

FIG. 5.7 Basis and levels in statistical polymer. Indices mean the following: 1 = first level (the basis); 2 = second level; 3 = third level; 4 = fourth level.

The total number of levels in the statistical N-mer is N, and the number of units on some levels can be below 1. Each level is characterized by the number of monomer units $R_k(N)$ [$R_k(N)$ do not need to be integer and can be less 1] and vacancies $V_k(N)$ meaning the number of units that can be *yet* accepted to $(k+1)$-th level. These parameters are found from the following recursive equations [7,10]:

$$R_k(N+1) = R_k(N) + \frac{V_k(N)}{V_\Sigma(N)} \tag{5.25}$$

$$V_k(N) = mR_{k-1}(N) - R_k(N) \tag{5.26}$$

$$R_1(N) = 1 \qquad V_1(N) = 0 \tag{5.27}$$

Equation (5.25) means that the probability of the acceptance of an additional unit onto kth level is proportional to the number of vacancies on $(k-1)$-th level $V_k(N)$ per the total number of vacancies $V_\Sigma(N)$. Equation (5.26) brings the relationship between the numbers $V_k(N)$ and $R_k(N)$. Equation (5.27) illustrates the obvious fact that there is no level "below" the basis.

Equations (5.25)–(5.27) allow the introduction of the notion of the characteristic size ('diameter') of the statistical N-mer, that being defined as [7]

$$Z(N) = \sum_{k=1}^{N} W_k(N) \tag{5.28}$$

$$W_k(N) = \begin{cases} 1 & \text{if } R_k(N) \geq 1 \\ R_k(N) & \text{if } R_k(N) \leq 1 \end{cases} \tag{5.29}$$

The definition of the characteristic size of statistical polymer by Eqs. (5.28) and (5.29) means that the size is normally measured in spots where at least one unit is found. For other spots we have a problem: of course, they influence the total size, but their contribution cannot be equivalent to that of spots really containing one or more units.

Let us notify that the characteristic size of statistical N-mer depends only on the number of branching m.

Figure 5.8 presents the dependence of the characteristic size of statistical polymers of different weight on the degree of polymerization,

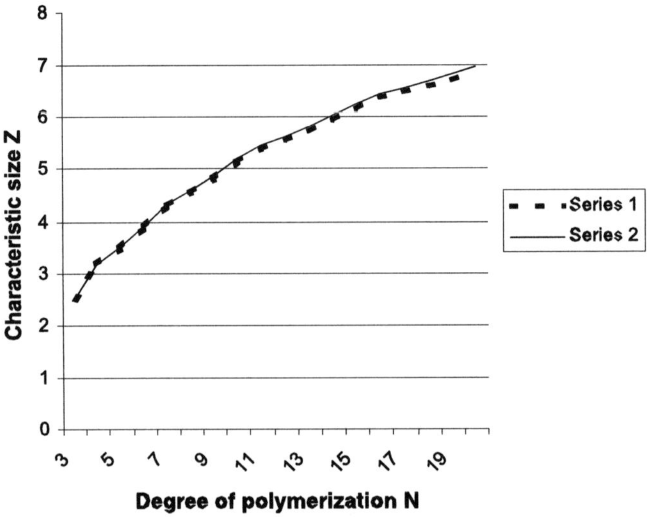

FIG. 5.8 Characteristic size of statistical N-mer: series 1: $m=5$; series 2: $m=3$.

for branching numbers 3 and 5: these numbers are frequently found in polymeric and aggregated systems (corresponding to functionality or coordination number 4 or 6, respectively).

As follows from Fig. 5.8, the value of branching m does not influence much the characteristic size: the curves for $m=3$ and $m=5$ are very close, while the curve $Z(N)$ for $m=3$ is over. Hence, errors in the evaluation of m for $m\geq 3$ do not influence much the estimated value of $Z(N)$.

The above-introduced notion of the characteristic size of statistical polymer allows us to solve many problems related to the structure of polymers. Let us illustrate that with calculating the inertia moment of statistical polymers. Rotation of a statistical polymer is characterized by two following parameters (first moment of inertia and the relationship between first and second moments of inertia):

$$I_1 = \frac{\sum_{k=1}^{N} kR_k(N)}{N} \tag{5.30}$$

$$I_n = \frac{\sum_{k=1}^{N} k^2 R_k(N)}{NI_1} \tag{5.31}$$

Figure 5.9 presents I_1 and I_2 as functions of N at $m=7$.

As follows from Fig. 5.9, the inertia moments increase monotonically with the degree of polymerization. We also notify that second moment of

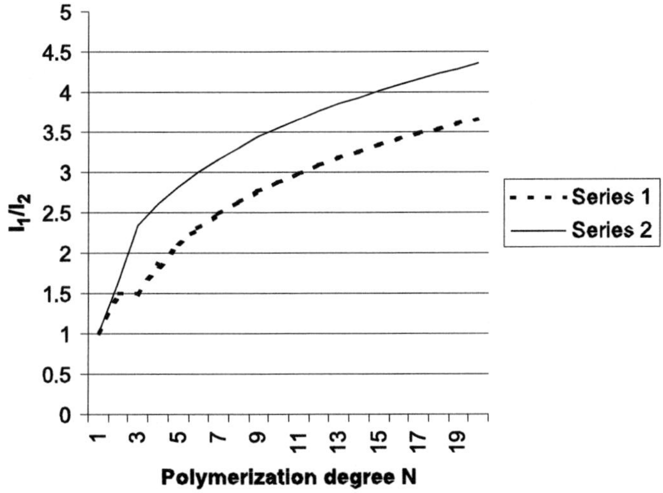

FIG. 5.9 Relationship between the degree of polymerization and the value of moments of inertia: series 1: I_1; series 2: I_2.

inertia increases very fast: though Fig. 5.9 presents not second moment of inertia but its relation to first moment of inertia, this relation increases even faster than first moment of inertia.

As the characteristic size of statistical N-mer, I_1 and I_2 depend only on m.

B. Influence of Cross-Linking on the Characteristic Size of the Statistical Polymer

Cross-link formation between monomeric units belonging to k_1th and k_2th levels ($k_1 < k_2$) is equivalent to the transfer of the unit from k_2th level onto ($k_1 + 1$)-th one. Hence:

Cross-linking between kth and ($k + 1$)-th levels does not change the characteristic size.

If $k_2 > (k_1 + 1)$, all units having level more k_2 and connected with the linked unit on k_2th level "are moved" onto the corresponding levels.

Of course, the assumption of nonchanging of the characteristic size because of cross-linkages inside kth level or between kth and ($k + 1$)-th levels is an approximation, but the fraction of such cross-links is so low that the caused error can be neglected.

The probability of getting cross-linkage between k_1th and k_2th levels is

$$\Pr(k_1,k_2) = \alpha_c R_{k_1}(N) R_{k_2}(N) \tag{5.32}$$

$$\alpha_c = \frac{1}{\sum_{k_1=t}^{N} \sum_{k_2=t}^{N} R_{k_1}(N) R_{k_2}(N)} \tag{5.33}$$

where α_c is the normalization coefficient.

The influence of cross-linking on the characteristic size is equivalent to the change of the values of $R_k(N)$:

$$\delta R_k(N) = u_{k+} - u_{k-} \tag{5.34}$$

$$u_{k+} = \sum_{k_1=K_2+2}^{N} \nu_{k_1} \tag{5.35}$$

$$u_{k-} = \sum_{k_2=1}^{k-2} \nu_{k_2} \tag{5.36}$$

where $\nu_{k_1} = \Pr(k,k_1)$ *and* $\nu_{k_2} = \Pr(k,k_2)$.

The change of the characteristic size because of formation of C_l cross-links in a statistical polymer is equivalent to C_l times repeated transformation given by Eqs. (5.32)–(5.36).

The change of the characteristic size because of cross-linking is given by Fig. 5.10.

We notify that, as follows from Fig. 5.10, the number of cross-links is not so important as their existence itself (meant even the existence of one single cross-link): even one cross-link changes the characteristic size much more than the formation of second and third links. This result allows us to neglect the formation of second, third, fourth, etc. cross-links in evaluations of characteristic size and related parameters of SPM and simplify computer calculations.

C. Characteristic Size of Multicomponent Statistical Polymer

Equations for the characteristic size of multicomponent statistical polymer are very similar to those for the one-component case (considered above), but there are some changes related to the multicomponent composition.

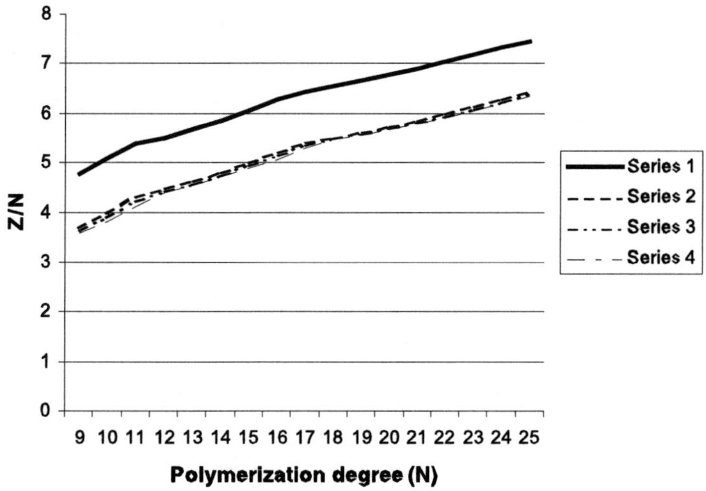

FIG. 5.10 Influence of cross-linkage onto the characteristic size of statistical polymer: series 1: non-cross-linked statistical polymer; series 2: statistical polymer with one (single) cross-link; series 3: statistical polymer with two cross-links; series 4: statistical polymer with three cross-links.

Equation (5.25) is written in the multicomponent case as follows:

$$R_{ik}(N+1) = R_{ik}(N) + \frac{y_i V_k(N)}{V_\Sigma(N)} \tag{5.37}$$

where R_{ik} is the number of units of ith component on kth level, y_i is the concentration of ith component in the polymeric phase (that may easily differ from the concentration in the bulk solution, because of preferences in interactions between various components). As R_k, the value of R_{ik} does not need to be integer and may be below 1.

The number of vacancies on kth level provided by ith component is found from the following equation equivalent to Eq. (5.26):

$$V_{ik}(N) = m_i R_{i(k-1)}(N) - \sum_{j=1}^{M} R_{jk}(N) \frac{V_{j(k-1)}(N)}{V_k(N)} \tag{5.38}$$

As in Chapter 3, M is here the number of components able to polymerize, aggregate, or associate.

Equation (5.27) for the multicomponent case is transformed to a less obvious form:

$$R_{i_1}(N) = Y_i \qquad V_{i_1}(N) = 0 \tag{5.39}$$

The first part of Eq. (5.38) is because the basis unit is eventually chosen in the polymeric structure; hence, its value is proportional to the concentration of ith component in the polymeric phase.

Now, we just take into account that

$$R_k(N) = \sum_{i=1}^{M} R_{ik}(N) \tag{5.40}$$

and all mechanical characteristics for multicomponent statistical polymers are written as in the one-component case.

Let us notify that the above-presented method of the estimation of the characteristic size of statistical polymer allows also the evaluation of the radius of gyration and correlation function of branched macromolecules [7].

III. ANALYSIS OF THE RANDOM FRACTAL MODEL

A. Random Fractals as Structures

Some definitions related to random fractals, their formation conditions and properties were considered in Chapters 1 and 3. We notified that random fractals can be considered as normal objects (in many cases having high symmetry), structure of which corresponds to non-three-dimensional space. Now, let us consider the relationships between various parameters characterizing random fractals.

The main idea of the formalism of random fractals consists in the numerical description of extensive parameters of objects identified as random fractals through a scaling-law relation [11]:

$$\text{Parameter} \sim \text{scale}^{\beta} \tag{5.41}$$

where β is a coefficient, value of which is determined by the dimensionality of the fractal d_f.

One may divide the space of random fractal into little cells, each having size l_f, and count their number intercepted by the curve of the surface. This number N_{cell} allows the estimation of the length or surface area of the fractal [11]:

$$L_f(l_f) = N_{\text{cell}}(l_f)l_f \tag{5.42}$$

$$A_f(l_f) = N_{\text{cell}}(l_f)(l_f)^2 \tag{5.43}$$

where $L_f(l_f)$ is the measured length (characteristic size), and $A_f(l_f)$ is the surface area of the considered fractal [11]. For a line and a surface one obtains, respectively [11],

Line:

$$N_{\text{cell}}(l_f) \sim (l_f)^{-1} \tag{5.44}$$

Surface:

$$N_{\text{cell}}(l_f) \sim (l_f)^{-2} \tag{5.45}$$

In the general case, for a fractal having dimensionality d_f:

$$N_{cell}(l_f) = \Theta_f l_f^{-d_f} \tag{5.46}$$

where Θ_f is the coefficient determined by the geometry of the fractal (its symmetry). Hence, one obtains from Eqs. (5.42)–(5.46):

$$L_f(l_f) = \Theta_f l_f^{1-d_f} \tag{5.47}$$

$$A_f(l_f) = \Theta_f l_f^{2-d_f} \tag{5.48}$$

As it was shown in Chapter 3, random fractals can be also consistently characterized by volume and surface area [8]:

$$A_f = \alpha_f V_f^{1-1/d_f} \tag{5.49}$$

where the coefficient α_f characterizes the geometry of the fractal; particularly, for cubes and spheres $\alpha_f = 2d_f$. In estimations of α_f it is always useful to remember that (as it was shown in Chapter 3) the symmetry of the resulting structure is the same as the initial one: cluster growth inside cube brings cube fractal, inside ball brings ball fractal, etc.

In most of problems, fractal formation is considered for discrete models (otherwise the above model of cell counting would be invalid).

In any form of the characterization of fractals, the main problem consists in the evaluation of the fractal dimensionality d_f or both volume and surface area of the coexisting continuous and empty phases. Let us consider some problems of this kind.

B. Fractal Formation by the Casual Mechanism

The casual mechanism of fractal formation means that the growth of the cluster inside the primary structure does not depend on the prehistory of the system. Let us consider a discrete system having the total volume V_0 and consisting of cells having (each) volume $v_0 = 1$. The volume balance is given by

$$dV_1 = -dV_2 \tag{5.50}$$

where V_1 and V_2 are the empty (resulting) and the filled (initial) substructures. Of course, $V_1/V_0 = \xi$ (the porosity). Each cells has $n_n = 6$ neighbor cells. Among the neighbor cells, there are m_e empty cells (belonging to the porous cluster) and $m_f = (n_n - m_e)$ filled cells belonging to the initial (filled) substructure. If the amount $(dV/v_0) = dV$ of filled cells are transformed to empty cells, the change of the internal surface area is

$$dA_i = (n_n - 2m_e)A_0 \frac{dV/n_n}{V_0} \tag{5.51}$$

The total surface area of the porous cluster includes internal and external areas:

$$A_{p\Sigma} = A_i + A_{pe} \tag{5.52}$$

$$A_{pe} = \frac{A_0 V_1}{V_0} \tag{5.53}$$

In accordance with the Monte Carlo approach, $m_e/n_n = V_1/V_0$, hence, Eq. (5.51) is transformed to the following simple form:

$$A_{p\Sigma} = A_0 V_1 \frac{1 - V_1/V_0}{V_0} = A_0 \xi (1 - \xi) \tag{5.54}$$

From Eqs. (5.49) and (5.54) we obtain the value for the fractal dimensionality of the filled substructure:

$$d_f = d_{f0} - \xi(d_{f0} - 1) = 3 - 2\xi \tag{5.55}$$

and for $\xi = 1$, $d_f = 1$ (dimensionality of line).

C. Structural Characteristics of Microporous Cluster Formed Randomly Under Thermodynamic Limitations

The physical aspect of the growth of microporous cluster randomly under thermodynamic limitations was considered in Chapter 3. Now, let us use the equations derived in Chapter 3 for the evaluation of structural characteristics of microporous cluster.

The principal equations describing the evolution of microporous cluster under thermodynamic limitations are written as (see Chapter 3):

$$W_r(n_e) = N_p(n_e)\exp\left(\frac{-n_n - n_e)\sigma a_0}{R_g T}\right) \tag{5.56}$$

$$dN_{f0} = -dV\psi\frac{N_{f0}n_n}{\left(\sum_{K=0}^{n_n-1}(n_n - K)N_{f(n_n-K)}\right)} \tag{5.57}$$

$$dN_{f_{nn}} = -W_r(n_n)dV + \frac{\psi dV}{\sum_{K=0}^{n_n-1} KN_{f(n_n-K)}} \tag{5.58}$$

$$dN_{fne} = -W_r(n_e)dV + \frac{\psi dV}{[N_{f(n_e-1)}(n_n - n_e + 1) - N_{fn_e}(n_n - n_e)]/\sum_{K=0}^{n_n-1}(n_n - K)N_{fk}} \tag{5.59}$$

$$dV = \frac{dN_f}{\rho_f} \tag{5.60}$$

where N_{fne} is the number of filled cells having (each) n_e empty neighbors, n_e is the number of empty neighbor cells, Ψ is the normalization coefficient, and dV is the change of the volume of the porous cluster.

Figures 5.11–5.16 below present the change of the numbers of N_{fne} ($n_e \leq 5$) vs. the volume fraction of the porous phase for varied value of $\sigma^* = (\sigma a_0)$ and $T = 500\,\mathrm{K}$. The parameter σ^* has the sense of the surface tension per 1 mol of cells forming the continuous solid phase. Of course, N_{fne} do not need to be integer numbers.

We can notify, based on Figs. 5.11–5.16, that the surface tension (cohesive force) in the continuous phase influences much the structural parameters of the product of synthesis of microporous material. The tendency of the increase of the fraction of filled cells having less empty neighbors with the increase of the surface tension (N_{f0}) on Fig. 5.11 and N_{f1} on Fig. 5.12) is obvious. Respectively, the fractions of other filled cells (having more empty neighbors-N_{f2}, N_{f3}, N_{f4}, N_{f5}) decreases. Physically,

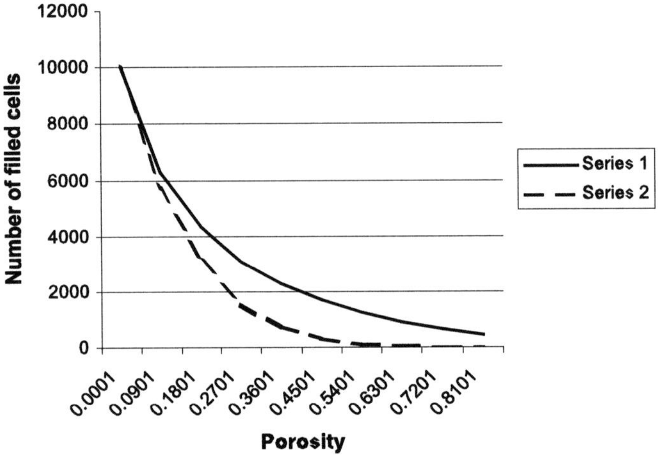

FIG. 5.11 Number of filled cells having filled neighbors only (N_{f0}) as function of porosity, for varied surface tension in the continuous phase: series 1: $\sigma^* = 10{,}000$ J/mol; series 2: $\sigma^* = 1000$ J/mol.

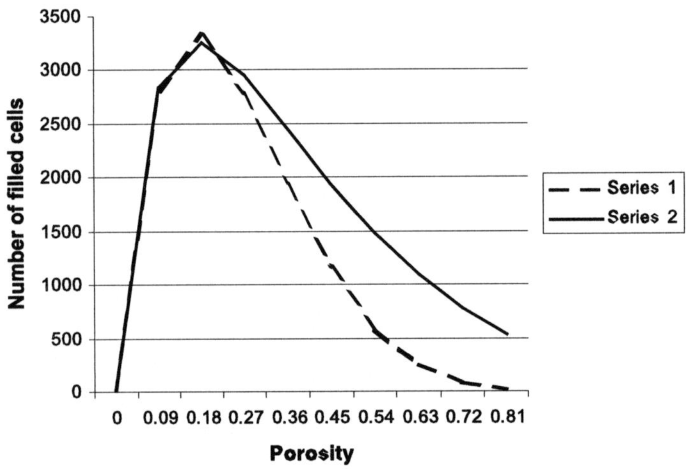

FIG. 5.12 Number of filled cells having (each) one empty neighbor only (N_{f1}) as function of porosity, for varied surface tension in the continuous phase: series 1: $\sigma^* = 10{,}000$ J/mol; series 2: $\sigma^* = 1000$ J/mol.

one can explain this fact by the tendency of avoiding the growth of the internal surface with the increase of the cohesive force. One of the sequences of this phenomenon consists in the increase of the connectedness of the resulting microporous structure.

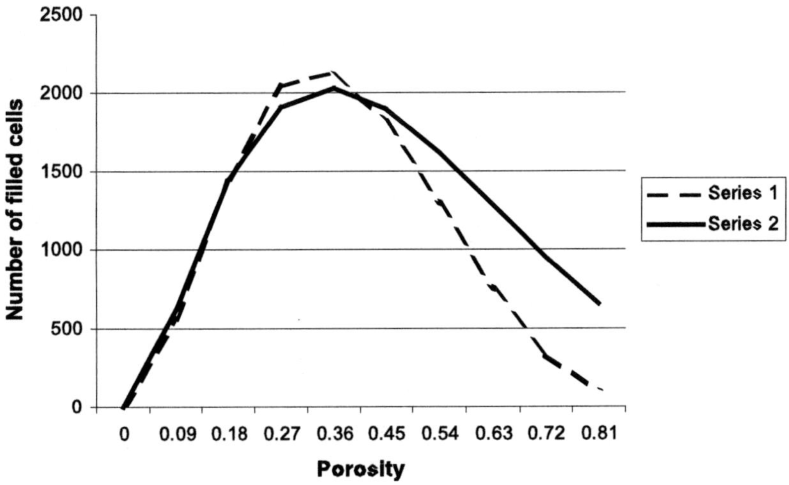

FIG. 5.13 Number of filled cells having (each) two empty neighbors only (N_{f2}) as function of porosity, for varied surface tension in the continuous phase: series 1: $\sigma^* = 10{,}000$ J/mol; series 2: $\sigma^* = 1000$ J/mol.

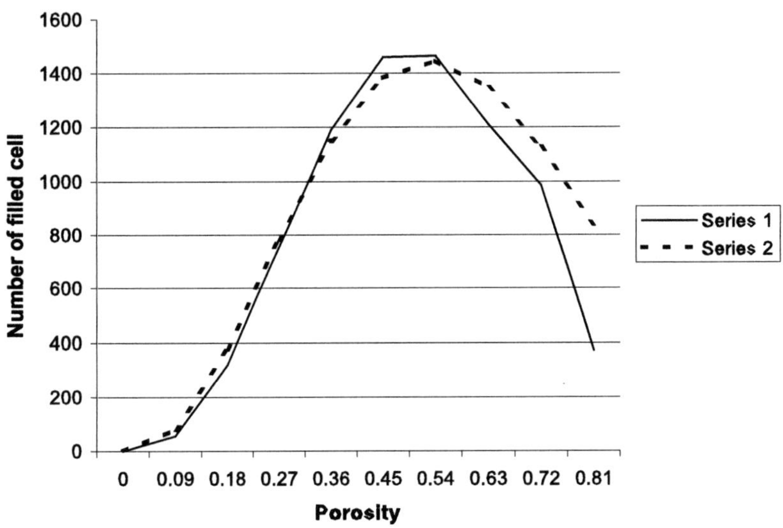

FIG. 5.14 Number of filled cells having (each) three empty neighbors only (N_{f3}) as function of porosity, for varied surface tension in the continuous phase: series 1: $\sigma^* = 10{,}000$ J/mol; series 2: $\sigma^* = 1000$ J/mol.

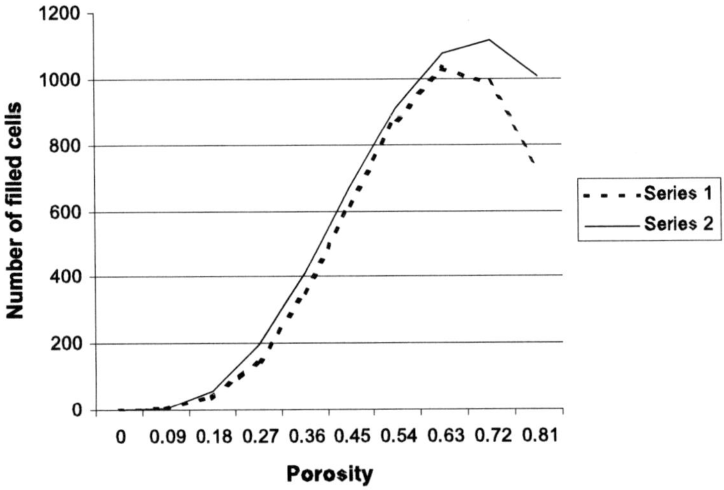

FIG. 5.15 Number of filled cells having (each) four empty neighbors only (N_{f4}) as function of porosity, for varied surface tension in the continuous phase: series 1: $\sigma^* = 10{,}000$ J/mol; series 2: $\sigma^* = 1000$ J/mol.

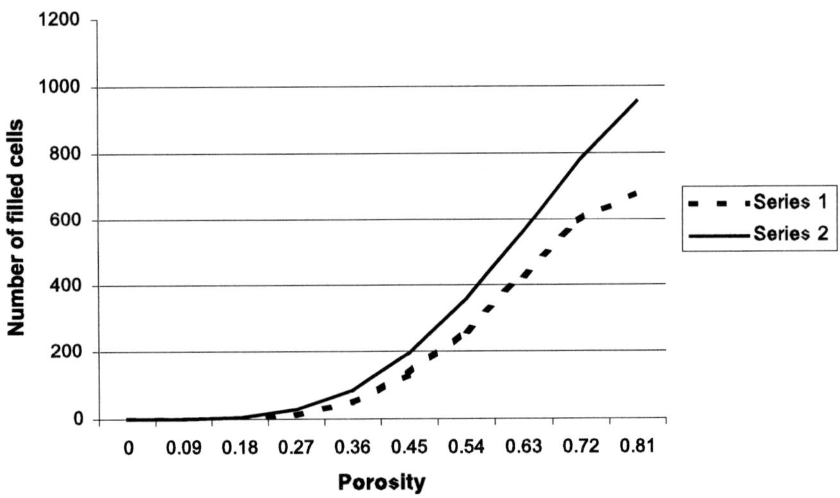

FIG. 5.16 Number of filled cells having (each) five empty neighbors only (N_{f5}) as function of porosity, for varied surface tension in the continuous phase: series 1: $\sigma^* = 10{,}000$ J/mol; series 2: $\sigma^* = 1000$ J/mol.

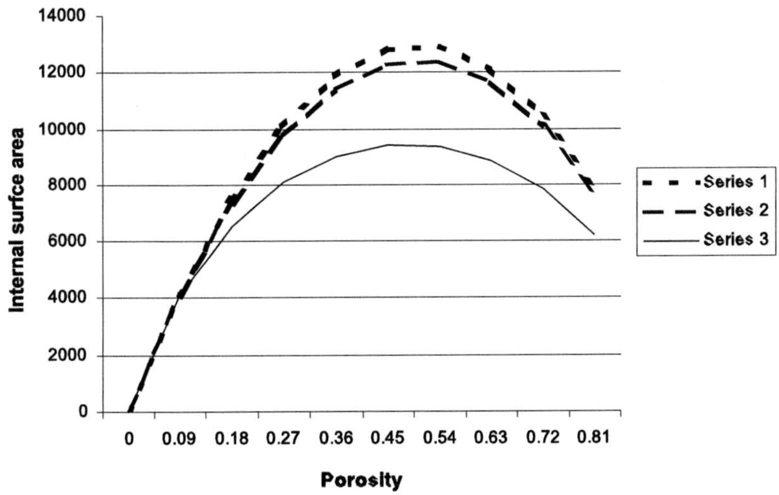

FIG. 5.17 Internal surface area in system with microporous cluster: series 1: $\sigma^* = 100$ J/mol; series 2: $\sigma^* = 1000$ J/mol; series 3: $\sigma^* = 10{,}000$ J/mol.

Figure 5.17 presents the behavior of the internal surface area.

It is important to notify that the free energy of the microporous cluster is always negative for low values of porosity, due to the entropy factor.

IV. CONCLUSIONS

1. Elementary assumptions about the limit range of available energy for micropores allow the obtainment of the consistent system of equations for characteristic parameters of microporous media formed in steady-state conditions. The error of the mentioned assumptions can be neglected for micropores.
2. In the thermodynamic aspect, specific properties of microporous media can be formulated in terms of chemical potential introduced for micropores. Most of properties of the function of the chemical potential of micropores are very similar to those of traditional substances.
3. The driving force for aging of micropores and many kinds of hysteresis is the high chemical potential of micropores.
4. The characteristic size of statistical polymer, needed for the calculation of structural parameters of branched macromolecular systems, is introduced in the probabilistic sense for statistical polymer itself.

5. All structural characteristics of non-cross-linked branched macromolecule depend only on its number of branching (related to its functionality).
6. The effect of cross-linkage reduces significantly the characteristic size of statistical polymer, while the number of formed cross-links is not so important.
7. Fractal properties of microporous substructures (clusters) can be characterized in one of the following form:
   a. Just the value of the fractal dimensionality
   b. The values of the internal surface and volume of both empty and continuous phases
8. Calculations on the base of the discrete model show that structural properties of microporous clusters formed due to a mechanism of removal of filled cells depend much on the surface tension (cohesive forces) in the continuous phase. The increase of the value of the surface tension leads to the increase of the connectedness of the empty substructure.
9. Evaluations of the free energy of microporous substructure show that not all microporous media have specific (first of all adsorptive) properties. The conditions of the appearance of adsorptive properties for a microporous substructure comprise:
   a. Sufficient value of the surface tension in the continuous phase per the thermal energy (determined by the current temperature)
   b. Sufficient value of the porosity

REFERENCES

1. McKetta, J.J.; Cunningham, W.A. (Eds). *Encyclopedia of Chemical Processing and Design*; v. 41, p. 130. Marcel Dekker: New York, 1976.
2. Tsekouras, G.A.; Provata, A. Fractal properties of the lattice Lotka-Volterra model. Phys. Rev. E: Statistical, Nonlinear, and Soft Matter Physics **2002**, *65*(1–2), 16204/1–16204/8.
3. Cerofolini, G.F. Model which allows for the Freundlich and the Dubinin–Radushkevich adsorption isotherms. Surf. Sci. **1975**, *51*(1), 333–335.
4. Romm, F. Thermodynamic description of formation of microporous adsorbents. (linear steady-state approximation). J. Coll. Interface Sci. **1996**, *179*(1), 1–11.
5. Romm, F. Thermodynamic description of formation of microporous adsorbents. (non-steady-state approximation). J. Coll. Interface Sci. **1996**, *179*(1), 12–19.

6. Romm, F. Derivation of the equations for isotherm curves of adsorption on microporous gel materials. Langmuir **1996**, *12*(14), 3490–3497.
7. Romm, F. Thermodynamics of microporous material formation. In: *Surfactant Science Series "Interfacial forces and fields: Theory and applications"* (Monographic series, Ed. Hsu, J.-P.), Marcel Dekker: New York, 1999, Chapter 2; pp. 35–80.
8. Romm, F. Modeling of internal structure of randomly formed microporous material of illimited volume following thermodynamic limitations. J. Coll. Interface Sci. **2000**, *227*, 525–530.
9. Romm, F. Modeling of random formation of microporous material following thermodynamic limitations. J. Coll. Interface Sci. **1999**, *213*(2), 322–328.
10. Romm, F.; Figovsky, O. Theoretical modeling of mechanical resistance and stability and related characteristics of polymeric systems with branching/cross-linking. J. Solid State Chem. **2002**, *164*(2), 237–245.
11. Gutfraind, R.; Sheintuch, M. Scaling approach to study diffusion and reaction processes on fractal catalysts. Chem. Eng. Sci. **1992**, *47*(17/18), 4425–4433.

6
Engineering Applications of the Concept of Microporous Systems

I. ORGANIZATION OF ENGINEERING CALCULATIONS: USING TOOLS OF MICROPOROSITY THEORY

A. General Procedure of Simulations of Synthesis, Structure, and Properties of Microporous Materials

In this chapter we will consider some examples of the practical application of the above-developed tools for the theoretical study of microporosity. These examples will comprise simulations of processes of synthesis of a microporous material, structural studies of the resulting product and estimations of its properties having technical merit. We will also analyze the question about the initial data needed on each stage of calculations. The procedure will include

1. The formal definition of the problem (by the user)
2. The division of the problem entire into some stages of calculations and the scientific definition of calculation problems for each of these stages
3. Theoretical tools used on each stage of calculations
4. Input and output on each stage of calculations
5. Results of calculations and their comparison with the formal definition of the problem

B. Formal Definition of the Problem

The formal definition of the problem of a theoretical (calculative) study of microporosity is formulated in traditional industrial, domestic, and commercial terms. The formal definition of the problem aims to explain to the researcher (person or group responsible for theoretical studies and computer simulations of microporosity) the practical needs of the user. The formal definition of the problem comprises:

The request of the user (a consumer of microporous materials, using them for various needs) about the theoretically estimated properties of some material that is planned to be prepared from several raw materials, with using some technological process

Initial data for the calculations (if those are not widely available or the user prefers, for such or such reasons, to suggest his own initial data)

Intermediate results of calculations (e.g., structural data) in which the user is interested in addition to the final results

For example, if the user is interested in forecasting adsorptive–desorptive properties of some material prepared by a novel technology, the user may order

The information about the forecasted structural characteristics and thermodynamic properties of the expected product

Some measurable structural parameters—for the comparison with the available experimental data

The forecasted adsorption isotherm curves and hysteresis

The forecasted aging of the hypothetical material

As follows from above, the user may be interested not only in the final results about the measurable properties of the material but also in the reproducible information about its structure—for example, for the estimation of the possible changes in the preparation technology, for improving the resulting properties.

The user may also order calculations not related directly to microporosity but using some results from microporosity simulations. In such case, the user is responsible also for bringing to the researcher (performer of calculations) of all additional tools necessary for the accomplishment of such additional job.

The user is responsible for the initial data he brings to the researcher. If the user does not provide some initial data necessary for the solution, he must suggest any source for finding them. If these data differ in different sources, he should choose the sources he prefers.

C. Stages of Calculations

The division of the formal problem into some stages of calculations is carried out by the researcher and aims to prepare "the translation" of the formal definition of the problem into the language of the microporosity theory. For example, for the above formal definition of the problem, the stages of solution will comprise

The computer simulation of the synthesis of the mentioned material
Getting its structural properties as output of the previous stage
Building of the theoretical isotherm of adsorption, based on initial data provided by the user and the output of the previous stage
Estimations of hysteresis and aging

D. Scientific Definition of Calculation Problems

The scientific definition of the calculation problem(s) aims to present the formal definition of the problem (given by the user) in the language of microporosity theory. The scientific definition is required for all stages of calculations (see above) separately.

For example, for the above formal definition of the problem and specification of stages of calculations, the scientific definition of problems will comprise

The class of the mentioned material (polymeric or not), the regime of its treatment (raw materials, treatment close to equilibrium or not, level of temperatures, principal chemical reactions, etc.), and the resulting structural parameters needed for bringing to the user and/or for the further calculations
Estimation of the needed model for the simulation of the adsorption isotherm (VFMT or Langmuir-like model, etc.)
Analysis of the possible mechanisms for aging and hysteresis and the choice of their most appropriate model, in the case of the considered material

E. Choice of Theoretical Tools

The scientific definition of the problem determines the list of theoretical tools needed for its solution. For the considered example, such theoretical tools will comprise

The appropriate model for the process of synthesis of the considered material
The appropriate model for the adsorption on the considered material
The appropriate model for aging and hysteresis for the considered material

The list of the relevant models determines automatically the list of the needed computer programs for the numerical solution of above-defined scientific problems.

F. Input and Output

The lists of needed theoretical models and corresponding computer programs determine the input and the output on each stage of calculations. The information sources for building input may comprise

Initial data provided by the user
Physical constants
Handbook data allowed by the user
Output from the previous stage(s) of calculations

The output will comprise the final and/or intermediate results: characteristics of the obtained microporous structure, its thermodynamic parameters, standard graphs (like the adsorption isotherm), visible percolation threshold (if necessary), etc.

For the considered example, the input and output will comprise

1. On the stage of the simulation of synthesis: thermodynamic parameters of raw materials (from handbooks or brought by the user), constants of rate of chemical reactions (if the process is far from equilibrium), parameters of the regime (temperature, pressure, etc., determined by the user), necessary constants (e.g., gas constant R_g, used in this book), and expected structural parameters of the product (e.g., the coordination number for its minimal fragments).

   The output on the same stage is the resulting porosity, the internal surface area, the amounts of filled cells having some number of empty neighbors, and the value of the free energy for the microporous substructure. The output may comprise also special data, though not needed for the further calculations, but inquired by the user: the moments of inertia, the gyration radius, the correlation function, the fractal dimensionality, etc.
2. On the stage of building the theoretical isotherm of adsorption: Input is the free energy of the microporous substructure (from the previous output), the micropore energy distribution function (from the previous output), the porosity (from the previous output), and constants of adsorption (from the user or allowed handbooks).

   Output is the theoretical isotherm of adsorption (as a graph).
3. On the stage of estimations of aging and hysteresis: Input is the free energy of the microporous substructure (from the output of

stage 1), the micropore energy distribution function (from the output of stage 2), the porosity (from the previous output), and constants of the rate of aging and hysteresis (from the user or allowed handbooks).
Output is curves of aging and hysteresis (as graphs).

Let us notify that the main output used in the calculations on stages 2 and 3 comprises structural parameters calculated on stage 1. That is a normal thing: the structural information is the base for all estimations in the concept of the joint consideration "synthesis–structure–properties" proposed in this book.

G. Results of Calculations

After all the above stages accomplished, the needed calculations are performed and the required results (output) obtained. These results can be of three kinds:

1. Intermediary results not only needed for the further calculations but also requested by the user
2. Intermediary results needed only for the further calculations
3. Results needed only for the user

Among these three kinds of obtained results, the group of results 2) is the interior affair of the performer of calculations (the researcher), but others need to be presented in the appropriate form before transferred to the user. Such presentation means the use of widely adopted terms, graphs, etc., with emphasizing the relationship between the obtained results and the formal definition of the problem by the user.

For the above-considered problem formally defined by the user, the results will comprise:

The values of the calculated structural parameters requested by the user
The forecasted curves of isotherms of adsorption and desorption
The forecasted curves for aging and hysteresis

Now, let us consider some examples of the formal definitions of problems related to microporosity, and the ways for their solution. We will consider problems of three kinds:

1. Estimation of structural parameters and measurable properties from preparation conditions of a microporous material, defined by the user;

2. Estimation of structural parameters indirectly from experimental data about measurable properties of microporous material, provided by the user;
3. Estimation of structural parameters and some measurable properties indirectly from experimental data of measurements of other properties, provided by the user.

II. EXAMPLE 1: ESTIMATION OF STRUCTURAL CHARACTERISTICS AND ADSORPTIVE AND PERMEABILITY PROPERTIES FROM PREPARATION CONDITIONS OF POLYMERIC MATERIAL

A. Formal Definition of Problem by User

The user is interested in forecasting of the some structural parameters (porosity and internal surface area) of a polymeric material prepared by the polymerization of a special monomer solution (within 10 s at 400 K, about 50% of monomers got conversion to macromolecules; at 450 K, the time of semi-conversion of monomers was 4 s) under the normal pressure 1 ata. In 2 h, the polymerization degree was 71 at 400 K and 52 at 450 K. The monomer functionality $f_0 = 6$. The surface tension per surface area of one mole of the products is about $\sigma^* = 2000$ J/mol. The chemical potential of monomers is about 10 kJ/mol. The synthesis should be carried out at the temperature 380 K. The conversion time allowed for the synthesis can be 10 s, 15 s, or 20 s. Please estimate the weight distribution for each option of conversion duration time. The process of the synthesis can be approximately considered as steady state, and the temperature influence on the energy dissipation approximately satisfies Arrenius equation. The formation of cross-links can be neglected. The product is planned for the following uses:

1. In the Separation of Binary Gas Mixture (Gases A and B) by Fast Adsorption (with the Thermal Recovery of the Adsorbent) in Counter Current Mixing Reactors

The cascade of such reactors is adequately described by the matrix presentation of counter current reactors with mixing (see Appendix 1). The price of each reactor is U.S. $ 3000. The constants of adsorption by the Langmuir-like model are for 298 K

For gas A, $\gamma_p = 0.5$ and $K_{\text{ads}}(\varepsilon_0) = 0.6$.
For gas B, $\gamma_p = 0.2$ and $K_{\text{ads}}(\varepsilon_0) = 0.6$.

At 400 K and normal pressure, the amounts of adsorbate are 20 times less (for both gases) than for 298 K. Adsorption of gas A does not influence the adsorption of gas B, and versus. The saturation pressure for both gases should be 10 ata. The efficiency of pumps and compressors is about 90%. Heat capacity of the polymer is about 33 J/mol/K.

The aging and hysteresis for the considered material can be neglected. The energetic estimations of the processes of adsorption and desorptive regeneration of the adsorbent can be carried out by the Carnot cycle.

2. As Molecular Sites in Which Monomer–Polymer Mixture will be Separated

The polymer molecules are so large their permeability can be neglected, while molecules of monomers are little, and their characteristic size is below the expected sizes of the fragments of the molecular site.

In addition to the above-mentioned structural parameters, please evaluate

The theoretical isotherms of adsorption of A and B on the products of the synthesis of the considered material.

The optimal number of the reactors in the cascade, in which enough high efficiency of separation will not require large investments to the equipment.

Which among three considered options for the studied material allow the permeability over the visible percolation threshold? The permeability can be estimated in any form appropriate for the researcher, but the evaluations for all three samples must be in the comparable conditions, meaning that the result may depend only on the internal structure of each sample.

Thus, that is an example of the formal definition of the problem of estimating structural characteristics and some measurable properties of polymeric material. The information provided by the user is, in principle, complete, but needs some treatment before used in computer programs.

B. Stages of Calculations

The stages of calculations for the formally defined problem above will comprise the following:

1. The presentation of the initial data provided by the user in the form appropriate as input in the computer programs. The data about the kinetics and equilibrium of polymerization must be presented as constants of the rate of polymerization, activation

energy of polymerization, etc. The matrix presentation of countercurrent cascade of reactors of mixing must be formulated as a computer program.

2. The computer simulation of the polymerization process of the mentioned material, with the evaluation of thermodynamic parameters. Since porosity and internal surface area are requested (not only as intermediate but also final results), the simulation of the synthesis itself is not sufficient (because both these parameters depend not only on the weight distribution function but also on inter-penetration), and one needs to use also a program for the evaluations of porosity and internal surface area.
3. Computer building of theoretical isotherms for gases A and B and evaluation of the maximal relative distance between them, which corresponds to the best separation.
4. Computer simulation of the cascade of countercurrent reactors, in which gases A and B are separated, with the estimation of the costs for such cascade.
5. Computer simulation of the thermal regeneration of the adsorbent.
6. Computer evaluation of the theoretical permeability of three kinds of the polymeric product, aiming to compare that to the permeability corresponding to the visible percolation threshold.

Thus, after the above formulation of the formal problem defined by the user as a chain of logically related computer calculations, the problem is ready for the scientific definitions in terms of microporosity.

C. Scientific Definition of Calculation Problems

The scientific definition of the theoretical and calculation problem(s) for the considered formally defined problem can be formulated as a number of acts of application of theoretical tools of microporosity concept and related computer calculations, as follows:

1. Solution of the problem of equilibrium non-cross-linked branched polymerization of monomer, functionality of which is 6 and the number of branching of which is 5. Assume that the polymerization process is fast enough; hence, the data about the degree of polymerization practically correspond to the equilibrium weight distribution of the polymers. The solution should correspond to the averaged degree of polymerization: 71 for the temperature 400 K and 52 for the temperature 450 K. The output should contain the changes of enthalpy and entropy in the process of polymerization.

2. Using the results from the first stage of calculation, estimate the constants of the rate of the direct and reverse reactions of polymerization; evaluate the activation energy of polymerization.
3. Using the results from first and second stages of calculations, estimate the evolution of the free energy of the polymerized system with time. Find the values of the free energy after 10 s, 15 s, and 20 s of synthesis duration.
4. Using the results obtained on third stage of calculations, evaluate the internal surface area and the porosity of microporous material, coordination number in which is 6, while the free energy is equal, respectively, to the values obtained on third stage. Evaluate also the distribution of filled cells in the number of their empty neighbors for the further solution of the permeability problem.
5. Provide the values of porosity and internal surface area to the user.
6. Build the theoretical adsorption isotherm curve for gases A and B using the Langmuir-like model of adsorption on micropores (Chapter 4). Forward the adsorption isotherm curves (as one of ordered results) to the user. Comparing both adsorption isotherm curves for A and B, estimate the maximal relative distance between them. Estimate the corresponding pressure, that is, the optimal pressure for the separation of A from B with the use of the reference material. Forward this recommendation to the user.
7. Try some options for the number of reactors in the countercurrent cascade, using the computer program based on the matrix presentation of the cascade. Let us assume that their maximal number is 10. Evaluate the separation efficiency. Forward the results to the user.
8. Evaluate the optimal temperature of the thermal regeneration of the adsorbent, using the Carnot-cycle approximation.
9. Using the structural results obtained on fourth stage (distribution of filled cells in the number of their empty neighbors), evaluate the permeability for three requested situations. Compare the result to one corresponding to the visible percolation threshold. Forward the results of the comparison to the user.

After the accomplishment of all ten stages of calculations and bringing the requested results to the user, the formal problem is solved.

D. Choice of Theoretical Tools and Computer Programs

The above scientific definition of the problem determines the following list of theoretical tools in microporosity theory needed for the obtainment of the solution:

The equilibrium version of the statistical polymer method
The nonequilibrium version of the same method
The discrete model of the formation of microporous material
The Langmuir-like model for adsorption on micropores
The percolation trajectory method for the evaluation of permeability

Each of the above theoretical tools is related to a single computer program. However, in addition to the above list, one needs the following supporting programs not related to microporosity:

The computer program of the matrix presentation of countercurrent cascade of reactors of mixing (based on the information presented in Appendix 1)
The computer program on the evaluation of the energetic efficiency of the thermal regeneration of the adsorbent

If all the above programs are available, the researcher possesses all necessary technique for the solution of the above-defined problem.

E. Input and Output

The input and the output of the calculations on all the stages given above are determined by the available data and the expected results. Let us present input and output for each stage of the calculations as Table 6.1.

As follows from Table 6.1, the above-suggested procedure of calculations can be accomplished on the available theoretical tools and obvious computer programs. The computer programs have the simple structure: just arithmetic calculations, no necessity of object-oriented programming. In most of cases, the order of calculations cannot be changed, because output of previous stage of calculations is usually necessary for the input in further stages.

F. Calculations and Results

Now, let us illustrate the performance of the operations described in Table 6.1.

1. Stage 1. Equilibrium in Non-Cross-Linked Branched Polymerization of Monomers

The functionality 6 corresponds to the branching number 5. The results of the calculations are given on Fig. 6.1.

TABLE 6.1 Input and Output on Each Stage of the Computer Solution of the Problem Defined in Example 1

Stage N	Name of the stage	Model used	Input needed	Output available
1	Simulation of equilibrium polymerization	Statistical polymer (equilibrium version), Chapter 3, Eqs. (3.157)–(3.162)	Number of branching, averaged degree of polymerization	Changes of enthalpy ΔH^0 and entropy ΔS^0 of equilibrium polymerization
2	Simulation of nonequilibrium polymerization	Statistical polymer (nonequilibrium version), Chapter 3, Eqs. (3.170)–(3.176)	ΔH^0, ΔS^0 of polymerization; duration time of semiconversion of monomers	Parameters of nonequilibrium polymerization ΔS_e^0, ε_0
3	Free energy with time	Free energy of insulated system, Chapter 3, Eq. (3.29)	Duration time of semiconversion of monomers, Parameters ΔS_e^0, ε_0	Values of free energy within 10, 15, and 20 s of conversion duration time
4	Estimation of structural parameters	Discrete growth of microporous cluster, Chapter 5, Eqs. (5.56)–(5.60)	Free energy, coordination number, surface tension $\sigma^* = \sigma a_0$, temperature	N_{f0}, N_{f1}, N_{f2}, N_{f3}, N_{f4}, N_{f5}, N_{f6}; porosity ξ; internal surface area A_{int}
5	Forwarding ξ and A_{int} to the user	–	–	–
6	Building of theoretical isotherms of adsorption	Langmuir-like model, Chapter 4, Eqs. (4.34)–(4.36)	γ_p, $K_{ads}(\varepsilon_0)$ for gases A and B; ξ; temperature T	Isotherms of adsorption for the user

7	Estimation of optimal pressure, distribution of gases in phases	Comparison of the theoretical isotherms	Theoretical isotherms	Optimal β_i for gases A and B, the optimal pressure
8	Optimization of cascade of reactors	Matrix presentation of countercurrent cascade, Appendix 1, Eqs. (A1.3) and (A1.4)	β_i for gases A and B	Optimal number of reactors, resulting concentrations of gases in both phases
9	Optimization of the regeneration regime	Arrenius equation for desorption	Initial concentration of A and B in the adsorbate; change of this concentration with temperature	Optimal temperature of desorption, for the user
10	Simulation of permeability	Percolation trajectory method, Chapter 4, Eqs. (4.63)–(4.70)	Coordination number; N_{f0}, N_{f1}, N_{f2}, N_{f3}, N_{f4}, N_{f5}, N_{f6}; porosity ξ.	Permeability vs. thickness for each sample

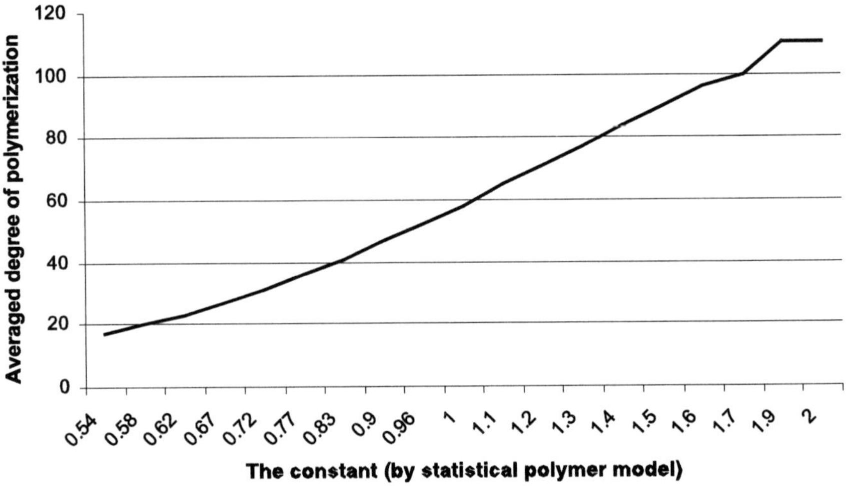

FIG. 6.1 Averaged degree of polymerization for various values of equilibrium polymerization constant K_0 [Eq. (3.161)].

As follows from Fig. 6.1, the polymerization degree 71 corresponds to the constant $K_0 = 1.2$, while the polymerization degree 52 corresponds to the constant $K_0 = 0.96$. That corresponds to the enthalpy of polymerization $\Delta H^0 = -66.8$ kJ/mol and change of entropy $\Delta S^0 = -15.2$ J/mol. Respectively, for 380 K, the value of K_0 will be 1.334; the free energy of the polymeric mixture in equilibrium will be about 2 kJ/mol.

2. Stage 2. Nonequilibrium Polymerization

Now, let us consider the nonequilibrium aspect of the same problem. We will use the negentropy approach: negentropy in the quasi-insulated system changes in accordance to the following law:

$$U(t) = U(t = 0)\exp(-\sigma_d t) \tag{6.1}$$

where the initial negentropy $U(t = 0)$ is equal to the standard potential of pure monomers $\mu_1(t = 0)$, and the dissipation parameter σ_d should be found from these kinetic evaluations. When $t = \tau$ (τ is the period of semiconversion of monomers), the contribution of monomers is

$$U_1(\tau) = C_1(\tau)\mu^0(\text{monomers}) + C_1(\tau)R_g T \ln C_1(\tau) \tag{6.2}$$

where $C_1(\tau) = 0.5$. When equilibrium is gained, the contribution of monomers in the total (minimal) negentropy is

$$U_1(\infty) = C_1(\infty)\mu^0(\text{monomers}) + C_1(\infty)R_g T \ln C_1(\infty) \tag{6.3}$$

According to the superposition principle (Chapter 3), the following correlation is valid:

$$\frac{U_1(0) - U_1(\tau)}{U_1(0) - U_1(\infty)} = \frac{U_\Sigma(0) - U_\Sigma(\tau)}{U_\Sigma(0) - U_\Sigma(\infty)} \tag{6.4}$$

The parameters regarding equilibrium ($t = \infty$) are found from the equilibrium solution. Hence, the only unknown parameter in Eq. (6.4) is $U_\Sigma(\tau)$:

$$U_\Sigma(\tau) = U_\Sigma(0) - [U_\Sigma(0) - U_\Sigma(\infty)]\frac{U_1(0) - U_1(\tau)}{U_1(0) - U_1(\infty)} \tag{6.5}$$

We obtain from Eq. (6.5): for 400 K, $\sigma_d = 0.091$ Hz; for 450 K, $\sigma_d = 0.102$ Hz. The visible activation energy for the energy dissipation by polymerization is 3.415 kJ/mol. Using the Arrenius form for the dissipation parameter, we obtain $\sigma_d = 0.1202$ Hz for 380 K. The resulting weight distributions of polymers after 10 s ($U_\Sigma = 4.405$ kJ/mol), 15 s ($U_\Sigma = 3.563$ kJ/mol), and 20 s ($U_\Sigma = 2.723$ kJ/mol) of conversion, respectively, are given on Fig. 6.2.

Results on Fig. 6.2 are forwarded to the user.

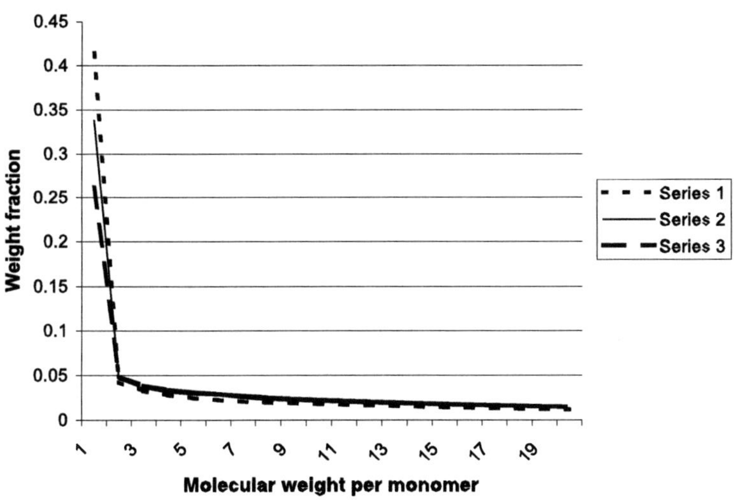

FIG. 6.2 Weight distribution of polymers (Example 1): series 1: after 10 s of conversion; series 2: after 15 s of conversion; series 3: after 20 s of conversion.

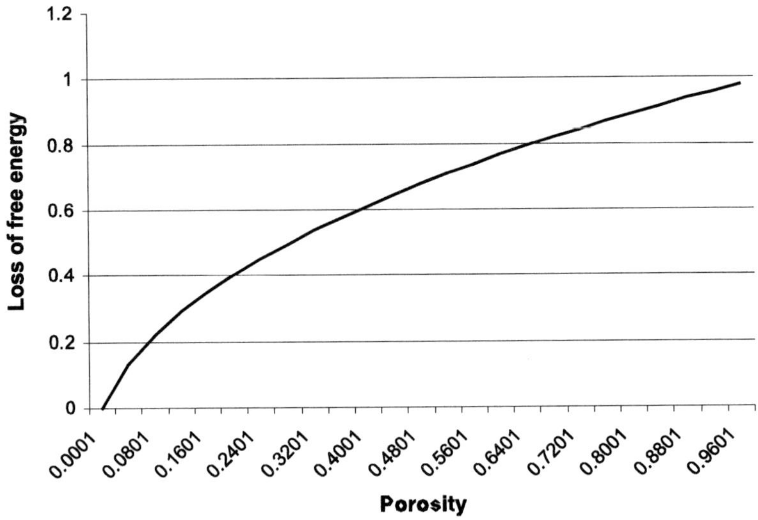

FIG. 6.3 Energy loss per 1 mol of gas phase vs. porosity.

3. Stage 3. Evaluation of the Loss of the Free Energy of the System

Following the results of stage 2, we find the loss of free energy corresponding to various periods of conversion: for 10 s, 44%; for 15 s, 75%; 20 s, 82%. This loss of free energy corresponds to the formation of free surface. The theoretical equilibrium corresponds to the loss of free energy R_gT per mole of the gas phase. Figure 6.3 shows the relationship of the free energy loss per 1 mol of the gas phase.

As follows from Fig. 6.3, the energy loss for 10 s (44%) corresponds to porosity about 0.32, 15 s (energy loss 75%) to porosity = 0.63, and 20 s (82%) to 0.75.

4. Stage 4. Evaluation of the Distribution of Filled Cells in the Number of Their Empty Neighbors, the Internal Surface and Energy of the System

For the considered system, the behavior of N_{f0}, N_{f1}, N_{f2}, N_{f3}, N_{f4}, N_{f5}, N_{f6}, and the internal energy and internal surface of the system (per 10,000 cells) with porosity rising is given on Figs. 6.4–6.13, respectively.

Let us notify that the function on Fig. 6.11 is the direct line, that is because both energy and the volume of the porous phase are additive parameters.

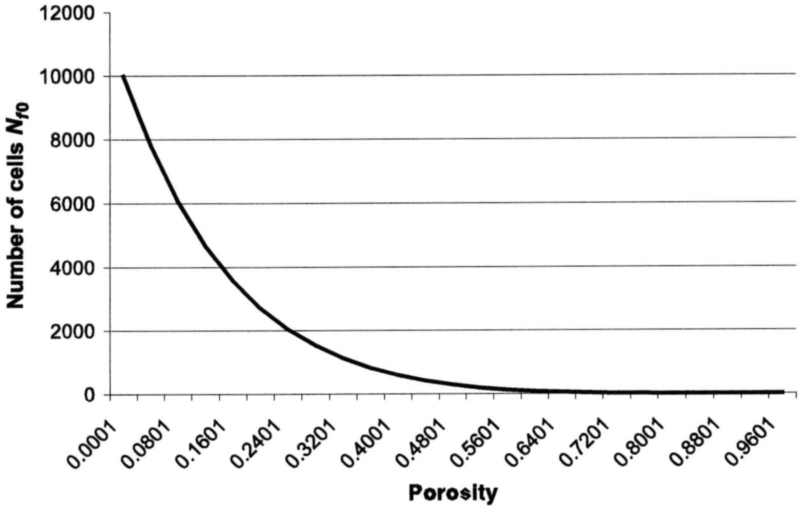

FIG. 6.4 Number of filled cells having no empty neighbors N_{f0} vs. porosity.

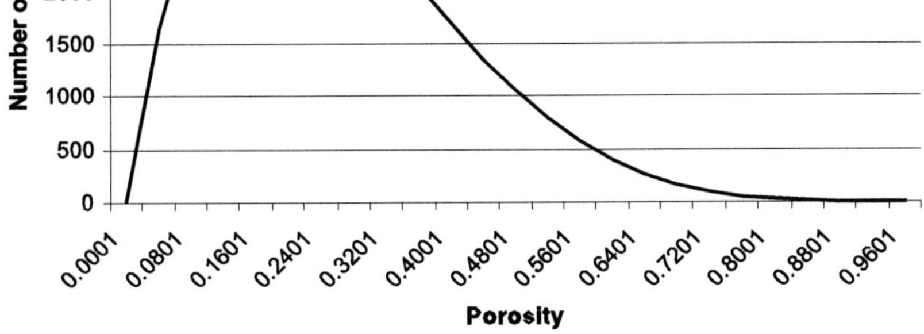

FIG. 6.5 Number of filled cells having (each) one empty neighbor only, N_{f1}, vs. porosity.

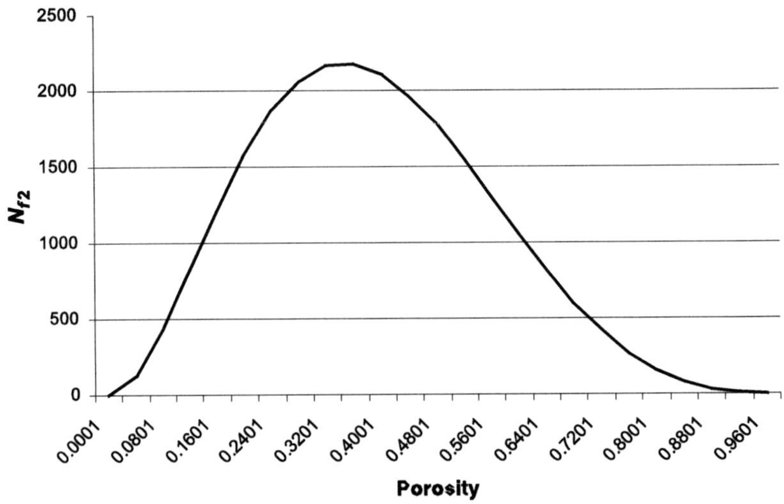

FIG. 6.6 Number of filled cells having (each) two empty neighbors only, N_{f2}, vs. porosity.

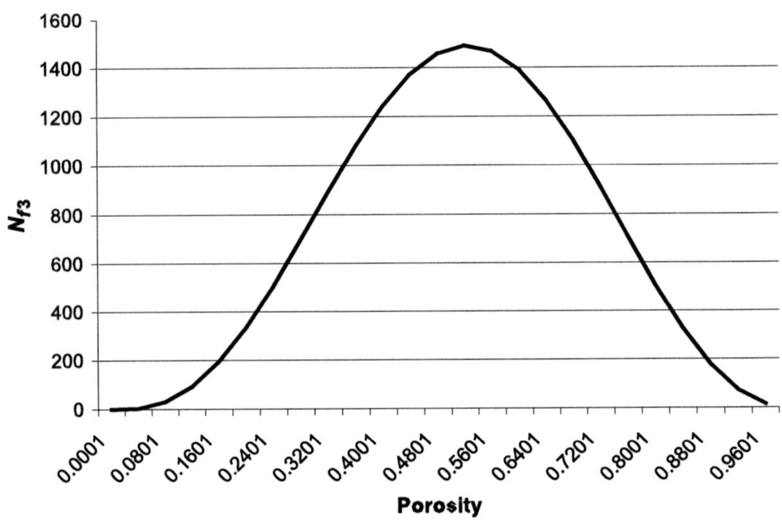

FIG. 6.7 Number of filled cells having (each) three empty neighbors, N_{f3}, vs. porosity.

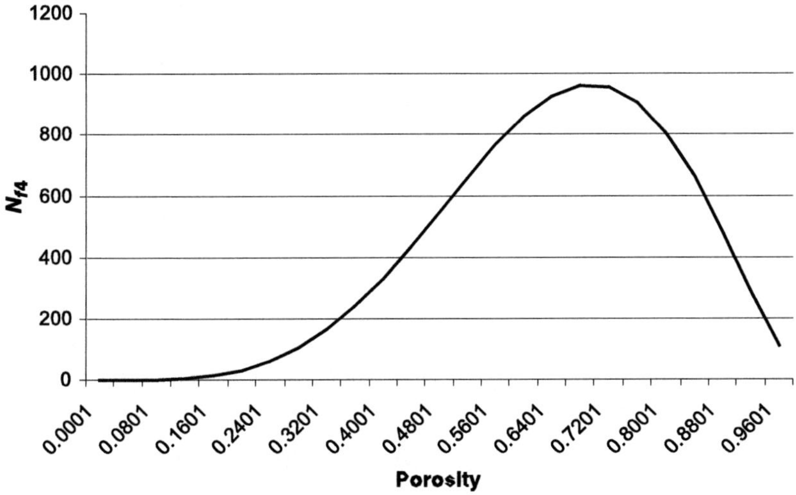

FIG. 6.8 Number of filled cells having (each) four empty neighbors, N_{f4}, vs. porosity.

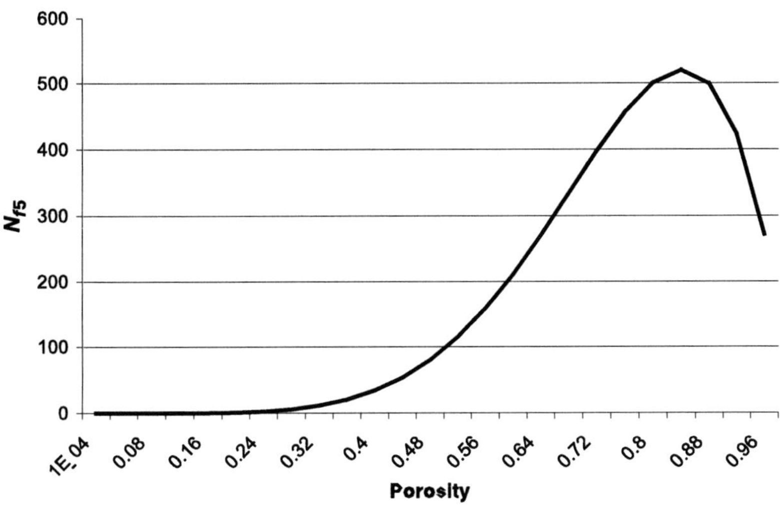

FIG. 6.9 Number of filled cells having (each) five empty neighbors, N_{f5}, vs. porosity.

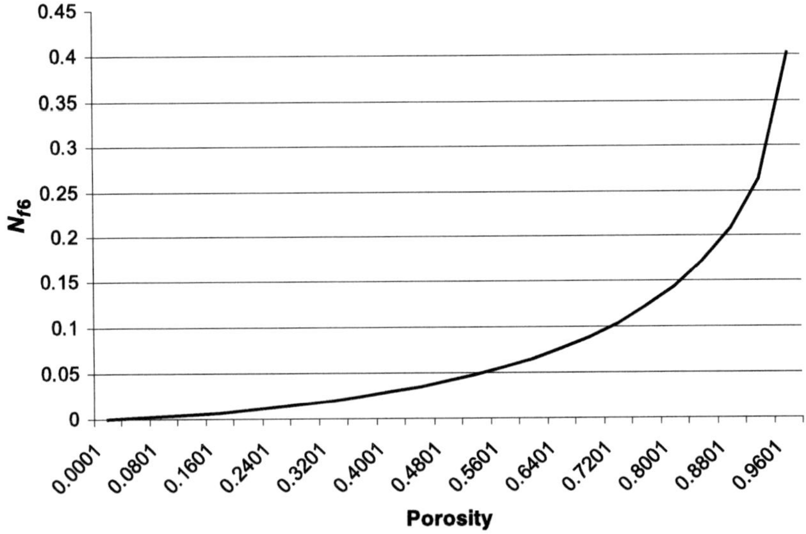

FIG. 6.10 Number of filled cells having (each) six empty neighbors, N_{f6}, vs porosity.

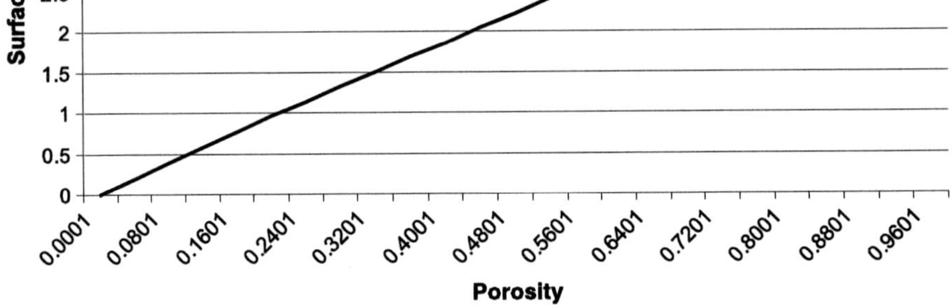

FIG. 6.11 Internal surface vs. porosity.

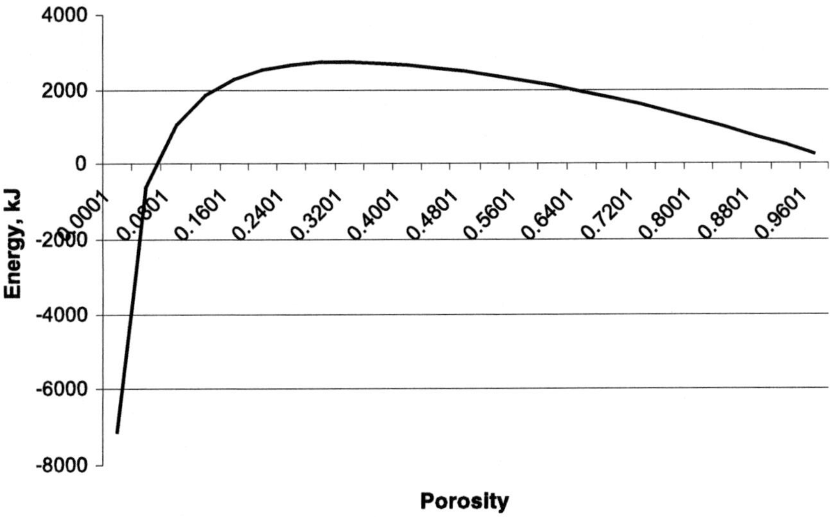

FIG. 6.12 Internal energy per amount of empty phase vs. porosity.

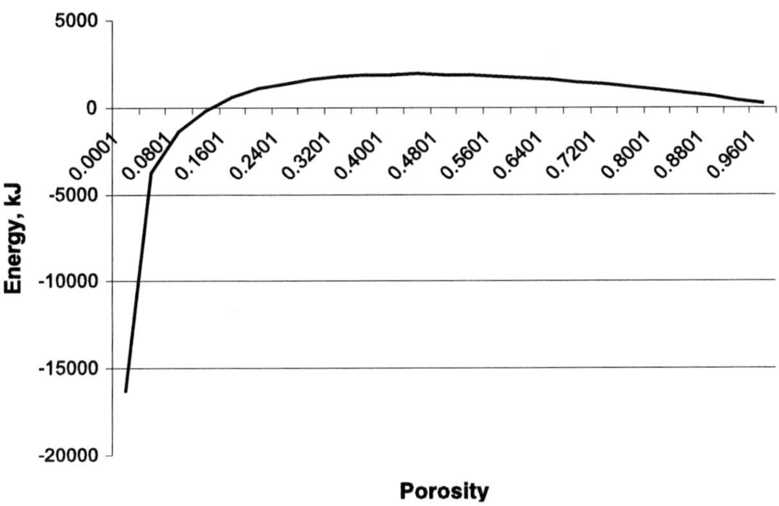

FIG. 6.13 Internal surface per amount of the continuous phase vs. porosity: 10 s ($U_\Sigma = 4.405$ kJ/mol), 15 s ($U_\Sigma = 3.563$ kJ/mol), and 20 s ($U_\Sigma = 2.723$ kJ/mol) of conversion.

5. Stage 5. Preliminary Results (the Values of Porosity and Internal Surface Area) to the User

The preliminary results of calculations are given below in Table 6.2.

6. Stage 6. Theoretical Adsorption Isotherms

For gas A, $\gamma_p = 0.5$ and $K_{ads}(\varepsilon_0) = 0.6$.
For gas B, $\gamma_p = 0.2$ and $K_{ads}(\varepsilon_0) = 0.6$.

Let us notify that the adsorption isotherms in all these cases are very similar (see Figs. 6.14–6.16). Therefore, in the below analysis we neglect their divergence.

7. Stage 7. Comparison of Adsorption Isotherm Curves for A and B

Now, let us consider the effectiveness of the considered species for the separation of gases A and B, on the base of their adsorption isotherms. As we wrote above, all three samples have similar curves of adsorption isotherms for A and B, and the researcher recommends the user to use the cheaper product among three considered options. The separation effectiveness of this product is determined by the relative divergence in the adsorption of A and B (Table 6.3), and one finds three opportunities for A–B mixture separation:

After the treatment in the adsorber cascade, the concentration of A in the adsorbent is much higher than that of B.

TABLE 6.2 Preliminary Results of Calculations of Structural Characteristics of the Considered Materials

Parameter	Conversion 10 s	Conversion 15 s	Conversion 20 s
Porosity	0.32	0.63	0.75
Internal surface area per amount of continuous phase, units/mole[a]	1.5	2.7	3.5
N_{f0}	1900	60	10
N_{f1}	2433	270	60
N_{f2}	2167	850	280
N_{f3}	892	1300	750
N_{f4}	162.8	920	910
N_{f5}	11.34	260	440
N_{f6}	0.02	0.074	0.118

[a]The surface area equal 6 units corresponds to the situation when all filled cells are isolated.

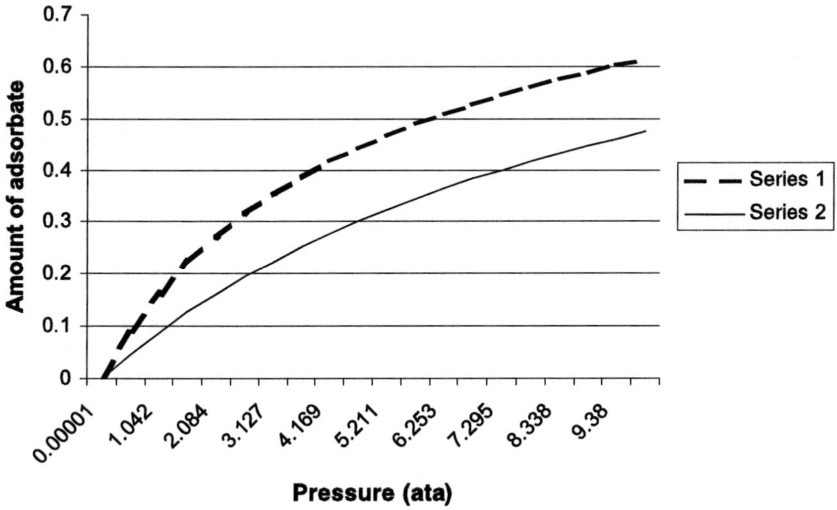

FIG. 6.14 Theoretical isotherms for adsorption of A and B on the adsorbent (10 s of conversion): series 1: gas A; series 2: gas B.

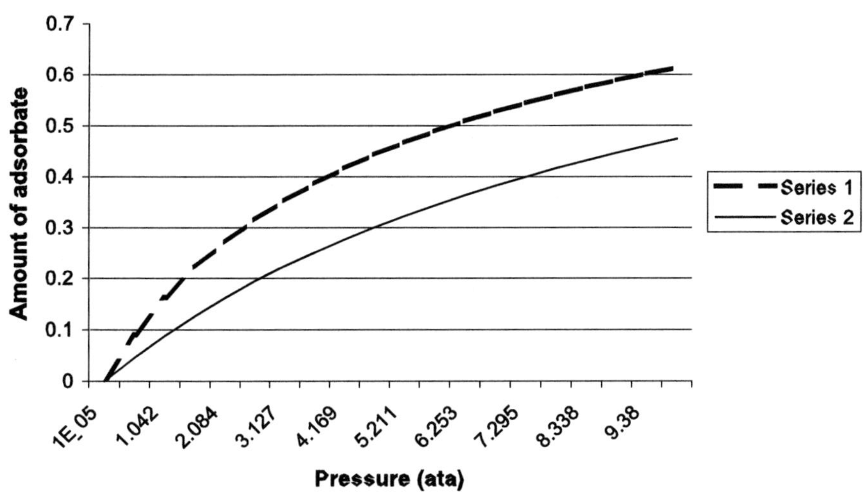

FIG. 6.15 Theoretical isotherms for adsorption of A and B on the adsorbent (15 s of conversion): series 1: gas A; series 2: gas B.

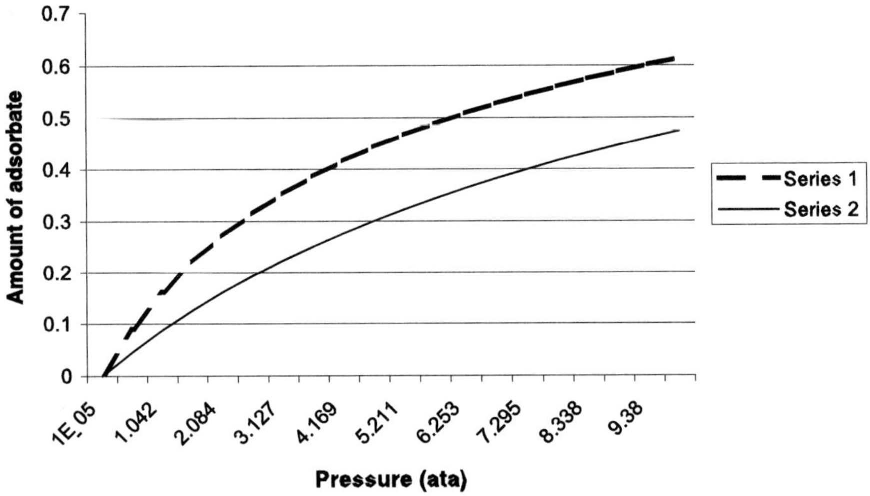

FIG. 6.16 Theoretical isotherms for adsorption of A and B on the adsorbent (20 s of conversion): series 1: gas A; series 2: gas B.

TABLE 6.3 Parameters for Adsorptive Separation of Gases A and B on Different Samples Considered

Conversion duration time, s	β for gas A	β for gas B
10	0.162	0.0882
15	0.162	0.0882
20	0.162	0.0882

After the same treatment, the concentration of B in the gas phase is much higher than that of A.

After the same treatment, the concentrations of A and B in both components differ greatly.

In the first case, the separation is due to the high concentration of A in the solid phase; in the second case, it is due to the high content of B in the gas phase; in the third situation, it is due to high concentration of each component in some phase.

The calculations on the base of the matrix presentation of counter-current cascade of adsorbers show that A is well removed from the gas phase, while the content of both components in the solid phase is high, that is the second case (see Fig. 6.17).

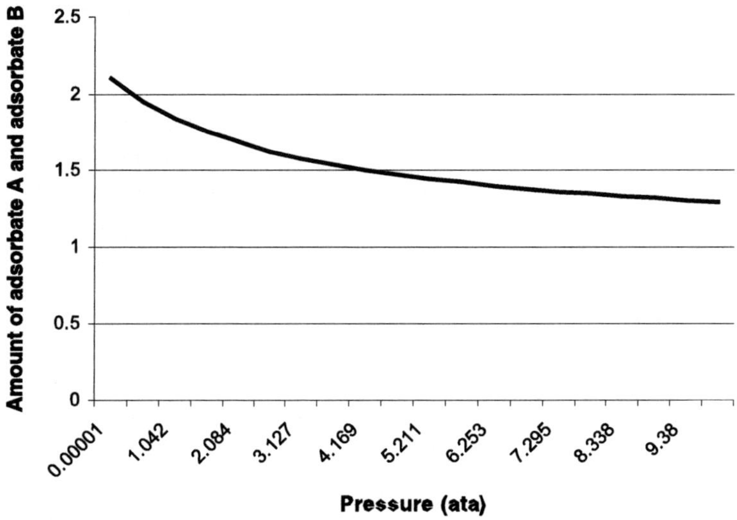

FIG. 6.17 Relative divergence between adsorption isotherms of A and B on the adsorbent.

As follows from the above estimations, after the adsorption the gas phase is enriched with B, while the mixture of adsorbates in the solid phase can be delivered due to adsorbent regeneration and treated again, until the content of A becomes enough high, and the separation is completed.

The researcher recommends the user to work at atmospheric pressure, because the pressure increase just reduces the separation effectiveness. Pumps and compressors should be used only for the compensation of friction of the interacting phases.

Conclusion: All three samples have the same ability to separate A and B, and the user needs to choose for this aim the cheaper option.

8. Stage 8. Modeling of Cascade of Adsorbers

The countercurrent cascade of adsorbers is modeled in accordance with the method of matrix presentation of countercurrent cascade described in Appendix 1. The criterion for the separation is formulated as follows: concentration of A or B in one phase (or both phases) is much more than the concentration of the second component. See Fig. 6.18.

As follows from Fig. 6.18, after the cascade of 10 adsorbents in countercurrent, the concentration of A in the gaseous phase is much less than that of B; hence, the separation has place, and the removed gas phase contains mostly B.

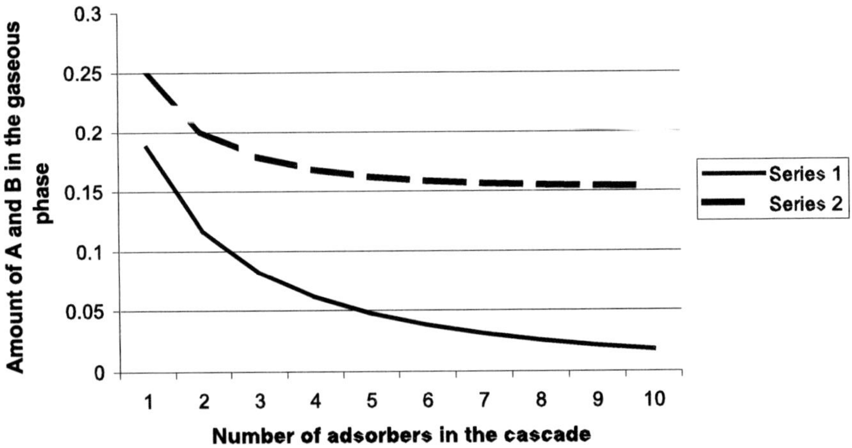

FIG. 6.18 Rest concentration of A and B components in the gaseous phase after the countercurrent cascade of adsorbers: series 1: rest of A in the gaseous phase; series 2: rest of B in the gaseous phase.

9. Stage 9. Optimal Temperature of the Thermal Regeneration of the Adsorbent

The criterion for the optimal regeneration is

$$\eta_r = \frac{(S_0 - S_{\text{cur}})\eta_c}{C_p} \tag{6.6}$$

where η_r is the energetic efficiency, η_c is the Carnot coefficient, S_0 is the initial entropy, S_{cur} is the current entropy of the system, C_p is the heat capacity of the adsorbent. The entropy of the system is found from the following condition:

$$S_x = R_g(Q_{\text{ads}} + Q_{\text{gas}})[x_{\text{gas}} \ln(x_{\text{gas}}) + (1 - x_{\text{gas}}) \ln(1 - x_{\text{gas}})] \tag{6.7}$$

$$x_{\text{gas}} = \frac{Q_{\text{gas}}}{Q_{\text{ads}} + Q_{\text{gas}}} \tag{6.8}$$

where Q_{ads} is the amount of the adsorbent, Q_{gas} is the amount of the gaseous components, and x_{gas} is the concentration of the gaseous components in the solid phase. The functions of x_{gas} and η_r vs. temperature are given on Figs. 6.19 and 6.20, respectively.

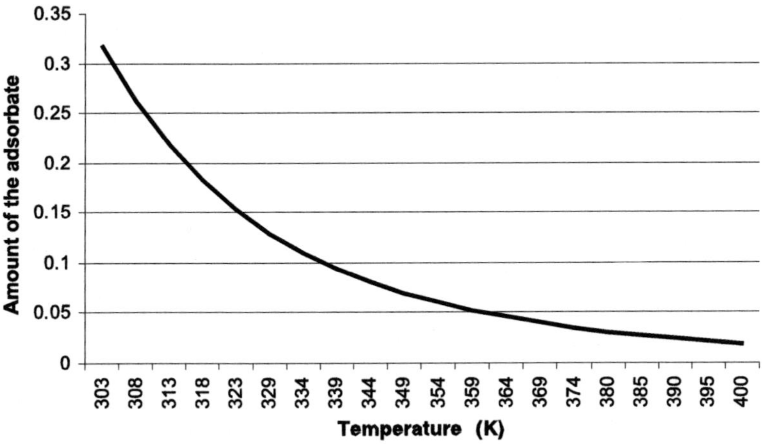

FIG. 6.19 Rest concentration of the adsorbate (gaseous components) in the solid phase after regeneration.

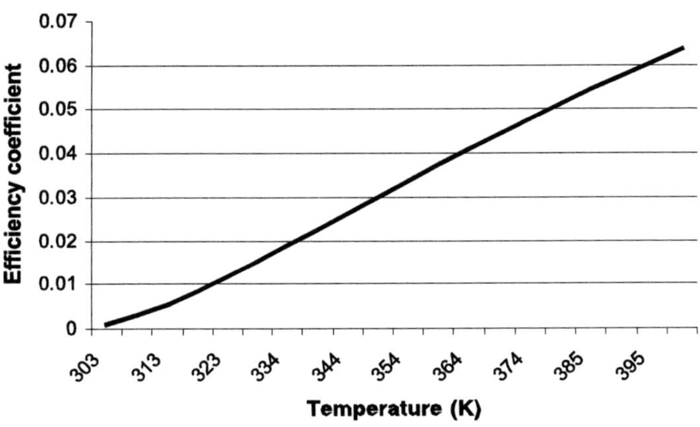

FIG. 6.20 Energetic efficiency of the thermal regeneration of the adsorbent.

10. Stage 10. Permeability Evaluations

The permeability was evaluated as described in Chapter 4 for the case of slab symmetry. The results of calculations are given in Table 6.4.

As follows from Table 6.4, the permeability of first sample (product of 10 s, polymerization) is negligible, this sample cannot be used as molecular sieve. Two other samples have good characteristics regarding their permeability, and the user may choose between them the cheaper option.

TABLE 6.4 Estimated Permeability of Considered Polymeric Samples

Conversion duration time	10 s	15 s	20 s
Permeability estimated	0.59223×10^{-7}	0.41990×10^{3}	0.16018×10^{7}

III. EXAMPLE 2: EVALUATION OF STRUCTURAL PARAMETERS (SURFACE TENSION OF FORMATION σ^*) FROM ADSORPTION ISOTHERMS

A. Formal Definition of Problem by User

The user is interested in forecasting of surface tension of some material on the frontier with the delivered gas at the moment of pore formation. One may assume the preparation conditions of the reference material close to equilibrium. One may assume that the energy range for the micropores is from zero to infinity. The result can be presented as σ^*. The porosity of the material, being estimated from density measurements, is about 25%. The coordination number of the material is 6. The initial data for the estimation of structural parameters of the solid phase comprise three isotherms of adsorption for three different gases (Fig. 6.21).

This time, we deal with the problem of the estimation of structural characteristics (more concretely—σ^*) from measurable properties: adsorption isotherms.

B. Stages of Calculations

The calculations for the formally defined problem above will comprise two stages only:

1. The treatment of the initial data provided by the user in order to obtain the value of the internal energy of the adsorbent.
2. The evaluation of the surface tension parameter σ^* corresponding to the porosity value brought by the user and the internal energy calculated in Stage 1.

C. Scientific Definition of Calculation Problems

The scientific definition of the theoretical and calculation problem(s) for the considered formally defined problem can be formulated as a number of acts of application of theoretical tools of microporosity concept and related

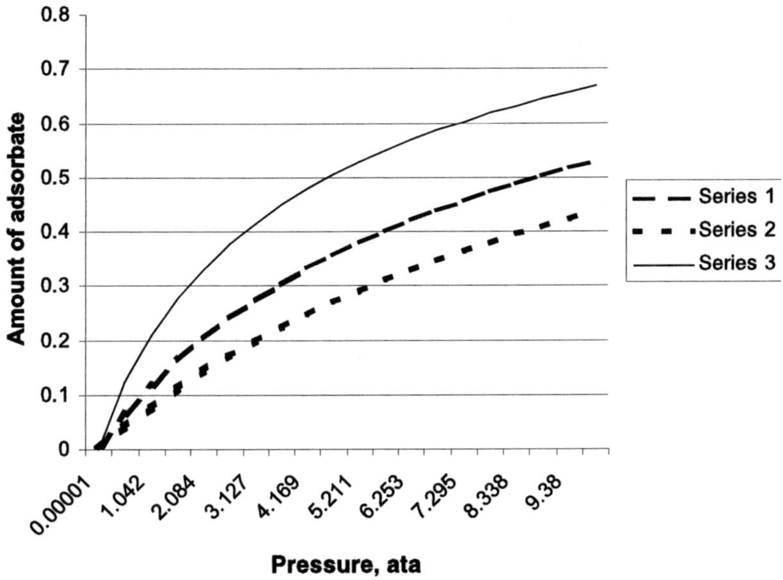

FIG. 6.21 Experimental adsorption isotherm for gases A, B, and C: series 1: gas A; series 2: gas B; series 3: gas C.

computer calculations, as follows:

1. The computer treatment of the known adsorption isotherms by least square method, with the estimation of the parameter α of the microporous adsorbent. This corresponds to the internal energy of the microporous system.
2. Using the results from the first stage of calculation, estimate the porosity of the system corresponding to the previously found value of the internal energy, for varied surface tension. The surface tension corresponding to the requested combination *porosity* + *internal energy* is the solution of the problem.

D. Choice of Theoretical Tools and Computer Programs

The above scientific definition of the problem determines the following list of theoretical tools in microporosity theory needed for the obtainment of the solution:

The program for the treatment of experimental adsorption isotherm curves by least square method

The program for the simulation of the growth of the porous phase in the discrete model

Each of the above theoretical tools is related to a single computer program. Only programs related to microporosity are needed.

E. Input and Output

The input and the output of the calculations on all the stages given above are determined by the available data and the expected results. Let us present input and output for each stage of the calculations as Table 6.5.

As follows from Table 6.5, the procedure of calculations suggested above can be accomplished on the available theoretical tools and obvious computer programs. The computer programs have the simple structure. The order of calculations cannot be changed, because output of first stage of calculations is necessary as the input in the second stage.

F. Calculations and Results

Now, let us illustrate the performance of the operations described in Table 6.5.

1. Stage 1. Treatment of the Experimental Isotherms of Adsorption

The equation for the minimized functional for the graphs treated is written as

$$\sum_{k_1=1}^{n_1}\left(\frac{y_{1t}}{y_{1e}}-\frac{y_{1e}}{y_{1t}}\right)^2+\sum_{k_2=1}^{n_2}\left(\frac{y_{2t}}{y_{2e}}-\frac{y_{2e}}{y_{2t}}\right)^2+\sum_{k_3=1}^{n_3}\left(\frac{y_{3t}}{y_{3e}}-\frac{y_{3e}}{y_{3t}}\right)^2=\text{minimum} \tag{6.9}$$

The treatment of the experimental isotherms of adsorption with least square method (eight fitted parameters, while we are interested mostly in the internal energy), we obtain the value of the internal energy: 4.157 kJ/mol. Thus, we have the input for stage 2:

Internal energy = 4.157 kJ/mol (by stage 1).
Porosity = 0.25 (by the user).
Coordination number = 6.

2. Stage 2. Evaluation of Surface Tension from the Internal Energy and the Porosity

The value of the surface tension satisfying the imposed conditions is $\sigma^* =$ 2824 J/mol. The resulting value of $\sigma^* = 2824$ J/mol is forwarded to the user.

TABLE 6.5 Input and Output on Each Stage of the Computer Solution of the Problem defined in Example 2

Stage N	Name of the stage	Model used	Input needed	Output available
1	Treatment of experimental adsorption isotherm curves	Langmuir-like model of adsorption in micropores	Experimental points	Parameters in Eqs. (4.34)–(4.36)
2	Simulation of the growth of the porous phase	Discrete model of pore growth	Internal energy, porosity, coordination number	σ^*

IV. EXAMPLE 3: ESTIMATION OF PERMEABILITY FROM ADSORPTION ISOTHERMS

A. Formal Definition of Problem by User

The user is interested in forecasting of permeability characteristics of some material. One may assume the preparation conditions of the reference material close to equilibrium. One may assume that the energy range for the micropores is from zero to infinity. The result can be presented as the visible percolation threshold. The porosity of the material, being estimated from density measurements, is about 15%. The coordination number of the material is 6. The initial data for the estimation of structural parameters of the solid phase comprise three isotherms of adsorption for three different gases (Fig. 6.22).

This time, we deal with the problem of the estimation of some measurable characteristics [more concretely, the visible percolation threshold (VPT)] from other measurable properties (adsorption isotherms).

B. Stages of Calculations

The calculations for the formally defined problem above will comprise three stages:

1. The treatment of the initial data provided by the user in order to obtain the value of the internal energy of the adsorbent
2. The evaluation of the number of filled cells having m filled neighbors (N_{fm}) for m varied from 0 to 6

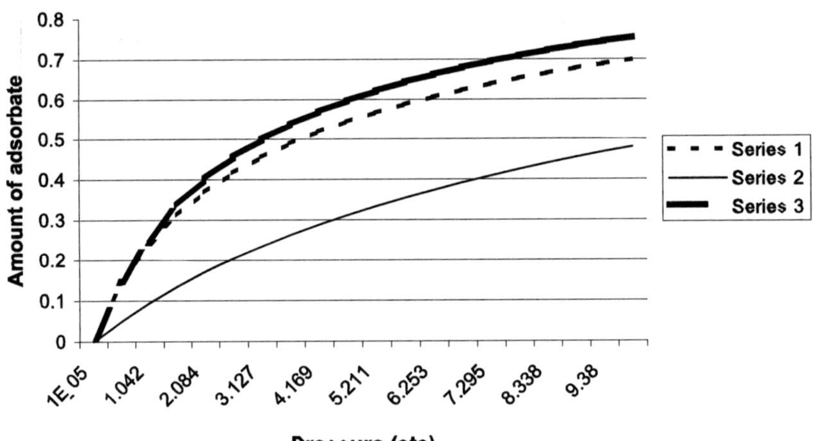

FIG. 6.22 Experimental adsorption isotherm for gases A, B, and C (Example 3).

3. The evaluation of the visible percolation threshold from the numbers of N_{fm}

C. Scientific Definition of Calculation Problems

The scientific definition of the theoretical and calculation problem(s) for the considered formally defined problem can be formulated as the following application of theoretical tools of microporosity concept and related computer calculations:

1. The computer treatment of the known adsorption isotherms by least square method, with the estimation of the parameter α of the microporous adsorbent. This corresponds to the internal energy of the microporous system.
2. Using the results from the first stage of calculation, estimate the porosity of the system corresponding to the previously found value of the internal energy for varied surface tension. The complex of solutions satisfying the conditions porosity $= 15\%$ and internal energy $= \alpha$ found in the previous stage is the needed result. Then, the values N_{fm} are found automatically.
3. Using the results obtained in the previous stage, estimate the visible percolation threshold in accordance with the usual procedure (Chapter 4). The obtained result is that requested by the user.

D. Choice of Theoretical Tools and Computer Programs

The above scientific definition of the problem determines the following list of theoretical tools in microporosity theory needed to obtain the solution:

The program for the treatment of experimental adsorption isotherm curves by least square method
The program for the simulation of the growth of the porous phase in the discrete model
The program for the evaluation of the permeability flow from structural parameters N_{fm}

Each of the above theoretical tools is related to a single computer program. Only programs related to microporosity are needed.

E. Input and Output

The input and the output of the calculations on all the stages given above are determined by the available data and the expected results. Let us present input and output for each stage of the calculations as Table 6.6.

TABLE 6.6 Input and Output on Each Stage of the Computer Solution of the Problem Defined in Example 3

Stage N	Name of the stage	Model used	Input needed	Output available
1	Treatment of experimental adsorption isotherm curves	Langmuir-like model of adsorption in micropores	Experimental points	Parameters in Eqs. (4.34)–(4.36)
2	Simulation of the growth of the porous phase	Discrete model of pore growth	Internal energy, porosity, coordination number	N_{fm}
3	Evaluation of permeability flow	Model of permeability	N_{fm}, coordination number, porosity	Visible percolation threshold

As follows from Table 6.6, the procedure of calculations suggested above can be accomplished on the available theoretical tools and obvious computer programs. The computer programs have the simple structure. The order of calculations cannot be changed, because output of each stage of calculations is necessary as the input in the next stage.

F. Calculations and Results

Now, let us illustrate the performance of the operations described in Table 6.6.

1. Stage 1. Treatment of the Experimental Isotherms of Adsorption

The equation for the minimized functional for the graphs treated is identical to Eq. (6.9) in Example 2 (Section III.F). The treatment of the experimental isotherms of adsorption with least square method (eight fitted parameters, while we are interested mostly in the internal energy), we obtain the value of the internal energy: 6.651 kJ/mol. Thus, we have the input for stage 2:

Internal energy = 6.651 kJ/mol (by stage 1).
Porosity = 0.15 (by the user).
Coordination number = 6.

2. Stage 2. Evaluation of Surface Tension from the Internal Energy and the Porosity

The resulting value of $\sigma^* = 3620\,\text{J/mol}$ is intermediary and needed just for the evaluation of N_{fm}. The output of this stage is: $N_{f0} = 0.456$, $N_{f1} = 0.399$, $N_{f2} = 0.128$, $N_{f3} = 0.0163$, $N_{f4} = 0.000741$, $N_{f5} = 0.00000959$, and $N_{f6} = 0.000000289$;

$$\sum_{m=0}^{n_n} N_{fm} = 1 - \xi$$

3. Stage 3. Evaluation of Visible Percolation Threshold

Involving the above result into the program for the evaluation of the permeability flow, we obtain: the resulting flow is 9.123×10^{-14}. Hence, the considered material has the negligible permeability, and its parameters are very far of the visible percolation threshold.

V. CONCLUSIONS

1. The practical (engineering) application of the concept of microporosity needs, as an intermediary stage, the evaluation of structural characteristics, the knowledge of which allows the estimation of all needed parameters.
2. The steps in the theoretical solution of problems related to microporosity comprise:
   a. The formal definition of the problem by the user and the definition of initial data
   b. The division of the entire problem into some stages of calculations
   c. The scientific definition of calculation problems
   d. Choice of theoretical tools for each stage of calculations
   e. Definition of input and output on each stage of calculations
   f. Results of calculations and their comparison with the formal definition of the problem
3. Among the facilities available in the theory of microporosity, one notices the opportunity of the solution of the following problems:
   a. Direct estimation of structural parameters from preparation conditions
   b. Direct estimation of measurable parameters from structural characteristics
   c. Indirect estimation of structural parameters from experimental data on some measurable characteristics
   d. Indirect estimation of some measurable parameters from experimental data on others
4. The mathematical model most employed for the evaluation of intermediary structural parameters is the discrete model of porous cluster growth. The corresponding computer program for the simulation of such growth is used in all examples considered in this chapter.
5. In addition to theoretical and calculative tools of the microporosity concept, the solution of practical problems may also need the use of other tools, e.g., the matrix presentation of countercurrent cascade of adsorbers used for gas separation. If such tools are not widely available, they must be brought by the user in the stage of the formal definition of the problem.

7

Perspectives for Further Development of the Concept of Microporous Systems

In this Chapter, we will analyze the perspectives of the further development of the concept of microporosity and microporous materials. We will analyze the principal existing obstacles hindering the fast progress in this concept and provide some recommendations about canceling these obstacles, comprising some initiatives aiming to attract investors to all phases of microporous material elaboration.

Our analysis will be based on methods of the analysis of computer systems. This approach allows the optimization of information systems, maximization of their effectiveness and usefulness.

The illustrations were prepared with using CASE Studio 2 (for DFD and ERD) and Access (for DSD).

I. METHODS OF ANALYSIS USING COMPUTER SYSTEMS

Below, we will use the following methods of analysis using computer systems:

A. Entity-Relationship Diagram (ERD)

This method presents the *entities*—"participants" (subjects) in the information exchange process, the objects on which the action is performed, and the principal operations allowing their relationship. The *relationship* can be presented as one entity related to one entity (one to one) or one entity related to many entities (one to many). The presentation *one to one* means that there is the single relationship between two different entities (e.g., an individual and his or her assurance number), while the presentation *one to many* means that there are two or more options of relationships of entity 2 to entity 1 (e.g., first name and individual: the same first name can be found for many individuals, that is the relationship "one (first name) to many (persons)"). The relationship "many entities to many entities" is also possible but, for technical reasons, is presented through intermediate entities as a series or relationships "one entity to many entities". An example of the relationship "many to many": the same author may write two or more books, but the same book may be written by two or more coauthors, hence, the relationship "authors-books" needs an intermediary entity "author-book" formed for each situation when a concrete person is one of the coauthors in a concrete book.

The drawback of ERD presentation consists in the absence of a logical relationship between events; such relationship is allowed by "data flow diagrams" method.

B. Data Flow Diagrams (DFD)

This method allows understanding of the relationship between objects, databases and events (functions). DFD comprises three types of components:

Subjects: some interested persons or groups actively participating in some events

Events (functions): processes and their results changing the situation for the subjects

Databases: storage of information about some phenomena, technique, persons, etc.

Combined with ERD, DFD allows the presentation of both relationships between the subjects and acted objects, but the system of information storage is not optimal. The optimization of the information storage is allowed due to using of data structure diagrams.

C. Data Structure Diagrams (DSD)

This method allows the optimal systematization and presentation of information databases, comprising the logical relationship between them. That is usually done in the form of tables, including:

The code and the principal characteristics of the object (or subject) described

The codes of principal related objects or subjects

The relationships between different objects and subjects

Combinations of ERD, DFD, and DSD allow the presentation of relationships between the subjects and acted objects, the optimized systems for storage and furnishing information about events related to changes in the considered system.

More detailed information about such methods of computer system analysis as ERD, DFD, and DSD is given in Refs. 1–4.

II. PRINCIPAL PROBLEMS IN THE DESIGN OF NEW MICROPOROUS MATERIALS

The principal problems in the design of new microporous materials are NOT related to the scientific knowledge or talent of explorers but only to the system of their financial support.

The existing system of the financial support of the research of microporosity comprises two principal subsystems working almost independently and related mostly by the publication of results. These subsystems are

The academic system of fundamental theoretical and experimental studies of microporosity in universities, specialized colleges, and governmental institutes. These institutions are supported by governments and/or private philanthropic sponsors, while the relationship with the industry is minor. The researchers working in this subsystem are not very interested in the financial results of their explorations, and their interest is mostly morale (nice to contribute to the science and the perspectives of mankind). They publish many papers and sometimes get patents, but the real output of their efforts is dubious.

Private and some governmental (mostly military) research centers and institutes. They depend much more on the results of their work, and their interest to obtain really important scientific results is minor. They solve the current problems related to materials science, and they are informed about the achievements of their colleagues working in the subsystem above mostly due to scientific publications and conferences. The worst thing is that they are not interested even in forecasting the global and environmental sequences of the realization of their ideas, unless that is imposed by existing laws.

Is the existing system (described above) good enough? In principle, it allows the obtainment of new scientific results and the solution of urgent technical problems—at least, today. However, this system is not optimal of all. The whole previous part of this book (Chapters 1–6) presented the power of the concept of the joint consideration of synthesis, structure and measurable properties of microporous materials. Nevertheless, all scientific results presented above cannot attract workers in the existing system of the research of microporosity, because

Workers in sponsored institutes, universities, and colleges are not interested in rejecting their previous suggestions and plans without getting the serious financial stimulation.
Workers in private and military research centers are not interested in accepting the risk of the realization of novel organizational ideas, effectiveness of which cannot be approved in few months.

On the other hand, *there are* subjects really interested in the increase of the effectiveness of studies of microporosity—those financing the sponsored sector and investing in the private sector. They are really interested in getting maximum scientific and technical effect per each dollar spent.

Is it possible to suggest any initiatives stimulating the sponsored researchers and reducing the risk of private researchers in the case if they accept the methodology of the joint consideration "synthesis–structure–properties"? The below scheme suggests a way for the solution of this problem.

A. Traditional Scheme of the Elaboration of New Microporous Materials

In the previous chapters, we mentioned sometimes the specificity of the existing system of study of microporosity. Its ERD scheme is presented on Fig. 7.1.

As follows from Fig. 7.1, all begins by the user of material, needing a new material with special properties that are not found for existing materials. The user of material request this new material from the supplier (in principle, the normal situation is: *many* users request materials from *many* suppliers, which is relationship ERD many to many simplified to a pair of relationships one to many through orders of materials). The supplier orders the needed research to the researcher of material (or many suppliers order research to numerous researchers), this one synthesizes the material with the needed properties, forwards the technical information to the supplier, and publishes the scientific results. The patent rights are transferred, in most cases, to the supplier.

Thus, we find in the considered scheme the following subjects:

User of material
Supplier of material
Researcher of material

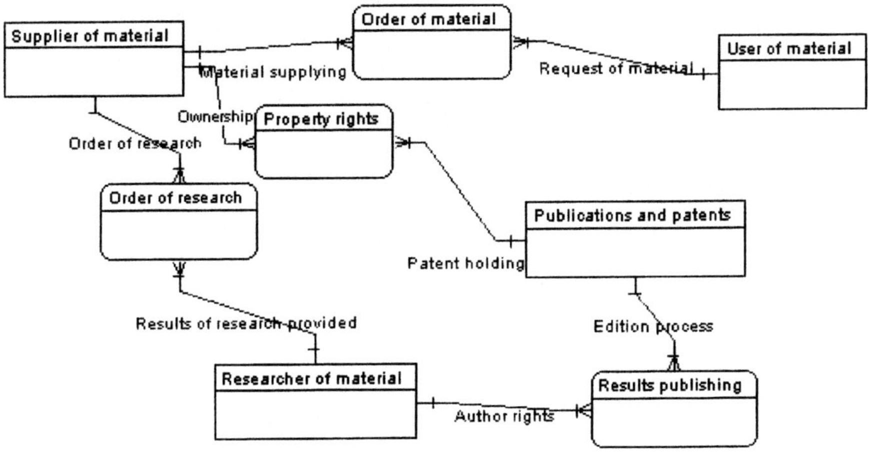

FIG. 7.1 The existing scheme ERD of relationships of subjects in the existing system of studies of microporosity.

Of course, talking about user of material, supplier of material, and researcher of material, we mean not only concrete persons but related collectives, institutes, companies, ministries etc.

Now, let us consider the DFD scheme of the same system. That is presented by Fig. 7.2.

As follows from Fig. 7.2, the most problematic phase of the new material creation is its elaboration by the researcher. First of all, this one tries the existing materials and finds them inappropriate for completing the received order. However, the researcher finds some materials, properties of which get closer to those of the requested material. Based on these materials, the researcher tries some combinations of known materials, modifies them, and, after all, maybe prepares the material with the desirable properties. But—maybe not! The main disease of the existing scheme is the impossibility to forecast whether or not the requested material can be ever prepared—just theoretically! Well, let us assume that the order will be completed, but the resources needed for the solution of the defined problem may be too expensive. After all, the material is synthesized, the supplier and user are satisfied, the novel scientific data are published, the special technical information is patented, and the existing database on microporous materials, their synthesis, structure, and properties is updated.

In addition to the active participants in the traditional scheme (subjects already listed above) we find also the database on microporous materials, touched twice by the material researcher: the first time when he

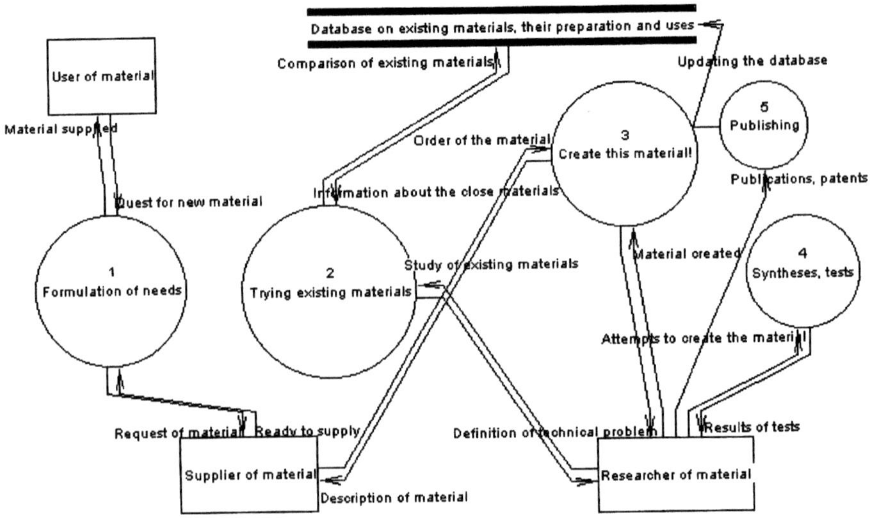

FIG. 7.2 The DFD scheme of the traditional system of microporous material design.

seeks for the appropriate or close materials and the second time when he updates this database with papers and patents and, respectively, enriches the human knowledge about microporosity.

The financial aspect of the traditional scheme of the synthesis of new material with desired properties is illustrated by Table 7.1.

As follows from Table 7.1, the most problematic phase of the creation of the novel material with the desired properties—empirical synthesis by the researcher of material—corresponds to the expenses ($n_{s1} \times \mathrm{COST}_{s1}$). If the number of attempts n_{s1} becomes infinite, that corresponds to one of two regretful options: the initial order of the material is incorrect and requires impossible things, or the researcher of material does not possess the needed resources. In both cases, the result is negative, and the research fails. If the researcher of material belongs to a sponsored institute, his damage is just moral; otherwise, his losses are much more material.

B. Suggested Scheme of the Elaboration of New Microporous Materials

The suggested scheme of the elaboration of new microporous material with desired properties is based on the concept of the joint consideration

TABLE 7.1 Expenses for the Synthesis of Novel Material with Desired Properties by the Traditional Scheme of Study of Microporosity

Operation or event	Performer of the operation	Performer of the payment	Number of the operations	Expenses per one operation
Formulation of order	Material user	Material user	1	$COST_{or11}$
Receipt of order	Material supplier	Material supplier	1	0
Order or research	Material supplier	Material supplier	1	$COST_{or12}$
Choice of close materials	Researcher	Researcher	1	$COST_{ch11}$
Synthesis and tests of new materials	Researcher	Researcher	n_{s1}	$COST_{s1}$
Comparison with the order	Researcher	Researcher	n_{c1}	$COST_{com1}$
Patenting, publishing, report preparation	Researcher	Researcher and/or supplier[a]	1	$COST_{pub1}$
Completing of the order	Researcher	Supplier	1	$COST_{matres11}$
Supplying material	Supplier	User	1	$COST_{matsup12}$
Total costs:				
Total11	—	User	—	$COST_{or11} + COST_{matsup12}$
Total12	—	Supplier	—	$COST_{or12} + COST_{matres11}$
Total13	—	Researcher	—	$COST_{ch11} + (n_{s1} \times COST_{s1}) + (n_{c1} \times COST_{com1})$

[a]We assume that in any case these expenses are paid by the supplier.

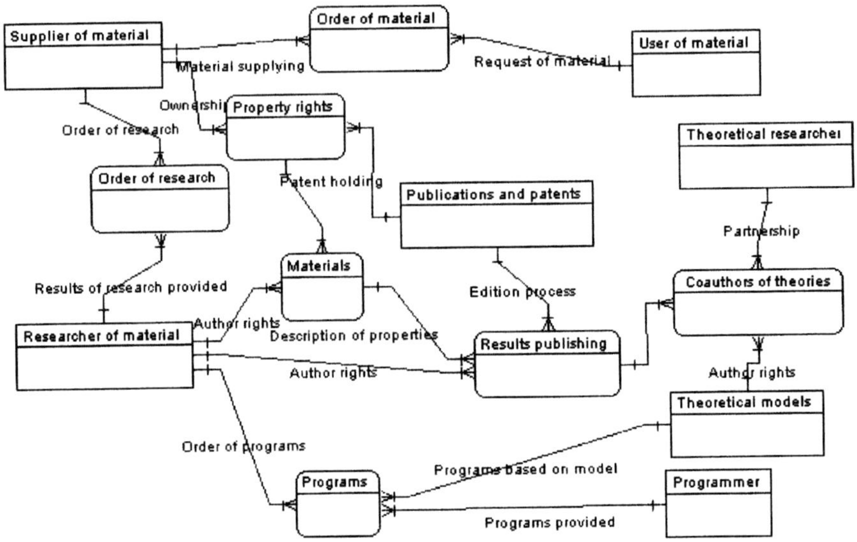

FIG. 7.3 The suggested scheme ERD of relationships of subjects in the system of studies of microporosity.

of synthesis, structure, and properties. Its ERD illustration is presented by Fig. 7.3.

As follows from Fig. 7.3, the suggested scheme of the relationships between the participants in the novel material elaboration is more complex than the traditional system. That includes not only user of material, supplier of material, and researcher of material, but also programmer and theoretical researcher, whose results are necessary for the realization of the joint concept. Also they are interested in the protection of their rights, especially in the case if the elaboration of the new material succeeds and results in a financial profit.

Now, let us consider the DFD scheme of the suggested system, which is presented by Fig. 7.4.

As follows from Fig. 7.4, there are no significant changes in the events related to the required microporous material synthesis, until the design itself is begun by the researcher of material. In the proposed scheme, the researcher of material, instead trying all close materials, first of all estimate the needed structure able to bring the required properties. This estimation is done on the base of computer calculations, using, as semiempirical parameters, experimental data from the database on existing microporous materials. The previously mentioned problematic phase of the new material

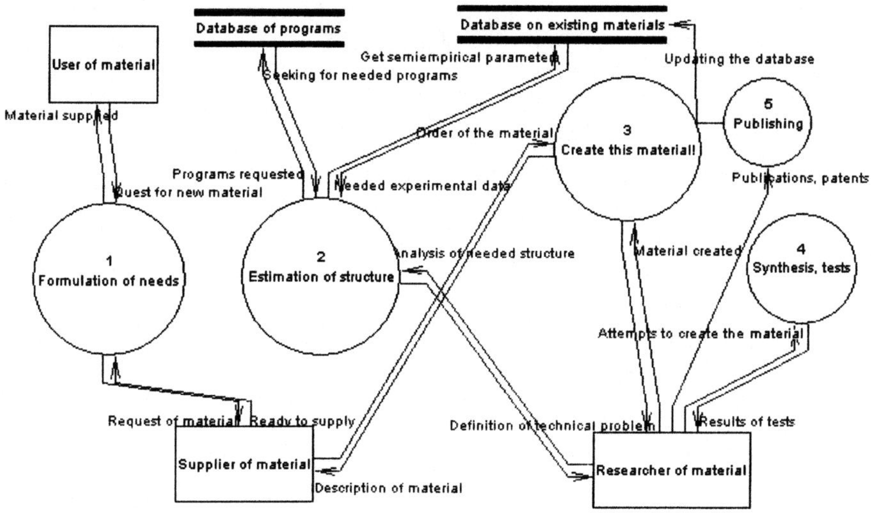

FIG. 7.4 The DFD scheme of the suggested system of microporous material design.

creation by the traditional scheme is cancelled already on the stage of the estimation of the structure: just after executing the calculating program, the researcher of material immediately knows whether the problem received from the user (through the supplier) is defined correctly. If not, the researcher of material informs the supplier that the problem has no solution or the needed resources are not available; then, the supplier decides what to do: ask the user to soften the requirements, or provide the needed resources to the researcher of material, or anything else. In any case, the researcher of material economizes efforts, time, and money.

In the case if the computer calculations do not bring any warning result about the chances of the solution of the defined problem, the researcher of material just takes the recommendations about the structure he needs to obtain and the preparation conditions he needs to assure. The synthesis is accomplished, in the normal situation, once only (unless the results of calculations are not enough clear).

In addition to the database on existing microporous materials, the proposed scheme comprises also the database on computer programs for estimating the structure allowing several (desired) measurable properties, and the parameters of the synthesis process for the preparation of the needed structure. During the realization of the proposed scheme, this database gets no change.

The theoretical researcher does not still appear in the suggested scheme, because he is not directly related to the researcher of material and the attempts to create the novel microporous material with the required properties.

Let us note also another difference of the proposed scheme from the traditional one: the database on the existing materials is used not for finding the closer materials but mostly for finding semiempirical parameters for the input in the program to run. The really obtained material with the desired properties may differ very much from "closer" materials! In the traditional scheme, such result cannot be ever obtained.

The financial aspect of the traditional scheme of the synthesis of new material with desired properties is illustrated by Table 7.2.

As follows from Table 7.2, the main financial effect of the realization of the proposed scheme consists in the fantastic reduction of the expenses for the stage of the synthesis and the tests. In the case if the calculating program is not expensive and its running does not require any expensive technique, $COST_{est} < COST_{sl}$ and the economy is assured almost always. The profit obtained due to the realization of the proposed concept can be used by the researcher of material in two ways (probably both):

Increase of revenues
Reduction of the price of supplying the results of the research to the supplier of material ($COST_{matres21} < COST_{matres11}$)

In the second case, also supplier of material gets an additional profit due to the use of the proposed scheme.

The profit obtained due to the realization of the proposed concept can be used by the supplier of material in two ways (compare to researcher of material):

Increase of revenues
Reduction of the price of supplying the results of the research to the user of material ($COST_{matres22} < COST_{matres12}$)

In the second case, also user of material gets an additional profit due to the use of the proposed scheme.

C. Suggested Scheme of the Elaboration of New Programs for Simulations of Structure and Properties of Microporous Materials

The suggested scheme of the elaboration of new microporous material with desired properties works okay in condition that the price required by the program owner does not ask too high price for the software. Otherwise, the

TABLE 7.2 Expenses for the Synthesis of Novel Material with Desired Properties by the Suggested Scheme of Study of Microporosity

Operation or event	Performer of the operation	Performer of the payment	Number of the operations	Expenses per one operation
Formulation of order	Material user	Material user	1	$COST_{or21} = COST_{or11}$
Receipt of order	Material supplier	Material supplier	1	0
Order or research	Material supplier	Material supplier	1	$COST_{or22} = COST_{or12}$
Estimation of needed structure, parameters of synthesis	Researcher	Researcher	1	$COST_{est}$
Synthesis and tests of the new material	Researcher	Researcher	1	$COST_{s2} = COST_{s1}$
Comparison with the order	Researcher	Researcher	2	$COST_{com2} = 2 \times COST_{com1}$
Patenting, publishing, report preparation	Researcher	Researcher and/or supplier[a]	1	$COST_{pub2} = COST_{pub1}$
Completing of the order	Researcher	Supplier	1	$COST_{matres21} \leq COST_{matres11}$
Supplying material	Supplier	User	1	$COST_{matres22} \leq COST_{matres12}$
Total costs:				
Total21	—	User	—	$COST_{or11} + COST_{matsup22}$
Total22	—	Supplier	—	$COST_{or12} + COST_{matres21}$
Total13	—	Researcher	—	$COST_{est} + COST_{s1} + 2 \times COST_{com1}$

[a]We assume that in any case these expenses are paid by the supplier.

condition $COST_{est} < COST_{s1}$ is not completed, and researcher of material is not interested in the novel way to material synthesis. However, in such case the program owner also loses the profit; hence, he is not interested in overestimating the real value of the programs.

Let us consider the process of the creation of novel computer program for modeling structure and properties of microporous material. DFD of this process is given in Fig. 7.5.

As follows from Fig. 7.5, the process of designing a program for simulating microporous materials is very similar to that for designing novel materials according to the proposed general scheme. This time, the role of the user is played by the researcher of material, and the order completing is performed by the programmer. This DFD includes two databases: database on programs (that is checked by program owner before ordering the new program writing, and later, when the program is ready, this database is updated) and database on theoretical models, among which the programmer finds the most appropriate mathematical tools allowing him to write the ordered program.

The resulting expenses for the program design are determined by two principal factors:

The price of licensing the appropriate theoretical model
The salary of the programmer

The theoretical researcher appears in this scheme indirectly, through the database on theoretical models.

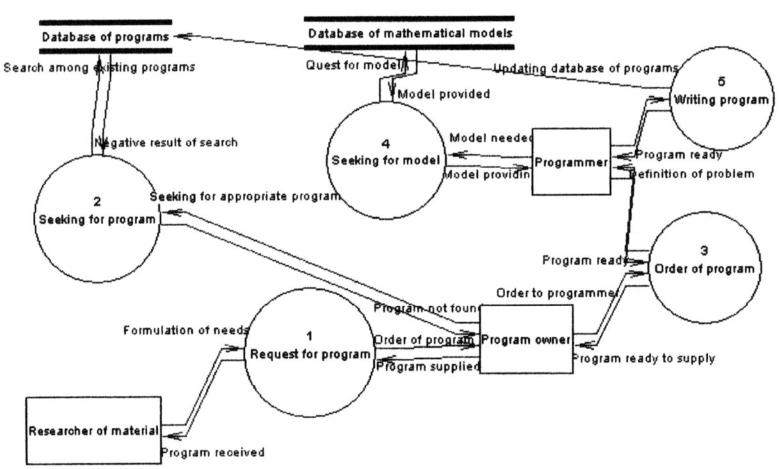

FIG. 7.5 The DFD scheme of the suggested system of design of computer programs for simulating structure and properties of microporous material.

The financial aspect of the traditional scheme of the synthesis of new material with desired properties is illustrated by Table 7.3.

As follows from Table 7.3, the condition of the financial reasonability of the design of a novel program is given by: $COST_{reqmod31} + COST_{wrpr31} < COST_{prog31}$, $COST_{prog31} < COST_{prog32}$; hence, too high price for licensing the needed theoretical model can destroy all financial interest: not only for the programmer and the program owner but also for the researcher of material and owner of material.

Today, everyone can take an existing theoretical model for free and write a related computer program, but, as we noted above, this system destroys all interest in theoretical studies and investments in theories; below we suggest some ways to compromise between the interests of programmers to get the theoretical models for minimal price and the interests of theoretical researchers in maximal royalties.

In the case if licensing the theoretical model is not expensive, $COST_{reqmod31} + COST_{wrpr31} < COST_{prog31}$, and the profit is obtained by both programmer and program owner. As in the case of material synthesis, the profit obtained by the programmer in two ways (probably both):

Increase of revenues
Reduction of the price of supplying the results of the research to the program supplier [$(COST_{reqpr32} + COST_{reqpr33} + COST_{prog31}) < COST_{prog32}$].

In the second case, also program supplier gets an additional profit due to the use of the proposed scheme.

If program supplier gets an additional profit due to the reduction of price by the programmer, that one can use his profit material in two ways (compare to the programmer and the supplier of material):

Increase of his revenues, and/or
Reduction of the price of supplying the results of the research to the researcher of material (reduction of $COST_{prog32}$).

In the second case, also the researcher of material gets an additional profit due to the use of the proposed scheme.

Now, let us consider the proposed scheme for the theoretical researcher.

D. Suggested Scheme of the Joint Elaboration of Novel Theories, Models, Computer Programs, and Materials

As we noted earlier, the existing scheme of the financial support of theoretical researchers does not stimulate all of them. Independently of

TABLE 7.3 Expenses for the Design of Computer Program for Simulating the Structure of Novel Material with Desired Properties

Operation or event	Performer of the operation	Performer of the payment	Number of the operations	Expenses per one operation
Request of program	Researcher	Researcher	1	$COST_{reqpr31}$
Query for the program	Program owner	Program owner	1	$COST_{reqpr32}$
Order of the novel program	Program owner	Program owner	1	$COST_{reqpr33}$
Quest for the model	Programmer	Programmer	1	$COST_{reqmod31}$
Writing program	Programmer	Programmer	1	$COST_{wrpr31}$
Order completing	Programmer	Program owner	1	$COST_{prog31}$
Program supplying	Program owner	Researcher of material	1	$COST_{prog32}$
Total costs:				
Total31	—	Researcher of material	—	$COST_{reqpr31} + COST_{prog32}$
Total32	—	Programmer	—	$COST_{reqmod31} + COST_{wrpr31}$
Total13	—	Program owner	—	$COST_{reqpr32} + COST_{reqpr33} + COST_{prog31}$

the novelty and level of proposed theories and models, their correctness and capability to treat the maximum of experimental data with minimum of fitted parameters, the theoretical researchers get some specified salary—or do not get anything. Their results can be used by everyone, just if copyright is not violated. There exists no system of protection of propriety rights of theoretical researchers or institutions paying them salary.

The system proposed above makes sense only on the condition that propriety rights of theoretical researchers and their employers will be respected in the same manner as the rights of patent holders. The organizational aspect of this problem will be considered below. Now, let us consider only the DFD and financial aspect of theoretical research, assuming that the proposed system makes impossible the traditional use for free of the original and effective theoretical tools. We also introduce the notion of owner of theoretical model (university, college, academy, private research company, etc.), meaning the employer of the theoretical researcher, defining the general problem for the theoretical research and possessing all propriety rights onto the obtained results.

The above-suggested scheme of the elaboration of new microporous material with desired properties and related computer programs works okay if the price required by the owner of theoretical propriety, while offering some significant profit, is not too high for the theoretical models he possesses. Otherwise, the programmer and the researcher of material are not interested in the novel way to material synthesis. However, in such case the owner of theoretical models also loses the profit and, hence, is not interested in overestimating the real value of the models.

Let us consider the process of the creation of novel theory and related models allowing the solution of particular problems in microporosity. DFD of this process is given below on Fig. 7.6.

As follows from Fig. 7.6, the process of designing a new theory of the synthesis, structure, or measurable properties of microporous materials is similar to that for designing novel materials and related computer program according to the proposed general scheme. This time, the role of the user is played by the programmer, and the order completing is performed by the theoretical researcher. This DFD includes two databases: database on existing microporous materials (that is treated by the theoretical researcher on all stages of creating theory and later, when mathematical models are applied to known experimental data to estimate the correlation between theoretical models and the experimental data) and database on theoretical models, to which the new theory is compared (aiming to estimate its novelty and place in the hierarchy of theoretical concepts of microporosity).

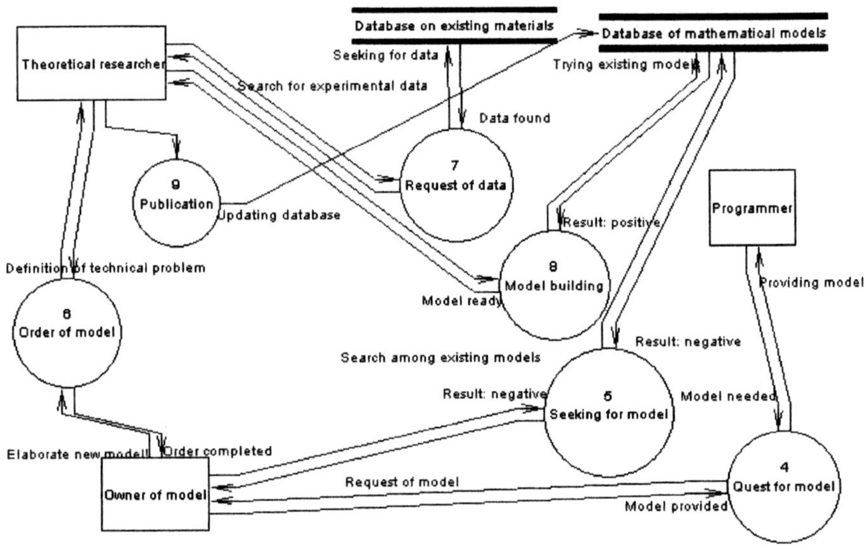

FIG. 7.6 The DFD scheme of the suggested system of design of theories and theoretical models of microporous materials.

After the receipt of the model order from the owner of theoretical model, the theoretical researcher:

1. Requests access into the database of existing theories and models.
2. Analyzes the existing theories—their advantages, principal assumptions and drawbacks.
3. Concludes about the necessary changes of the assumptions, in order to improve the theoretical base.
4. Builds novel theoretical models based on the "fresh" theoretical concept, using the appropriate mathematical tools applicable to various particular problems in microporosity science.
5. Compares the fresh models to these existing in the database; estimates their advantages and shortcomings.
6. Applies the fresh models to the treatment of real experimental data.
7. Concludes about the effectiveness of the proposed models.
8. Forwards the results to the owner of theoretical model.
9. Decides together with the owner of theoretical model about the relevant publications and protection of the created intellectual propriety.

The resulting expenses for the program design are determined by two principal factors:

The price of access into the database of theoretical models
The salary of the theoretical researcher

The financial aspect of the traditional scheme of the synthesis of new material with desired properties is illustrated by Table 7.4.

As follows from Table 7.4, the condition of the financial reasonability of the design of a novel theoretical concept is given by

$$\begin{aligned} &\mathrm{COST}_{\mathrm{theoran41}} + \mathrm{COST}_{\mathrm{theoran42}} + \mathrm{COST}_{\mathrm{theorimp41}} \\ &\quad + \mathrm{COST}_{\mathrm{newmod41}} + \mathrm{COST}_{\mathrm{compmod41}} + \mathrm{COST}_{\mathrm{compmod42}} \\ &\qquad < \mathrm{COST}_{\mathrm{matmod41}} \end{aligned} \tag{7.1}$$

and

$$\mathrm{COST}_{\mathrm{matmod41}} < \mathrm{COST}_{\mathrm{matmod42}} \tag{7.2}$$

Let us note that

The theoretical researcher may need writing computer programs on the base of the models, aiming to apply them to known experimental data but should do it without any assistance of the programmer (to avoid a sharp increase of the expenses), and the costs of writing such temporary programs are counted in $\mathrm{COST}_{\mathrm{compmod42}}$;

Since the theoretical researcher needs the database on existing theoretical models, the theoretical researcher is not interested in the increase of the price of the access to this database, which also limits profit;

Since the theoretical researcher needs the database on existing materials, the interest is in the collaboration with the researcher of material, aiming to maintain the prices of the access to all databases in limits allowing both of them to get a significant profit but preventing too sharp an increase of expenses.

III. REGISTRATION AND PROTECTION OF INTELLECTUAL PROPRIETY

As we noted above, one of the principal causes of the loss in the effectiveness of studies of microporosity is the absence of such protection of

TABLE 7.4 Expenses for the Design of Theoretical Models of Microporous Materials—Their Synthesis, Structure, and Properties

Operation or event	Performer of the operation	Performer of the payment	Number of the operations	Expenses per one operation
Request of new model	Programmer	Programmer	1	$COST_{reqmod41}$
Query for the new model	Owner of model	Owner of model	1	$COST_{reqmod42}$
Order of the novel model	Owner of model	Owner of model	1	$COST_{reqpr41}$
Access to database of models	Theoretical researcher	Theoretical researcher	1	$COST_{theoran41}$
Analysis of existing theories	Theoretical researcher	Theoretical researcher	1	$COST_{theoran42}$
Improving assumptions	Theoretical researcher	Theoretical researcher	1	$COST_{theorimp41}$
Building new models	Theoretical researcher	Theoretical researcher	1	$COST_{newmod41}$
Comparison of new models with existing ones	Theoretical researcher	Theoretical researcher	1	$COST_{compmod41}$
Application of new models to experimental data	Theoretical researcher	Theoretical researcher	1	$COST_{compmod42}$
Completing the order	Theoretical researcher	Owner of model	1	$COST_{matmod41}$
Completing the request	Owner of model	Programmer	1	$COST_{matmod42}$
Total costs:				
Total41	—	Programmer	—	$COST_{matmod42} + COST_{reqmod41}$
Total42	—	Owner of model	—	$COST_{reqmod42} + COST_{reqpr41} + COST_{matmod41}$
Total13	—	Theoretical researcher	—	$COST_{theoran41} + COST_{theoran42} + COST_{theorimp41} + COST_{newmod41} + COST_{compmod41} + COST_{compmod42}$

interests of the theoretical researcher that would allow him a good profit in the case of building an effective theory, solving classical problems of microporosity and applicable to numerous experimental data. This problem may have a solution due to the registration of novel theories, theoretical models, computer programs, and novel materials (even having no patent novelty) by some registering institutions. These institutions can be

International scientific or technological committees, centers, associations (e.g., UNESCO, IUPAC, NACE)
Governmental institutions (e.g., U. S. Department of Energy)
National institutions or associations (e.g., American Chemical Society, U. K. Royal societies)
Regional institutions or organizations (comprising universities)

The procedure of the registration should be automated, simple, and fast. The registration fees should be minimal (e.g., below U.S. $100 in the United States). The registration must officially assure

Author rights and owner rights
Limiting the similarity between different intellectual solutions of the same kind
The unification of the form of the presentation of different intellectual solutions of the same kind
Making easier the solution of possible conflict situations

The list of the data appearing in the registration forms for intellectual solutions (theories, theoretical models, computer programs, and microporous materials) is given in Table 7.5.

The simplest database system for storing information, allowing the automated regulation of the proposed scheme, is given on Fig. 7.7 as DSD relationship.

As follows from Fig. 7.7, the data structure for researcher of material and theoretical researcher is the same: of course, they are colleagues, and they differ only in the research experience and the profile of job!

All relationships on Fig. 7.7 are given through the codes only; such form allows the minimization of repeating data, which is typical for DSD.

However, the most important detail on Fig. 7.7 is the absence of any divergences between owners of materials, programs, models and theories. That means that the main function of owners is not the research itself but the financial support and control of research. Moreover, the same juridical person is allowed to be owner of materials, programs, and theoretical models. That is a very important aspect related to investments in studies of microporosity.

TABLE 7.5 Information in the Registration Forms for Intellectual Propriety Related to Studies of Microporosity

Data	Theory	Theoretical model	Computer program	Microporous material
Code	Code of theory	Code of model	Code of program	Code of material
Author(s)	Code(s) of theoretical researcher	Code(s) of theoretical researcher	Code(s) of programmer	Code(s) of researcher of material
Owner	Code of owner of model	Code of owner of model	Code of program owner	Code of owner of material
Name	Theory name	Model name	Program name	Material name
Designation	Sphere of applicability	Problems available to solve	Output available	Technical uses
Specificity	Assumptions	Equations	Language, algorithm	Most important properties
Information needed for use	—	Fitted parameters	Input	Restrictions
Hierarchic position	Class of theories	Initial theory (code)	Similar programs (codes)	Similar materials (codes)
Priority date	Date of first publication	Date of first publication	Date of registration	Date of registration, first publication, or patent
Publications available before the registration	Publications (codes)	Publications (codes)	—	Publications (codes)
Base	—	Theory code	Model code	Program code(s)

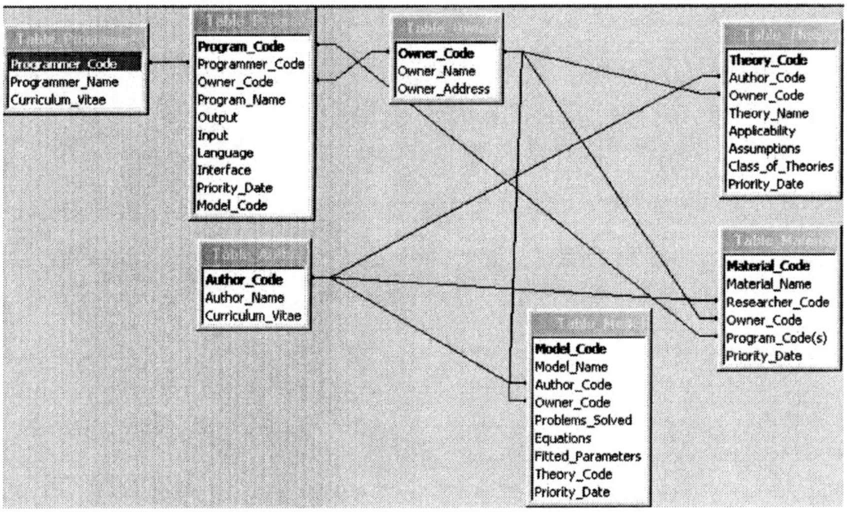

FIG. 7.7 DSD scheme for database on registered theories, theoretical models, computer programs, and new materials in the field of microporosity.

IV. INVESTMENTS IN STUDIES OF MICROPOROSITY

Some aspects of the investments in the existing system of studies of microporosity were considered above: nonprofit sponsorship by government or private philanthropic foundations and private investments aiming to get money back with the maximal profit as soon as possible. Of course, the scheme of investments in the proposed system of joint studies of materials, programs, and theory should be absolutely different.

First of all, all stages of study of microporosity can be realized on the profit base. Money can be invested not only in material synthesis and tests (as in the existing system) or programs but even in "pure" theory, because this becomes a product in the form of information from database. Moreover, investments in theoretical studies may have serious advantages: the expenses are related mostly to the salary of researchers that can return to the investor very fast (see Table 7.6: the expenses by the traditional scheme are compared with those by elements of the suggested scheme).

As follows from Table 7.6, the expenses for the suggested scheme are always much less than those for the traditional scheme; hence, in the aspect of costs the suggested scheme is definitely preferable for investments.

However, not only costs determine the effectiveness of systems. Let us compare the proposed system to the traditional system in various aspects

TABLE 7.6 Structure of Expenses in Different Studies of Microporosity

Item	Equipment amortization, energy consumption	Salary	Chemicals	Intellectual property (use of databases)
Researcher of material (by Table 7.1)	+++[a]	++[b]	+++	+[c]
Researcher of material (by Table 7.2)	+	++	+	++
Programmer (by Table 7.3)	+	++	–[d]	++
Theoretical researcher (by Table 7.4)	+	++	–	++

[a]+++ expenses are very high.
[b]++ expenses are high.
[c]+ expenses are moderate.
[d]– expenses are negligible.

TABLE 7.7 Various Aspects of Investments in Studies of Microporosity According to Different Schemes

Factor of comparison	Traditional scheme of research	Suggested scheme: only material study	Suggested scheme: only programming	Suggested scheme: only theory	Suggested scheme: all
Costs	Very large	Large	Minor	Minor	Large
Product to sale	Materials	Materials	Programs	Models	All
Relationship with material user	Direct	Direct	Indirect	Indirect	Direct
Dependence on partners	No	Yes	Yes	Yes	No

related to investments. We will consider three options for investments (see Table 7.7):

1. Investments in studies of microporosity according to the traditional scheme
2. Investments in theoretical research *or* programming *or* material synthesis
3. Investment in theoretical research *and* programming *and* material synthesis according to the suggested scheme

As follows from Table 7.7, only investing in all elements of microporosity studies can compete with the traditional scheme in all factors of comparison! Investing in theory and programming, although it seems very attractive financially, is too risky if the partners (for theoretician, programmer only; for programmer, both theoretician and material researcher; while for material researcher, programmer only) change the prices for their products. Moreover: theoretical researcher depends also on colleagues, whose intellectual products determine the prices in the database of theories and models, so necessary also for the theoretical study! The indirect relationship with the final consumer (material user) makes the additional risk for both theoretician and programmer.

Thus, the most effective in the aspect of the complex of characteristics of investing policy are investments in material synthesis, programming, and theoretical study together. Note that such investments are preferable also in the aspect of minimization of tax payments, because the products of theoretician and programmer are not always supplied to exterior consumers but first of all used inside the system.

V. CONCLUSIONS

1. The existing system of studies of microporosity does not stimulate the fundamental, especially theoretical research because theoretical results can be used by everyone without paying any fees to the author and/or investor in the fundamental research.
2. One of the sequences of the low efficiency of the existing system of studies of microporosity consists in the impossibility of forecasting whether a novel material with desired properties can be synthesized with the available resources.
3. The suggested system of studies of microporosity comprises the following:
   a. The protection of the propriety rights of owners of theoretical models of microporosity, with the registration of theories and models by competent organizations and/or institutions
   b. The registration of computer programs for forecasting properties of hypothetical materials in the same organizations and/or institutions
   c. The registration of novel materials not only in patent offices but also in the same organizations and/or institutions that register the novel theories, models, and computer programs on microporosity
4. The realization of the suggested system of studies of microporosity will allow:
   a. The financial stimulation of all aspects of studies of microporosity, comprising the elaboration of novel materials with

forecasted properties, writing computer programs for forecasting properties of new materials, and the creation of novel theories and theoretical models

b. The stimulation of investments not only in the elaboration of novel materials but also in new theories and computing programs linking theories to material synthesis

c. The increase of the effect of investments in studies of microporosity

REFERENCES

1. Shoval, P.; Shiran, S. Entity-relationship and object-oriented data modeling—an experimental comparison of design quality. Data Knowledge Eng 1997, 21(no issue), 297–315.
2. Shoval, P. Experimental comparisons of entity-relationship and object-oriented data models. Australian J Info Sys 1997, 4(2), 74–81.
3. Shoval, P.; Manor, O. Software engineering tools supporting ADISSA methodology for systems analysis and design. Info Software Technol 1990, 32(5), 357–369.
4. Shoval, P. ADISSA: Architectural design of information systems based on structured analysis. Info Sys 1988, 13(2), 193–210.

Appendix 1
Matrix Presentation of Cascade of Countercurrent Reactors of Mixing

It is a well-known fact that countercurrent interactions are preferable for the effective mass transfer between different phases to cocurrent technique. However, calculations of the efficiency of such interactions are fairly difficult. In the normal case, one assumes that the system of interactions comprises a cascade of some simple reactors of mixing, in which the interacting phases are mixed until getting to the local equilibrium. After that, the phases are separated and transported to other reactors, in which the procedure repeats some times.

Calculative simulation of such countercurrent cascade is allowed only if the number of reactors is specified, e.g., three or five. In such case, one can always write a system of equations for each reactor in the cascade; such system of equations is consistent and always has the single solution. However, if one is interested in the estimation of the optimal number of reactors in such cascade, the problem becomes too difficult.

The proposed matrix method for the presentation of cascade of countercurrent reactors of mixing aims to bring a tool for the description of mass transfer in such cascades in matrix form, that allows

- Simplification of writing systems of equations for their modeling
- Simplification of writing computer programs for the simulation of such cascades
- Simplification of the procedure of the optimization of cascades of countercurrent reactors.

The proposed solution is not applicable (without modifications) to systems, in which mass transfer is accompanied with heat transfer or chemical transformation.

I. DEFINITION OF PROBLEM

Let us consider a cascade of countercurrent reactors of mixing, in which fluid is mixed to high-dispersion solid phase or just two different fluids (in most cases, gas–liquid) are mixed, on purpose to carry out some mass-transfer processes, e.g., absorption, adsorption, desorption, degassing. The principal scheme of such cascade is given on Fig. A1.1 for the case of gas–liquid mixing.

The considered cascade comprises $N_c = 3$ reactors of mixing, in which gas interacts with liquid (for example, in the process of absorptive separation of gas mixture). The gas mixture is involved through a pump into the first reactor and there mixed to the liquid achieving to pass through the cascade. Then, the liquid leaves the cascade, while the gas comes to the next reactor, where the procedure repeats. After mixing N_c times, the gas is removed from the cascade system.

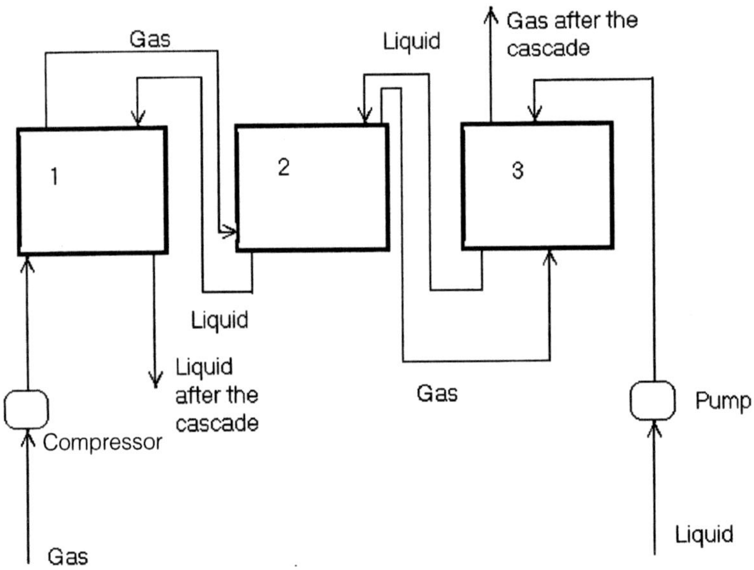

FIG. A1.1 Scheme of cascade of countercurrent reactors of mixing for gas–liquid interactions.

Of course, the real value of N_c can be 2, 3, 4, 5,..., while the interacting phases can be not only gas–liquid but also gas–solid powder and even liquid–solid or two immiscible liquids; the physics of the mass transfer is always very similar, while the technique for such processes becomes much more complex.

II. SOLUTION

Let us assume that the fluids contain K_v volatile components able to reversible transfer between the phases. Each reactor allows the obtainment of the equilibrium between the phases. The principal scheme of interactions in kth reactor is given on Fig. A1.2.

The liquid phase comes from $(k+1)$-th reactor to the head of kth reactor. The amount of ith component in the liquid is W_{+iLk}. However, the same component comes also from $(k-1)$-th reactor to the bottom of kth reactor in the gas phase, the amount of ith component being W_{+iGk}. Since no chemical transformation is allowed, ith component leaves kth reactor: from the bottom with the liquid phase (W_{-iLk}) and from the head with the gas phase (W_{-iGk}). We also notice that, as follows from Fig. A1.1, $W_{-iGk} = W_{+iG(k-1)}$ and $W_{-iLk} = W_{+iL(k+1)}$. The balance of ith component is given by the following equation:

$$W_{+iLk} + W_{+iGk} = W_{-iLk} + W_{-iGk} \tag{A1.1}$$

The distribution of ith component between both phases leaving kth reactor is given by the following equation:

$$W_{-iLk} = \beta_{ik} W_{-iGk} \tag{A1.2}$$

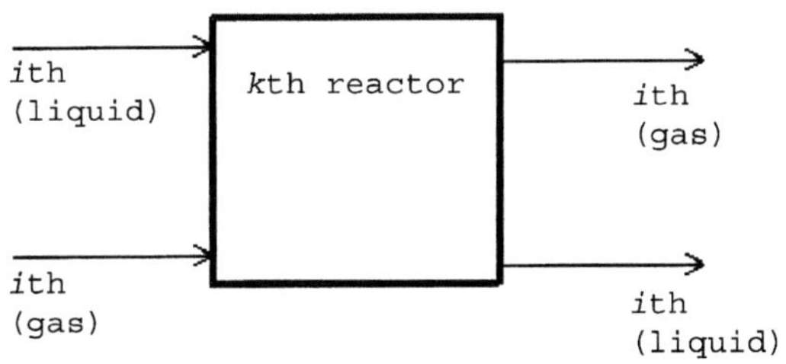

FIG. A1.2 The principal scheme of interactions in kth reactor.

Let us suggest that the process of interactions in kth reactor can be presented in the following matrix form:

$$\vec{W}_{i(k+1)} = B_{ik}\vec{W}_{ik} \tag{A1.3}$$

$$\vec{W}_{ik} = \begin{bmatrix} w_{+iGk} \\ w_{-iLk} \end{bmatrix} \qquad \vec{W}_{i(k+1)} = \begin{bmatrix} w_{+iG(k+1)} \\ w_{-iL(k+1)} \end{bmatrix} = \begin{bmatrix} w_{-iGk} \\ w_{+iLk} \end{bmatrix}$$

From Eqs. (A1.1)–(A1.3) we obtain

$$B_{ik} = \begin{bmatrix} 0 & \beta_{ik} \\ -1 & (1+\beta_{ik}) \end{bmatrix}$$

The process in the cascade entire is described as the following algebraic equation:

$$\vec{W}_{i(N+1)} = \left[\prod_{k=1}^{N} B_{ik}\right]\vec{W}_{i1} \tag{A1.4}$$

If $B_{ik_1} = B_{ik_2}$ (for all k_1, k_2)

$$B_i^N = G_i(N) = \begin{bmatrix} \gamma_{i11} & \gamma_{i12} \\ \gamma_{i21} & \gamma_{i22} \end{bmatrix}$$

$$\gamma_{i11} = -\beta_i \frac{1-\beta_i^{N-1}}{1-\beta_i} = -\beta_i \sum_{k=0}^{N-2} \beta_i^k$$

$$\gamma_{i12} = \beta_i \frac{1-\beta_i^{N}}{1-\beta_i} = \beta_i \sum_{k=0}^{N-1} \beta_i^k$$

$$\gamma_{i21} = \frac{-(1-\beta_i^{N})}{1-\beta_i} = -\sum_{k=0}^{N-1} \beta_i^k$$

$$\gamma_{i22} = \frac{1-\beta_i^{N+1}}{1-\beta_i} = \sum_{k=0}^{N} \beta_i^k$$

The proper eigenvectors for the matrix B_i are

$$\frac{1}{\sqrt{2}}\begin{bmatrix} 1 \\ 1 \end{bmatrix} \qquad \frac{1}{\sqrt{1+\beta_i^2}}\begin{bmatrix} \beta_i \\ 1 \end{bmatrix}$$

the corresponding eigenvalues, respectively, β_i and 1.

Appendix 2
"Big" and "Little" Phenomena by Self-Organization— Analogy of Pore Formation with Beginning of the Universe and Life (Chapter 3)

I. BEGINNING OF UNIVERSE BY SELF-ORGANIZATION

(a) General Aspect. Numerous models of the formation of the universe try to explain the beginning of the universe and predict general tendencies of its further evolution. The principal problems which needing a solution by astrophysics are

- What was the cause for beginning of universe?
- Does the universe expand? If yes, whether the expansion of the universe is unlimited or followed by collapse?
- Is there any cosmos outside universe?
- Since the principal physical laws are symmetric and do not distinguish past and future, what is the cause of the specified direction of the time arrow?
- What were the principal mechanisms of formation of heavy complex particles and nuclei?

(b) Historical Aspect. The problem of beginning of universe attracts not only scientists but also philosophers since ancient ages. The science before the 20th century assumed universe having no principal changes with time. The situation changed after the Einstein general theory of gravity was applied to universe. It was concluded that the universe parameters change with time. Particularly, about 12 billion years ago universe began due to an explosion (big bang) of a very massive singularity.

(c) Models of the Universe Formation. The main problem in studies of the universe formation consists in the impossibility of a direct experimental test of theoretically obtained results. Therefore, all studies in this field are theoretical, and the interpretation of available experimental data is very important.

(d) Big Bang: Closed Model of the Universe. In first microseconds of big bang, the temperature of the universe was so high that complex and massive elementary particles were formed. Then, the temperature decreased, and elementary particles formed nuclei of chemical elements (phase of *nucleogenesis*). The further expansion of the universe and the reduction of the temperature resulted in the appearance of nuclei of galaxies. The evolution of galaxies, accompanied by the further expansion of the universe and the decrease of the temperature of cosmos (background radiation temperature), led to formation of star systems, planets, and other cosmic bodies. Some researchers conclude that big-bang model of universe requires special initial conditions: the early universe was highly homogeneous and isotropic even though there existed causally disconnected regions (horizon problem) [1–6].

In principle, the theory of big bang answers most of above questions, but not all. First of all, the origin of the primary singularity is not clear. Was it a stage of cyclic evolution of the same universe (in the closed model of the universe) or a product of evolution of exterior cosmos (in the open model of the universe)?

Though the concept of big bang is the most widespread and recognized in astrophysics, not all specialists accept it. Some authors suggest alternative theories and deny the main assumptions of big bang theory and even general relativity, based on some results obtained due to observations of recently created galaxies, quasar–galaxy associations, etc. [7]. The existing theory of big bang cannot sufficiently solve some problems, comprising accidental superposition of galaxies and quasars with different red shift, problem of the density of the universe, and some others. Most of such arguments seem to concern first of all not the main idea of big bang but several of its sequences and related interpretations. Nevertheless, it seems obvious that experimental astrophysical data contain several contradictions complicating the image, and no existing model allows complete and clear solution of the mentioned problems. According to the opinion of several specialists, even if the theory of big bang is correct in principle, that stays in stagnation and needs principally new ideas [8,9].

(e) Closed Model. This model assumes that the space of the universe is closed; the universe has a finite lifecycle and should disappear again in no less than 100 billion years. Such assumption, whether or not applicable to the universe now, undoubtedly described some properties of the universe when this was a singularity and then, in the first period after big bang, because then the relation "mass–size" corresponded to black hole formation. However, black hole, though characterized by closed space, has energy and/or mass exchange with the environment. Is the same valid about the closed image of the universe? What about the time arrow? According to Hawking [3,4], the closed universe should have cycles of expansions and collapses, and the collapse phase might have the opposite direction of the time arrow (from future to past). If the time arrow does not change, does the entropy rising change the properties of the universe on each cycle? Can ever the cyclic evolution of the universe be transformed to irreversible expansion?

(f) Big Bang: Open Model. This model assumes that the space of the universe is open. As it is noted above, at least the first period of the existence of the universe was described by the closed model, hence, the open model has limited validity in any case. However, the primary singularity might be result of the evolution of exterior cosmos, and big bang could be accompanied by several interactions with exterior cosmos. In any case, the problem of exterior cosmos appears not only in the open model but also in the closed one.

We note that the analysis of the current mass characteristics of the universe gives preference to the open model.

(g) "No-Big-Bang" Theory. Though this concept is presented by minority of researchers, its versions are much more multiple than models of big bang. In the mathematical aspect, this is because the classical general relativity (GR) theory does not lead to some restrictions for admissible values of energy densities for gravitating systems. Different cosmological models and collapsing systems in GR have singular states with divergent energy density and curvature invariants [10].

Opponents of the big bang concept believe it to have shifted its ground frequently under observational constraints. Many of its present deductions are seen to be based on untested physics and unobservable events of the very early universe, while its beginning in a space–time singularity indicates its incompleteness as a physical theory. They may seem to claim sometimes that there is a conspiracy to hide these objects, but they themselves do not quote examples of galactic clusters made of normal galaxies and quasars all having the same red shift and therefore supporting the big bang.

(h) Self-Organization Theory. This theory can be considered as a development of the open model of the universe in the big bang theory. The main idea of the self-organization theory consists in the assumption that thermodynamics is applicable to cosmos before beginning of universe. Then big bang is considered as a process increasing the combinatorial entropy of cosmos [11], and the mechanism of bing bang is very similar to micropore formation (compare to Chapter 3).

Primary matter of cosmos before the universe formation is characterized by the following observed properties:

Primary matter consists of light particles that cannot be distinguished but for their energy only.

Particles of primary matter cannot be distinguished if the difference in their energy values is less than several threshold of distinguishability.

Primary matter is able to self-organization, which might result in the formation of the universe.

A self-organization process in a primary matter system leading to formation of a cell increases the combinatorial entropy of the system, and the additional entropy obtained due to it [11].

The self-organization theory assumes that the entropy rising is the main criterion determining the evolution of primary matter. During big bang, the nucleus of the future universe (singularity) overcomes a critical state characterized by minimum of the local entropy of the near part of the system, that is caused by formation of regular fluxes. Overcoming of the entropy minimum is due to certain fluctuations [11].

The first stage of big bang is related to formation of regular fluxes in form of a sphere, then transformed to the protection layer. This prevents processes of randomization able to destroy the network of self-organization. Near the protection layer containing the freshly captured high-energy particles of primary matter, processes of self-organization of matter are most intensive, while the central zones of universe are characterized by minimum of free energy; there, processes of self-organization are slow. The further expansion of the universe led to growth of the passive nucleus and decrease of its temperature because of various self-organization processes, comprising formation of galaxies, stars, etc. [11].

Now, let us consider the connection of the considered theories with known experimental facts.

(i) Background Radiation. This is considered as the main confirmation of the big bang cosmology. The background radiation (about 3 K) gets an exact explanation in each of considered versions of big bang as its sequence. The self-organization version of the open model of big bang supports such explanation but suggests that the temperature of the background radiation is different through the universe: minimum at the center, maximum at the frontier.

(j) Red Shift. The existence of the red shift in the spectrum of stars gets a similar explanation in all versions of big bang theory: galaxies "run off" the center of universe, as a sequence of big bang. Versions of no-big-bang concept sometimes suggest that the red shift is found because of observational constraints [6].

(k) Young Galaxies. This is one of most serious arguments against the big bang concept. In the closed model of big bang the explanation of this phenomenon is too difficult: if all universes formed in the big bang and galaxy nuclei appeared in the first period, young galaxies cannot exist.

In the open model of big bang, the existence of young galaxies may be explained as a result of the interaction with the exterior cosmos. Moreover, the self-organization version of the open model assumes that different parts of the universe should have different age.

In no-big-bang concept, young galaxies do not make problem.

(l) Time Arrow. The irreversibility of time gets a very simple explanation in the self-organization model of universe. As in classical thermodynamics, it is assumed that the time arrow is determined by rising of entropy resulting from self-organization processes and various interactions inside universe and with the exterior cosmos. In the traditional open model, the time arrow can be related to the irreversible expansion and interactions with the exterior cosmos. In the closed model, that is a problem: when the universe expands, the time arrow can be related to the expansion, but the stage of collapse, in such image, should turn back the time. Otherwise, the universe should get irreversible changes on each period expansion–collapse. In no-big-bang concept, this point is very problematic.

(m) Light Nuclei. For all versions of big bang theory, the abundance of light nuclei in universe is easily explained: those were preferably formed in first moments of big bang. In the no-big-bang concept, the explanation depends of the kind of the used version.

TABLE A2.1 Comparison of Existing Models of Beginning of the Universe with Known Experimental Facts.

Problem to solve	Big bang (closed model)	Big bang (open model)	Self-organization model	No big bang
Background radiation	+[a]	+	+	–[b]
Red shift	+	+	+	+
Density or mass	+/–[c]	+	+	+/–
Young galaxies	–	+/–	+	+
Time arrow	?[d]	+	+	?
Light nuclei	+	+	+	+/–

[a]+ connection is good.
[b]– connection is not good.
[c]+/– connection is problematic.
[d]? explanation is difficult.

(n) Methodological Aspect. As follows from the Table A2.1, the self-organization theory has several preference, due to having advantages of the open model without most of its drawbacks. In the experimental aspect, the self-organization theory predicts a significant divergence in values of density and background radiation temperature, which can be tested in the future. As most of existing models, the self-organization theory can be used for evaluation of fundamental constants.

II. BEGINNING OF LIFE BY SELF-ORGANIZATION

(a) General Aspect. Though the great importance of the problem of beginning of life, the existing theories did not still get a sufficient experimental support, meant a living organism formed artificially in a mineral environment.

The enormous importance of problems concerning the beginning of life makes necessary a review of existing theories on this matter, analysis of their advantages and disadvantages, their connection with known experimental facts, and their applicability to problems related to beginning of life.

(b) Beginning of Life on Early Earth and the History of the Problem. The problem of beginning of life interests not only researchers but also philosophers since ancient age. Attempts of its solution are found in all religions, all philosophies. Really scientific studies of this matter began the same time when basic principles of evolution theory were formulated: evolution from primitive and simple to developed and complex life meant also several minimum from which the evolution started.

Until recently, majority of biologists believed that the age of life is about 50% of the age of Earth. Recent discoveries in ancient terrains make us to conclude that life existed during at least 80% of the history of Earth. Microscopic fossils and fossils of microbial mats are found in sedimentary rocks in western Australian and southern African terrains (dated about 3.5 billion years ago) and Isua formation in Greenland (3.8 billion years ago) [12].

Thus, life began when conditions of existence were very strict, absolutely different from those in our habitat. Even the content of carbon in the environment was very low [12–15]. In addition to natural difficulties for life beginning, the conditions of the early Earth provided catastrophes that might seriously obstacle and even stop the development of life. Such catastrophes may be divided into three categories: "normal" events related to season changes (hurricanes, floods, etc.); geological cataclysms: volcano activity, earthquakes, their sequences; cosmic cataclysms: changes in the solar activity, impacts of cosmic bodies.

In many cases, catastrophes might be fatal for the first living organisms on the early Earth. Moreover: impacts of asteroids like Vesta, Pallas, or Hygiea might cause such increase of the temperature on the Earth that the ocean would be evaporated [12]. On the other hand, the catastrophes (especially cosmic ones) might stimulate amino acid formation and furnish biogenic elements [12,13].

Most of existing models of beginning of life are based on the astrophysical concept of evolution of cosmic bodies, according to which planets began as high-temperature massive objects with aggressive atmosphere, intense tectonic changes and large thunderstorm activity. For example, the atmosphere of the early Earth consisted mostly of hydrogen, methane, ammoniac, carbon dioxide, etc. According to experiments of S. Miller (1953), electrical discharges in atmosphere should cause formation of amino acids dissolved in water of the primary ocean. In 1957, Ph. Abelson showed possibility of production of more twenty amino acids and some proteins. Spontaneous polymerization allows formation of various macromolecules, comprising ribonucleic acid (RNA). In 1980s, it was demonstrated that RNA were capable to play role of enzymes (Dyson, 1985; Gilbert, 1986; Joyce, 1991; Cech, 1993). These results may be interpreted as RNA origin of the first living organisms [12,15].

(c) Creation Theory. The main idea of the creation theory consists in the suggestion that the life is created by upper forces. Historically, the creation theory was the first attempt of explanation of beginning of life, based on the Bible. In the general sense, the creation theory comprises also a version that life on the early Earth was due to a super civilization. Such a version does

not seem absurd, but that does not solve the main problem (because also super civilization should begin from a primitive life), and does not provide more information than the cosmic theory (see below). In the further analysis, we do not consider the creation version.

(d) Cosmic Theory. The cosmic theory considers the option of transportation of microorganisms due to an interplanetary motion of cosmic bodies. C. Sagan, C. F. Chyba, P. Thomas, and L. Brookshow (1990) modeled a collision of a bacteria-carrying comet with Earth and found that only exceptional conditions might allow organics to survive [12]. However, it was found possible that microorganisms could migrate between Earth and Mars in the debris ejected from the vicinity of giant impacts [12].

The cosmic theory does not solve the main problem: What was the mechanism of beginning of life? However, this theory enlarges the spectrum of conditions in which life might begin: in addition to early Earth, many of cosmic bodies that existed about 3.8 billion years ago and earlier.

(e) Oparin (Coacervate Droplet) Theory. The coacervate droplet theory suggests that the formation of first living organisms includes a step of separated droplets, as in an emulsion prepared by mechanical mixing of a complex solution with another, these being mutually insoluble or having limited mutual solubility. Such system is called *coacervate droplet* and characterized by developed surface. A coacervate drop could be formed, e.g., from a homogeneously constructed polymer consisting of a protein and a polyglycoside. Droplets have property of selective absorption and accumulation of several substances. Combined with the eventual formation of amino acids and proteins, the mass transfer in the droplets might become a mechanism for spontaneous evolution of nonliving structures to primitive life [15]. Coacervate drops can be considered as models for precellular structures [16].

In the Oparin model, the evolution of carbon compounds is postulated to be the basis of the origin of life on Earth and elsewhere. This process can be divided into three steps [14–20]:

- Formation of primitive organic substances, i.e., hydrocarbons, as in meteorites (carbonaceous chondrites), the abiogenic formation of primitive organic substances (such a situation is widespread in space)
- Formation of the primary soup: an aqueous solution of different and complex organic substances, monomers, and polymers, as well as polypeptides and polynucleotides
- Formation of complex macromolecular open systems, which are the primary forms for the origin of primitive creatures through evolutionary processes

In contrast to modern proteins, the high-molecular-weight polymers in the primary soup had only an accidental order of monomers in the chain. Coacervate droplets could have been formed in the primary soup from a simple mixture of randomly constructed polypeptides and polynucleotides. As soon as a certain level of polymerization had been reached in this mixture, the droplets could appear. These coacervate drops would be very concentrated and able to selective absorption of different low-molecular-weight substances from the external solution. If any of these substances are able to catalytic stimulation of the reactions in these drops, they become open systems. Such open macromolecular systems could easily have formed in the primary soup, owing to the incorporation of different organic and inorganic catalysts from the external medium; such systems are called *protobionts*. Prebiological selection may have ensued based on the growth rate and competition of the protobionts. Natural selection continually filtered out undesirable protobionts, which left only a few efficient ones, containing coenzymes we know today. Finally, the specific, strictly ordered association of mononucleotide residues in large molecules like DNA and RNA could have arisen only during prolonged evolution of living systems [21,22].

A serious disease of the coacervate droplet theory consists in difficulty of finding such droplets in nonliving nature, in which majority of liquids is presented by water or aqueous solutions, whereas non-aqueous liquids (e.g., petroleum) are mainly of organic origin and seem products of living organisms. As a result, the above-described process has a very low probability.

(f) Self-Organization Theory. The self-organization theory can be considered a development of the coacervate droplet theory. The principal difference between them consists in the suggested mechanism of the formation of the interface. Instead such event as mixing of two nonmiscible liquids, the self-organization theory suggests Benard-like mechanism that could easily realize, e.g., in the primary ocean near a volcano, or just because of intense solar radiation [23]. Introducing the self-organization mechanism into the explanation of beginning of life, this theory puts such exceptional phenomenon as life into the range of other processes related to self-organization.

The main idea of the self-organization theory is that beginning of life was thermodynamically favorable for mineral nature. The thermodynamic criterion of beginning of life in a planetary system consists in the following suggestion: due to formation of a living subsystem, the Earth system gets a profit in entropy. The formation of complex branched macromolecules contributes the rising of the entropy of the environment. Processes of amino acid polymerization and destruction do not need much energy and can take place not only due to electrical discharges but also due to solar radiation. Another thermodynamic factor stimulating beginning of life was formation of protein cells, which increased the combinatorial entropy of the ocean [23].

Notice the analogy with the formation of porous clusters in processes of micropore formation (see Chapter 3): in the role of the continuous phase is the primary ocean, while protein cells are very similar to micropores.

An intermediary step could comprise RNA-based cells.

After the formation of proteins, the next stage of life beginning comprised the formation of internal flows in protein systems. The change of free energy in polymerization processes was related to the formation of macromolecules and internal flows in the protein structures.

The self-organization theory suggests the following mechanism of evolution of amino acids to living systems [23]:

Reversible polymerization of amino acids
Formation of global protein heat engine
Involving oxygen in the system of global protein heat engine
Self-organization of the global protein structure
Internal self-organization of protein cells and their transformation to self-reproducing biocells

An experimental test for the proposed model of life beginning could be accomplished as follows [23]:

1. Exploration of the equilibrium amino acids RNA and DNA in conditions of various destructive factors supposed to exist on early Earth (solar and electromagnetic radiation, temperature changes, poison chemicals, etc.). Expected information obtained includes conditions of formation of periodical heat engine.
2. Study of destructive influence of the mentioned factors for the estimation of extreme conditions under which the system does not lose its main properties. Expected information includes changes caused by different doses of destructive factors.
3. Exploration of the formation of Benard-like cells in primary soup, in comparison with ordinary Benard cells formed in water. Expected result is the formation of protein cells and spectrum of parameters in which these are stable.
4. Behavior of protein cells treated with mentioned destructive factors. Expected result is the interior self-organization of protein cells and formation of elementary biocells.

TABLE A2.2 Connection of Existing Theories of Beginning of Life with Experimental Facts

Problem to solve	Oparin theory	Cosmic theory	Self-organization theory
Opportunity for RNA-based life	+[a]	+	+
DNA as the base for life on Earth	+/–[b]	–[c]	+
Signs of life in ancient rocks	?[d]	+/–	+
Stability of life	+/–	?	+

[a]+ connection is good.
[b]+/– connection is problematic.
[c]– the model contradicts to the fact.
[d]? the problem was not studied enough.

Now, let us compare the considered theories to known facts (see Table A2.2).

(g) Opportunity for RNA-Based Life. In all considered theories, the existence of RNA-based life is not impossible. In the coacervate droplet theory, the chemical composition of both interacting liquids is important but not decisive, and RNA-based mass-exchange is well possible [12,15,24]. In the cosmic theory, the base for an extraterrestrial life is determined by the conditions on the planet where this life began, and options for nonprotein life are well open. In the self-organization theory, thermodynamics of self-organization does not give absolute preference to proteins.

(h) DNA as the Base for the Life on the Earth. Though RNA-based life does not seem impossible, it is the fact that the life on the Earth now is based only on DNA. It is difficult to find any explanation of this fact in the coacervate droplet theory, while for the cosmic theory the situation is much worse: if the life came to Earth from another planet, why not from two or three different planets with different bases for life?

In the self-organization theory, the explanation of thermodynamic preference (not absolute but relative preference!) for DNA is given by the fact that macromolecules of DNA and their combinations in proteins are more complex and allow more microstates—hence, structures based on them have more entropy [23].

(i) Signs of Life in Ancient Rocks. The old age of the life on Earth does not get appropriate explanation in the Oparin theory [14,15,21,22,25–28]. Since the probability of coacervate droplet formation was very low, it is very

TABLE A2.3 Methodological Aspect: Usefulness of Existing Theories of Beginning of Life for Further Studies

Opportunity for reproduction of life by mankind	+/−[a]	+/−	+[b]
Recommendations for experiments for reproducing life	−[c]	−	+

[a]connection is problematic.
[b]connection is good.
[c]connection is not good.

strange that the life could begin almost immediately after the formation of the early Earth. In the cosmic theory, the situation is not much better: though bacteria might eventually get onto the early Earth, the probability of such event is too low. However, for the self-organization theory that is a normal result: as soon as the primary ocean was formed and amino acids appeared due to electrical discharges, all the mineral nature stimulated the formation of complex amino acid–based macromolecules (for the reason of rising entropy).

(j) Stability of Life. In the cosmic theory, this problem is not studied. In the Oparin theory, the stability of early life is probably determined by the stability of the environment, while this question is not analyzed. In the self-organization theory, the stability of life against little perturbations of the environment is described by the stability of the solution of thermodynamic equations for cell [23]. This result, being applied to a microorganism, means reflex; applied to a group of living organisms, it means evolution [23].

(k) Methodological Aspect. All three theories conclude positively about the principal possibility for an artificial reproduction of life (see Table A2.3). However, the theory of Oparin finds the probability of success of such experiment very low (at least, for conditions close to the early Earth). The cosmic theory can be tested, e.g., by the observation of other planetary systems [29–31]. The self-organization theory suggests an experimental way for the reproduction of life (see Ref. 23).

III. "BIG" PHENOMENA AND MICROPORE FORMATION

In Chapters 3 and 5 we considered the self-organization aspect of micropore formation. Above we have analyzed (brief) the same aspect for beginning of the universe and life.

TABLE A2.4 Analogy Between "Big" Phenomena (Big Bang and Beginning of Life) and Formation of Porous Clusters in Processes of Micropore Formation

Criterion for comparison	Big bang	Beginning of life	Micropore formation
Exterior phase	Big cosmos	Early Earth	Continuous phase
Mechanism of the process	Evolution of perturbed big cosmos	Evolution of macromolecular components	Evolution of metastable solid phase
Driving force for the process	Increase of the entropy of big cosmos	Increase of the entropy of the early Earth	Increase of the entropy of the solid phase
Resulting cluster of interior phase	The universe	Biological cells	Porous clusters

Let us notify the following factors of similarity between the above "big" phenomena and the "little" phenomenon of micropore formation:

For all these phenomena, the notion of the exterior phase is applicable: for the universe it is big cosmos; for life it is nonliving matter, and for micropores it is the continuous phase.

In all cases, the mechanism of the main process has the thermodynamic origin. For universe beginning it is the relaxation of eventually perturbed big cosmos; for life beginning it is the polymerization and destruction reactions of amino acids, having the tendency to form temporary-dissipative structure because of day-night changes, while for micropore it is the relaxation of the unstable solid phase.

Driving force for all three processes is the entropy increase for the exterior phase.

The result of all three processes consists in the formation of several clusters of the interior phase: universe, living organisms, or microporous cluster.

The above analysis is illustrated by Table A2.4.

REFERENCES

1. Moffat, J.W.; Vincent, D. The early universe in a generalized theory of gravitation. Can. J. Phys. **1982**, *60*(5), 659–663.
2. Borner, B. *The Early Universe: Facts and Fiction*; Springer-Verlag: Berlin, Heidelberg, New York, London-Paris, Tokyo, 1988.

3. Hartle, J.B.; Hawking, S.W. Wave function of the universe . Phys. Rev. D **1983**, *28*(12), 2960–2975.
4. Hawking, S.W. Arrow of time in cosmology. Phys. Rev. D **1985**, *32*(10), 2489–2495.
5. Wald, R.M. *Space, time, and gravity: the theory of the big bang and black holes*; University of Chicago Press: Chicago, 1992.
6. Kanno, S. Interaction between fermions and the gravitational field in the very early universe. Prog. Theor. Phys. **1988**, *79*(6), 1365–1377.
7. Arp, H.C.; van Flandern, T. The case against the big bang. Phys. Lett. A (Netherlands) **1992**, *164*(3–4), 263–273.
8. Schramm, D.C.N. Nuclear physics and cosmology. Proc. 1989 Intern. Nucl. Phys., Conf. V.2 Invited Papers, Sao Paulo, Brazil, 1989, 743–759.
9. Kierein, J.W. A criticism of big bang cosmological models based on interpretation of the red shift. Laser Part. Beams (UK) **1988**, *6*(3), 453–456.
10. Minkevich, A.V. Gravitating systems at extreme conditions and gauge theories of gravity. Avail. Nguyen van H., Tran T.V.J. Part. Phys. Astrophys., Proc. Rencontre Vietnam, 1st: Editions Frontieres, Gif-sur-Yvette, France, 1995, Volume Date 1993, 537–542.
11. Romm, F. Beginning of universe as result of self-organization of matter. Scientific Israel-Technological Advantages **2000**, *2*(1), 49–57.
12. Chyba, C. Origins of life. Buried beginnings. Nature **1998**, *395*(6700), 329–30.
13. Terzian, Y.; Bilson, E. *Carl Sagan's universe*. Cambridge University Press: Cambridge, 1997; 282 pp.
14. Oparin, A.I. Routes for the origin of the first forms of life. Subcell Biochem **1971**, *1*(1), 75–81.
15. Cech, T.R. The efficiency and versatility of catalytic RNA: implications for an RNA world. Gene **1993**, *135*(1–2), 33–36.
16. Gladilin, K.L.; Orlovskii, A.F.; Kirpotin, D.B.; Oparin, A.I. *Coacervate drops as a model for precellular structures*. Origin Life, Proc. ISSOL Meet., 2nd. Ed. Noda, H. 1978, Meeting Date 1977, Bus. Cent. Acad. Soc. Japan: Tokyo, Japan, 1978, 357–362.
17. Delsemme, A.H. Cometary origin of carbon, nitrogen and water on the Earth. Origins of Life and Evolution of the Biosphere **1992**, Volume Date 1991–1992, *21*(5–6), 279–298.
18. Dyson, F.J. A model for the origin of life. J. Molec Evol **1982**, *18*(5), 344–350.
19. Gilbert, W. Life after the helix: Watson and DNA: Making a Scientific Revolution, by Victor K. McElheny. Nature (London, United Kingdom) **2003**, *421*(6921), 315–316.
20. Sagan, C. The origin of life in a cosmic context. Origins of Life **1974**, *5*(3), 497–505.
21. Oparin, A.I. Chemistry and the origin of life. RIC Reviews **1969**, *2*(1), 1–12.
22. Oparin, A.I. Origin of life. Sci. Cult. **1968**, *34*(9), 4–9.
23. Romm, F. Beginning of life as result of self-organization of nature. Scientific Israel-Technological Advantages **2000**, *2*(1), 42–48.

24. Raff, R.A. *The shape of life*. The University of Chicago Press: Chicago–London, 1997.
25. Oparin, A.I. *The nature and origin of life*. Ponnamperuma, C., Ed. Comp. Planetol., [Proc. College Park Colloq. Chem. Evol.], 3rd Academic: New York, N.Y. 1978, Meeting Date 1976, 1–6.
26. Oparin, A.I. Modern concepts of the origin of life on the Earth. Scientia (Milan) 1978, 113(1-2-3-4), 7–25.
27. Oparin, A.I. Modern ideas concerning the pathways of the origin of life. Frank, G.M., Kayushin, L.P. Eds. Symp. Pap. Int. Biophys. Congr., 4th 1973, Akad. Nauk SSSR, Inst. Biol. Fiz.: Pushchino-on-Oka, USSR, Meeting Date 1972, 4, Pt. 2, 685–707.
28. Oparin, A.I. *Genesis and Evolutionary Development of Life*. Academic: New York, 1968.
29. Pilcher, C.B. Future of NASA's space science program. In: *Solar System Exploration*, NASA Headquarters: Washington, DC. Book of Abstracts, 218th ACS National Meeting, New Orleans, American Chemical Society: Washington, D.C. Aug. 22–26 1999, NUCL-208.
30. Chyba, C.F.; McDonald, G.D. The origin of life in the solar system: current issues. Ann. Rev. Earth Planetary Sci. **1995**, *23*(no issue), 215–249.
31. Sagan, C. On the origin and planetary distribution of life. Radiat. Res. **1961**, *15*, 174–192.

Notation

Notation is given only for parameters largely used in the book or in a given chapter. Locally used parameters are explained only in the location of the reference.

GENERAL SYMBOLS

A	surface area (m^2)
E	internal energy
$f(\varepsilon)$	distribution function of micropores in energy
G	free energy (Gibbs)
n_n	coordination number
m	number of empty or filled neighbors (changing from 0 to n_n or functionality f)
N_{fm}	number of filled cells having m empty neighbors
P	pressure (Pa or ata)
R_g	gas constant (J/mol/K)
S	entropy
T	temperature (K)
t	time (s)
U	negentropy
V	volume (m^3)
α	parameter of nonequilibrium of system
δB	very short variation of parameter B
ε	energy of pore
μ	chemical potential (J/mol)
ξ	porosity
σ	surface tension (N/m)

CHAPTER 1

$r_{\max}$	maximal distance between the closest opposite walls in the pore
$r_{\min}$	minimal distance between the closest opposite walls in the pore
$v(r)$	size distribution of the internal space (volume)
$\rho(r)$	size distribution of the internal surface

CHAPTER 2

c	velocity of light
Er_a	error of measurements by taking sample
Er_m	error caused by aging
W_a	initial amount of the treated material
W_s	maximal amount of the samples taken away
X	result of measurement
λ_{EM}	wavelength of electromagnetic rays
$\Theta(m)$	number of voids having (each) m empty neighbors
τ_m	period of measurements
τ_r	averaged relaxation period τ_r
τ_f	period of changing of the exterior intense parameters
P	measured parameter

CHAPTER 3

[A]	current local concentration of reagent A
C_N	weight fraction of N-mer
k_r	constant of reaction rate
$X\ (t, \varepsilon, T)$	volume fraction of micropores having energy ε at temperature T at moment t
$\varepsilon_{\min}, \varepsilon_{\max}$	limits of integration of ε
$E_a(\varepsilon)$	activation energy of formation of micropore having energy ε
M	monomer
$B_{l\Sigma}$	number of monomeric units having (each) l vacancies
$C_r(N)$	number of cross-links in N-mer
K_0	constant of polymerization (in the statistical polymer approach)
f	functionality
m	branching
Pol	polymer

SP(N)	statistical N-mer
$V_\Sigma(N)$	number of vacancies in statistical N-mer
$U_\Sigma(N)$	number of extreme units in statistical N-mer
$\mu(N)$	chemical potential of statistical N-mer
$\mu^\circ(N)$	standard chemical potential of statistical N-mer
$C(N)$	concentration of N-mers

CHAPTER 4

Ads	adsorbent
Ads · G	adsorbate
P_G	partial pressure of gas G
θ_t	amount of adsorbed fluid (adsorption isotherm)
l_+	number of steps in the percolation direction
l_-	number of steps back
l_0	number of steps in directions perpendicular to the percolation
l_Σ	total number of steps
C_n^m	number of combinations from n options by m manners
$\omega\ (l_\Sigma)$	linear velocity of the flow inside a path having the total length l_Σ
η	friction coefficient

CHAPTER 5

$\mu_{pore}(\varepsilon)$	chemical potentials of the micropore with energy ε per 1 mol of contained substances
μ^0_{pore}	standard chemical potentials of the micropore with energy ε per 1 mol of contained substances
T_f	temperature of pore formation (in equilibrium)
$R_k(N)$	number of particles on kth level of statistical N-mer
$V_k(N)$	number of vacancies on kth level of statistical N-mer
$Z(N)$	characteristic size of statistical N-mer

CHAPTER 6

$U_\Sigma\ (\tau)$	total negentropy of system at moment τ
$U_N(\tau)$	negentropy of N-mers at moment τ
σ_d	constant of rate of dissipation of negentropy

CHAPTER 7

COST_i	costs of *i*th kind
n_i	number of repeating costs of *i*th kind

APPENDIX 1

W_{+iLk}	amount of *i*th component incoming to *k*th reactor with liquid
W_{+iGk}	amount of *i*th component incoming to *k*th reactor with gas
W_{-iLk}	amount of *i*th component leaving *k*th reactor with liquid
W_{-iGk}	amount of *i*th component leaving *k*th reactor with gas
β_{ik}	distribution coefficient for *k*th reactor ($= W_{-iLk}/W_{-iGk}$)

Index